GUIDE TO FOODBORNE PATHOGENS

GUIDE TO FOODBORNE PATHOGENS

Edited by

Ronald G. Labbé
Department of Food Science
University of Massachusetts
Amherst, Massachusetts

Santos García
Department of Microbiology and Immunology
University of Nuevo Leon
Monterrey, Mexico

WILEY-INTERSCIENCE

A JOHN WILEY & SONS, INC., PUBLICATION

New York • Chichester • Weinheim • Brisbane • Singapore • Toronto

This book is printed on acid-free paper. ∞

Published simultaneously in Canada.

For ordering and customer service, call 1-800-CALL-WILEY.

Library of Congress Cataloging-in-Publication Data:

Guide to foodborne pathogens / edited by Ronald G. Labbé, Santos García.
p. cm.
"Proceedings of a seminar held in Monterrey, Mexico in 1998"—Pref.
Includes index.
ISBN 0-471-35034-6 (cloth : alk. paper)
1. Food—Microbiology—Congresses. I. Labbé, Ronald G., 1946– II. Santos García, Jose, 1961–
QR115 .G83 2001
615.9′54—dc21 2001026958

Printed in the United States of America.

10 9 8 7 6 5 4 3 2 1

"We can never tell what is in store for us."

Harry Truman
33rd President of the United States

CONTENTS

PREFACE

In recent years there has been increased public focus on microbial aspects of food safety. In part this is due to certain high-profile cases affecting the food service and food production industry resulting in deaths, illness, and massive product recalls. New, emerging, and well-established foodborne pathogens continue unabated in their attacks on unsuspecting and unprepared consumers.

The increasing globalization of the food supply will require a heightened awareness of food safety issues. To this end an international symposium entitled "Food Pathogens—Current Concerns and Approaches to Control" was held in 1998 in Monterrey, Mexico, with attendees from North, Central, and South America and the United Kingdom. This book is an outgrowth of the conference. It is intended to fill a need not readily filled by existing texts. While there are a number of monographs on the topic of foodborne diseases, most provide highly detailed and referenced treatments of each agent. We envision a need for a moderately priced, current source authored by experts in their respective fields that will appeal to the more general audience. Thus contributors here have avoided, for example, at-length discussions of molecular genetics of virulence factors and detailed mechanisms of pathogenicity. Some information on methodology is provided, but for detailed protocols readers are referred to dedicated sources such as the Compendium of Methods for the Microbiological Examination of Foods, the U.S. Food and Drug Administration's *Bacteriological Analytical Manual* (FDA BAM), the International Organization for Standardization (ISO), and similar publications from other governmental and non-governmental organizations. Discussions of main agents of foodborne illness are book-ended by chapters on globilization of the food supply, and physical methods for their destruction as well as role of hazard analysis critical contral point (HACCP) program.

RONALD G. LABBÉ
SANTOS GARCÍA

CONTRIBUTORS

David W. K. Acheson, Director, Food Safety Institute at New England Medical Center, 750 Washington Street, Box D41, Boston, MA 02111

Judd Aiken, Department of Animal Health and Biomedical Sciences, University of Wisconsin—Madison, 1656 Linden Drive, Madison, WI 53706

Jean Yves D'Aoust, Microbiology Research Division, Sir F. Banting Research Division, Postal Locator 220OH 4A2, Tunney's Pasture, Ottawa, Canada K1A OL2

Mafu Akier Assanta, Agriculture and Agri-Food Canada, Food Research and Development Center, 3600 Casavant Blvd. West, St. Hyacinthe, Quebec, Canada J2S 8E3

R. W. Bennett, Division of Microbiology, OSRS, Center for Food Safety and Applied Nutrition, Washington, DC 20204

Saumya Bhaduri, Eastern Regional Research Center, U.S. Department of Agriculture, Wyndmour, PA 19038

Deepak Bhatnagar, U.S. Department of Agriculture, ARS, Southern Regional Research Center, 1100 Robert E. Lee, New Orleans, LA 70124-4305

Lloyd B. Bullerman, Department of Food Science and Technology, University of Nebraska-Lincoln, 349 Food Industry Complex, Lincoln, NE 68583-0919

Dean O. Cliver, Department of Population Health and Reproduction, School of Veterinary Medicine, University of California, Davis, CA 95616-8743

Catherine W. Donnelly, College of Agriculture and Life Science, University of Vermont, 601 Main St., Burlington, VT 05401

Peter Feng, Division of Microbiology, OSRS, Center for Food Safety and Applied Nutrition, Washington, DC 20204

Eduardo Fernández-Escartin, Facultad de Ciencias Químicas, Universidad Autónoma de Querétaro, Apdo. 796, Querétaro, Mexico

Santos García, Fac. Ciencias Biologicas UANL, Lab. Bioquimica Microbiana, APDO Postal 124-F, San Nicolas N.L. 66451 Mexico

Robert B. Gravani, Department of Food Science, 8 Stocking Hall, Cornell University, Ithaca, NY 14853

Norma L. Heredia, Department of Microbiology and Immunology, Faculty of Biological Sciences, University of Nuevo Leon, Postal Box 124-F, San Nicolas, N.L. 66541, Mexico

James M. Hungerford, Seafood Products Research Center, 22201 23rd Drive, SE, Bothell, WA 98021

George J. Jackson, Center for Food Safety and Applied Nutrition, U.S. Food and Drug Administration, Washington, DC 20204

Ronald G. Labbé, University of Massachusetts, Department of Food Science, Amherst, MA 01003-1410

Barbara M. Lund, 8 The Walnuts, Branksome Road, Norwich NR4 6SR, United Kingdom

Michael W. Peck, Institute of Food Research, Norwich Research Park, Colney, Norwich NR4 7UA, United Kingdom

Denis Roy, Agriculture and Agri-Food Canada, Food Research and Developement Center, 3600 Casavent Blvd. West, St. Hyacinthe, Quebec, Canada J2S 8E3

Norman J. Stern, U.S. Department of Agriculture Russell Agricultural Research Center, 950 College St. Road, Athens, GA 30604-5677

Mark L. Tamplin, Eastern Regional Center, U.S. Department of Agriculture, Wyndmour, PA 19038

Ewen C. D. Todd, Bureau of Microbial Hazards, Health Protection Branch, Tunney's Pasture, Ottawa, Canada K1A OL2

Irene V. Wesley, U.S. Department of Agriculture, Agricultural Research Service, P.O. Box 70, Ames, IA 50010

GUIDE TO FOODBORNE PATHOGENS

CHAPTER 1

Epidemiology and Globalization of Foodborne Disease

EWEN C. D. TODD

1.1. INTRODUCTION

Infectious and toxigenic pathogens transmitted through food have been recognized for over 100 years, and by the 1950s the main pathogens of concern were *Salmonella, Staphylococcus aureus*, and *Clostridium perfringens* in the United Kingdom and the United States. Botulism had also been understood as a dangerous but rare disease related to canned food. Therefore, for most public health officials, foodborne disease, or food poisoning as it was called then, was generally considered to be an inconvenience for a day or two and more of nuisance than a threat to life. Not much was known in other countries because of a lack of any systematic reporting program. In fact, there was little interest or research being carried out on acute foodborne disease agents. We knew from outbreaks that most of the situations could have been avoided if there was proper time and temperature handling and storage of food. Once staff in food service establishments became better educated, these problems would resolve themselves. However, by the time that the 1980s came in, we were beginning to be a little more concerned with agents like *Campylobacter, Listeria monocytogenes*, and *Escherichia coli* O157:H7, but it took several years for health authorities to recognize that these had the potential to cause serious complications or death and could be transmitted by a variety of products and that there were limited control mechanisms in place to reduce such foodborne disease. Large outbreaks in the United States arising from *L. monocytogenes* in 1985 and *E. coli* O157 in 1993 resulted in changes to food safety policy specifically aimed at these organisms. Now, it is recognized that a variety of pathogens in many different types of foods can cause illnesses that may be life threatening.

Guide to Foodborne Pathogens, Edited by Ronald G. Labbé and Santos García
ISBN 0-471-35034-6

These include *Cyclospora, Cryptosporidium, E. coli* O157, *Salmonella, Shigella*, and small, round-structured viruses (SRSVs) in produce, dairy products, eggs, ice cream, and shellfish. Current surveillance systems are only capable of detecting a few of these pathogens.

1.2. SURVEILLANCE OF FOODBORNE DISEASE

There appears to be a general increase or at least a plateauing for foodborne disease cases throughout the world, even though new regulations and educational strategies are being adopted nationally and internationally. One of the reasons for this is that surveillance of foodborne and waterborne disease has been very limited in detecting cases. The traditional passive system of letting outbreak reports be sent to a central source in a very few countries has been the source of our knowledge for decades, but it is far from adequate. Outbreaks may only be investigated and written up if they are large enough. Many household illnesses are never documented. Therefore, we are more familiar with mass-catering outbreaks or those involving a processed or widely distributed product. Even with these limitations, we have learned a great deal about the types of implicated foods and can anticipate problems even if we are not ready to initiate control programs. The globalization of the food supply is another issue. Changes in farming practices, with larger operations and faster throughput, the drive to increase profit by recycling all animal materials, and the difficulty in disposal of manure, all lead to increased likelihood of contamination of raw animal products. There has been much more intensive rearing of animals that allow transmission of pathogens, even if the animals themselves are not affected. Treatment of flocks and herds with antibiotics is primarily for growth enhancement but leads to increased antimicrobial resistance in the gut flora, which is now a problem for the human population. Environmental sources of contamination such as rodents in barns and gulls in fields are being recognized as important links in the transmission chain in zoonotic diseases. New varieties of strains appear in human cases but may originate in environmental or animal-raising conditions where genes can be transferred from one organism to another. The development of large-scale aquaculture is another rapidly expanding industry in many parts of the world where fish and shellfish are raised in close proximity and are prone to carry enteric viruses or bacterial pathogens as well as seafood toxins if they are present in the aquaculture areas. We have also moved rapidly from local manufacturers producing our food to national industries and now international trade with wholesalers packaging products in ways appealing to local populations. The larger market size and wide geographic distribution of products mean that if problems occur, many people are at risk, and extremely large outbreaks with hundreds of thousands of cases have occurred. Trace-back of a product to its source becomes difficult, especially if the originating company is in a foreign country. Immigrants and travelers introduce people to new types of foods, and the public today wants a wide choice of products to eat. This leads to many companies trying out new products or modifying traditional ones, with each step a potential for the introduction or growth of a pathogen. The consumer wants more varieties of food and in packages that require the minimum of preparation. This means that many types of ready-to-eat packaged food may be stored in a refrigerator for long periods of time. This allows various organisms, such as *Listeria, Yersinia*, and *Clostridium botulinum*, if present, to grow slowly to reach numbers that can cause illness. Therefore, it is perhaps not surprising that we see an increase in foodborne disease throughout the world, especially as we are living in an age with a rapidly growing and migrating society. The demographics of the population is also changing because of our ability to live longer and improved public health services and medical treatment. However, as a result, there is an increase in the immunocompromised group, especially those who are aging. These people are at risk from infections by opportunistic pathogens, of which we have no or limited surveillance mechanisms and laboratory capabilities to detect.

Ways to improve the surveillance system is a linking of newer epidemiological approaches with fingerprinting of clinical and food isolates. In the United States, the PulseNet lab typing of strains by molecular techniques is possible by comparing isolates from across the country to look for identical strains and possibly detect common sources. The FoodNet program in several sites allow detailed investigative studies in consumer food hygiene and physician diagnostic practices, food preferences, and case–control studies when illnesses have occurred. For each outbreak there is much information that can be gathered if the incriminated food is available. Counts of pathogens in food, for example, are useful for determining doses and quantitative risk assessments. Since we know that surveillance data are very much underestimated, more realistic figures can be made, even if they are not precise, such as 76 million foodborne cases in the United States and 2.2 million cases in Canada. From these the economic burden of disease can be estimated. Although these are crude figures, they put the costs of foodborne disease and potential control measures into perspective. For instance, in the United States the cost of six foodborne bacterial diseases alone totaled $2.9–$6.7 billion, excluding industry losses and legal suits. For Canada, all foodborne disease cost $1.2 billion in 1989. The costs for salmonellosis in England and Wales were estimated to be between £263 million and £450 million. These types of figures allow cost–benefit studies of potential control measures to be evaluated. Surveillance must not only be national but also continental and eventually global. Enter-Net (formerly Salm-Net) allows tracking of *Salmonella* and other enteric pathogen isolates throughout Europe and a few other countries. There is logic for blocks of free-trade countries to compare systems and eventually adopt a common one, not to rapidly track international outbreaks and stop them, but to learn from the experience and understand the factors that led to them occurring.

In the following section on foods associated with outbreaks, mostly recent, we can see the wide range of problems that may be encountered in any country. Zoonotic diseases are the most difficult to control because there are many entry points for the pathogens in the food chain. In contrast, there is a better chance of reducing foodborne diseases spread by human carriers such as *S. aureus* where an organism must come from a food handler and grow to large numbers before the enterotoxin is produced. In fact, there has been a reduction in staphyloccocal food poisonings in the United Kingdom, the United States, and Canada. It is more difficult to prevent enteric virus transmission because the infectious dose is very low and no incubation time is necessary, but better hygiene by food workers and restricting infected workers to nonfood areas will help keep the contamination rate down.

1.3. MEAT AND POULTRY

It has been recognized that *Salmonella* has been transmitted by meat and poultry for many decades. Typically, limited food safety knowledge has led to undercooking, improper cooling, or cross-contamination. The following are some recent examples. Over 50% of guests at a church supper were ill after eating ham in Maryland in 1998 (750 cases and one death). The hams had been packed too tightly into cooking pots, creating uneven cooking conditions. Then, they were packed too tightly into the freezer, which prevented rapid cooling, to give the salmonellae opportunity to grow. One slicer was used to cut all the hams without any sanitation between the slicing process, thus allowing the transmission of salmonellae onto many of the ham pieces served. In Pasadena, California, salmonellosis occurred in about half of 90 guests gathered for a large family Thanksgiving meal in November 1997, and 13 were hospitalized. The kitchen was inadequate for catering to such a large number of people, and the turkey was left out because of insufficient refrigerator space. In the spring of 1998 there were about 50 cases of *Salmonella* Java associated with an outdoor pursuit center in England where many of the participants of paint-ball games ate roast pork from a whole pig roasted on a spit. The pig had been stored at room temperature for 38 h before roasting, and the inner parts may not have been completely cooked. In Texas from May to October 1995, 59 persons were ill from *Salmonella* Agona infection. All those who lived in San Antonio and Houston had eaten food in a San Antonio Mexican restaurant. The same pulse-field gel electrophoresis (PFGE) strain was isolated from machacado (air-dried raw beef), which had been shredded in a blender and cross-contaminated other foods like salsa. Austin residents had eaten machacado at a similar Mexican restaurant. Both restaurants had purchased the beef from the same supplier in Corpus Christi. Without PFGE typing of isolates, these outbreaks would not have been detected.

Acute illnesses are not the only problems arising from salmonellosis and other enteric diseases. For example, in 1984, 423 provincial police officers were infected with *Salmonella* Typhimurium after eating contaminated meat sandwiches provided to them as they lined a route in Ontario, Canada, for the Pope's visit; 6.4% had acute reactive arthritis (RA), one-third resolving within 4 months, and the majority continued to have mild symptoms in their joints for the following 5 years. Four had sufficient damage to force them to change jobs. The risk factors identified were severe diarrhea at the time of the outbreak and presence of specific genes that code for HLA-B27 and HLA-CREG antigens. In an Australian outbreak in March 1997, of those ill after eating pork rolls produced by a bakery, two died and 3% developed RA. One-half of those with arthritis (1.5%) continued to have it 12 months later. In another outbreak in August from the same product, 5% developed RA. The difference in percent affected may be due to the ethnic background of those ill who may have a differing proportion of susceptible genes (Vietnamese in March and Cambodian in August). Sequellae such as RA are more frequent and are more long lasting than previously recognized.

Salmonella Typhimurium DT 104 is a major problem in the United Kingdom, because it is resistant to a wide range of antimicrobial agents. It is associated mainly with cattle but also pigs, sheep, and poultry. No effective treatment of infected animals has yet been found. In humans, there is a relatively high mortality rate of 3%, especially in the elderly. In 1994, it accounted for 60% of *S.* Typhimurium and 30% of all *Salmonella* incidents in cattle. It is spread from farm to farm by water and is difficult to eradicate since it survives well in both dry and wet environments. It has been found in beef, pork, salami, chicken, and cereals. Foodborne disease outbreaks have been associated with hamburgers, sausages, and sausage rolls. This pathogen will be in poultry flocks and other meat animals and extend down the food chain from farm to fork. Because of its ability to colonize many farm animals and be in the environment, its impact will probably be worse that of *Salmonella* Enteritidis. In the United States and other countries there is evidence for an increase in multiresistant *S.* Typhimurium strains, and so this is going to become a global issue.

Campylobacter has been associated in outbreaks and epidemiological studies with undercooked chicken and meat as well as from other sources. A case–control study in four cities in New Zealand bears this out. Risk factors were raw or undercooked chicken, any chicken prepared in a sit-down restaurant, unpasteurized milk or cream, overseas travel, rain water, and contact with puppies and calves. Barbecued or fried chicken was associated with illness, whereas baked or roasted chicken was protective, although the reason for this was not determined, unless the former were more typically undercooked. Although cases of sporadic campylobacteriosis are frequent, large *Campylobacter* outbreaks from poultry or meat, however, are relatively rare. In the following example chicken was the source of the pathogen, but it was other foods that caused the illness through cross-contamination. In 1996, at least 14 patrons of a restaurant in Oklahoma were ill with campylobacteriosis after eating lasagna and lettuce. Raw chicken had been cut up on a cutting board and the lettuce and lasagna contaminated by the cook's unwashed hands, utensils, or counter top. Because the infectious dose of *Campylobacter* is low (about 500 cells), one drop of chicken exudate can be enough to infect a consumer. The worker had frequently used the same towel to dry hands. The investigation emphasized the importance of clean-up and sanitation of preparation areas and of hand-washing facilities. The institution now contracts for large quantities of meat, and hygiene practices are being improved.

Escherichia coli O157:H7 was first identified as a foodborne pathogen in

1982 from two outbreaks involving hamburgers served in fast-food restaurants of the same chain in Michigan and Oregon. Reports of other outbreaks were followed with undercooked ground meat as the most frequently associated vehicle. By 1985, the most serious outbreak so far reported occurred in a London, Ontario, home for the aged where 70 were ill and 17 died after eating ham sandwiches probably contaminated by an infected food worker. The largest incident occurred in 1993 when over 700 persons in four U.S. states were ill with 4 deaths in children after they ate hamburgers served in a fast-food chain. In subsequent years, the majority of outbreaks were associated with hamburgers, but some were from other meats, such as steak pies and roast beef. Factors contributing to outbreaks identified during the outbreak investigations were mainly undercooking, improper or inadequate cooling, infected food preparer, eating raw ground beef, and contaminated equipment and kitchen surfaces. The most serious outbreak of *E. coli* O157:H7 occurred in Scotland in November and December 1996, with over 500 cases and 20 deaths. Those ill had eaten cold cooked meats, meat sandwiches, and cooked steak in gravy prepared by the same butcher. As a result of this episode, numerous changes in food hygiene were initiated in the United Kingdom.

Ready-to-eat meat has also been a vehicle transmitting the pathogen, including fermented sausages in the United States, Canada, and Australia (salami and mettwurst). In 1995, salami contaminated with *S.* Typhimurium affected 83 persons in Italy; the main risk factor was a reduced ripening time of the salami because of a strong market demand for the product. In all these products, the pathogens survived any acidity developed during the fermentation processes. Eleven persons were ill after eating home-made jerky prepared from a deer that had been shot The hunters were not hygienic in storing or dressing the carcass and preserving the meat. *Escherichia coli* O157:H7 was found in the patients, jerky, uncooked deer meat, a saw used on the carcass, deer hide, and deer fecal pellets.

In 1990 in Western Australia, a cluster of 9 *L. monocytogenes* cases, including 6 stillbirths, was linked to a brand of pate. In 1998, 51 cases were reported in Australia in 1998, with 29 in New South Wales. As a result of these incidents, advice was given in 1999 for pregnant women to avoid soft cheeses, raw or smoked seafood, cold meats, and salad bars. In December 1991, goat meat had been home cooked and frozen in California and transported to Scarborough, Ontario, where some of the meat was defrosted, reheated in a microwave oven, and eaten. One woman suffering from lupus and receiving steroid treatment was infected. The same strain was isolated from the case and the meat. It is possible the meat thawed in transit, allowing *L. monocytogenes* to grow, and the meat was incompletely cooked in the microwave. Most recently, in late 1998, at least 100 people contracted listeriosis after eating hot dogs and other processed meat cold cuts in several U.S. states; 20 people died.

In an institution in England, 12 joints of lamb were thawed and roasted for 5 h and stacked on a single tray to cool before refrigeration. The next day, some of the meat was removed from the refrigerator, sliced, and left at room temperature for several hours before being served. This was repeated the next day with the rest of the lamb; 12 staff and residents were ill with cramps and diarrhea after an incubation period of 9–16 h. *Clostridium perfringens* was found in the stools of the cases and in the leftover lamb ($>10^5$ g^{-1}). Improper cooling of the joints and storage periods at ambient temperatures allowed growth of the pathogen. In 1998, 64 people in New Zealand were ill with *C. perfringens* food poisoning after eating a traditional Maori roast pork meal. The pig was roasted without a meat thermometer and cooled for 90 min before serving, allowing any spores of *C. perfringens* to germinate and grow rapidly.

A caterer with an SRSV infection transmitted it to 67 people who had attended an international acquired immunodeficiency syndrome (AIDS) conference in Wales. The food handler, who had boned the cooked chicken with bare hands, had been ill 2 days earlier.

1.4. FISH AND SHELLFISH

Cholera was first documented in Peru in 1991 and rapidly spread to other Latin American countries with millions of cases. *Vibrio cholerae* O1 was found in water, plankton, sewage, and seafood. Street-vended foods, crabs, shrimp, ceviche, and beverages containing ice were all vehicles of transmission. The economic loss from lost markets and tourism, absence from work, and medical care and the value of lives lost were high. In May 1998, 34 cholera cases in Hong Kong occurred after they had been to restaurants where hotpot (raw meat and seafood dipped into a communal boiling broth and eaten with a raw egg and sauces) was served. Live seafood from polluted water was probably the source of cholera, as was found in a 1994 outbreak when tank water for fish and lobsters contained *V. cholerae*. Mozambique has yearly catch of 20,000 tonnes, more than 50% exported to the European Union (EU). Cholera in Mozambique starting in August 1997 caused more than 30,000 cases and 780 deaths, and as a result the EU banned import of Mozambican fish, costing the country U.S. $240 million. In July–August, 1997, 209 persons were infected with *Vibrio parahaemolyticus* on the west coast from California to British Columbia after eating oysters and other shellfish. The water temperature was 1–5°C above normal. The last outbreak in the United States was in 1982, when 10 cases were confirmed; there never has been a *Vibrio* outbreak in Canada linked to food. The next year the problem occurred again. In July 1998, over 400 were ill in five states from *V. parahaemolyticus*-contaminated oysters harvested in Galveston Bay, Texas. The pathogen was identical to ones found in Southeast Asia and probably came from dumped ballast water. Texas had a heat wave with no water running into the bay, and any discharge would stay in the same area, allowing oysters to be contaminated. Sales were stopped if counts exceeded 10,000 g^{-1}, but this did not contain the outbreak, which continued in the United States until December. The same strain caused illnesses in September in Oyster Bay, Long Island, New York, affecting an oyster festival. The warmer water and less oxygen from fertilizer runoff may have allowed vibrios to multiply. At an army base in Peru, 279 cases of diarrhea caused by *V. parahaemolyticus* in recruits were traced to common meals served at a dining facility but no specific vehicle was identified. This *Vibrio* species is one of most frequently isolated enteric pathogens in Latin America. In 1994 in Sweden, the summer was unusually hot and water temperatures reached 17°C. That year *Vibrio vulnificus* was first observed infecting a bather. Since then, some summers have been just as warm and other cases have been reported, and one died. *Vibrio vulnificus* septicemias occur each year as a result of consuming oysters harvested in the Gulf of Mexico during the summer months when the water is warm enough to allow rapid growth of the pathogen. Illnesses and death have also been documented in Singapore, and probably all populations in tropical areas eating undercooked or raw seafood are at risk of infections from this and other vibrios. Dozens of people in Japan were infected with *E. coli* O157:H7 when they ate salmon roe. The owner knowingly sold 3.6 tons of contaminated roe to a food company that distributed it to many sushi shops. The company was liquidated, and the owner was arrested on criminal charges and faces up to 5 years in jail or a fine.

Listeriosis is occasionally linked to seafood, as in New Zealand in 1992, when two women contracted the disease, which resulted in the death of newborn twins after smoked mussels were eaten. The contamination was traced to the processor, and the product had been widely distributed; other cases were probably ill in England. In Sweden, nine persons suffered from listeriosis from June 1994 to June 1995, after they had eaten fermented or smoked rainbow trout produced by one manufacturer. In Canada in 1996, six cases were linked to imitation crab meat. Stored uneviscerated fish in various countries has led to botulism outbreaks, with one of the biggest in Egypt in 1991 resulting in at least 20 deaths from ingestion of locally made *faseikh*. Botulism is also well documented from consumption of fermented marine mammal parts in Inuit communities in Alaska and northern Canada and fermented salmon eggs in northwest coast Indian communities. Parasites have been found in shellfish, such as *Cryptosporidium* in oysters in Chesapeake Bay near river tributaries where there may be sewage discharge or farm runoff. No cryptosporidiosis cases have been linked to the ingestion of raw shellfish, but the potential exists.

Because for most countries much of the seafood eaten is imported (for the United States it is more than 50%), there are risks of contracting illnesses from pathogens in the originating country, like *L. monocytogenes* in the New Zealand mussels episode or from *Salmonella* derived from aquaculture raising practices in developing countries. Therefore, a vigorous inspection program by importing countries should be in place. Preferably this should be based on the hazard analysis critical control point (HACCP) program rather than random samples and microbiological testing.

Viruses are only beginning to be recognized as foodborne pathogens. In England and Wales, a study of SRSV infections from 1990 to 1995 showed that there were 707 outbreaks, mostly in the winter. There were 14,000 cases suspected to occur in hospital geriatric wards and homes for the aged as well as 17 SRSV-associated deaths. Foodborne transmission was reported in 97 of the 707 outbreaks (14%). Infected food workers were suspected in 30 of these outbreaks. In 37 outbreaks, implicated foods included oysters and other seafood, fresh salads, sandwiches, fruit, vegetables, cakes, and desserts; 96 outbreaks were associated with commercial food outlets (mixed person-to-person and foodborne transmission).

Seafood toxins are found locally in mollusks and fish throughout the world, and illnesses frequently occur but are not always documented. Most of these originate from naturally toxic plankton that are ingested by shellfish and include paralytic shellfish poison (saxitoxins and related toxins), diarrhetic shellfish poison (okadaic acid and dinophysistoxins), and amnesic shellfish poison (domoic acid). The first two affect fisheries in the Americas, Europe, Asia, and many tropical countries. Once illnesses occur or levels of toxins reach a certain limit, harvesting stops until the toxin level in the shellfish is below the limit or the toxic algae are no longer abundant. Domoic acid is found in shellfish worldwide but has only caused one outbreak. This was in Canada in 1987, when contaminated mussels affected at least 107 persons, several with permanent memory damage, and caused three deaths. Ciguatoxin is a seafood toxin present in many varieties of tropical fish, but currently there is no routine diagnostic test to detect it. Although its effects are mainly in regions where these fish are caught, consumption of exported fish in temperate regions has resulted in several outbreaks. For instance, in Texas, 17 crew members of a Norwegian ship suffered from ciguatera poisoning after eating a barracuda caught off the Bahamas, and medical help had to be sought. There is some evidence that toxic plankton are spreading throughout the world as a result of climate change affecting water temperatures and dumping of ballast water in harbors or increased aquaculture, and this may recently have happened in the Hong Kong area when the causative organism of ciguatera poisoning, *Gambierdiscus toxicus*, was found there for the first time in 1998. That same year, at least 71 suspected ciguatera cases occurred after they ate tiger grouper; most were hospitalized briefly. Authorities warned people not to eat fish larger than 1.5 kg.

Toxins are increasingly being found in new areas and new toxins are discovered. In New Zealand in 1993, illnesses from neurotoxic shellfish poisoning (NSP) and paralytic shellfish poisoning (PSP), previously unrecorded, were documented because of water temperature changes caused by the El Niño Southern Oscillation (ENSO) effect. In 1997, the first incident of diarrhetic shellfish poisoning (DSP) recorded in the United Kingdom with 49 cases may also be caused by water temperature changes. In Madagascar one incident of apparent ciguatera poisoning affected over 500 people who ate a shark in 1993 and caused 98 deaths. In fact, two heat-stable liposoluble toxins, carchatoxin A and B, were isolated from the shark. In Australia, 20 cases of mild foodborne disease were associated with consumption of small cockles from New South Wales beaches hundreds of miles apart. An unknown algal toxin was suspected, and a ban was placed on harvesting. Haff disease was first reported by people living by the Baltic Sea and associated with ingestion of burbot in the 1930s. The same disease affected people in the United States in 1984–1986 after eating buffalo fish. In 1997, six cases developed muscle aches, rigidness or stiffness, and weakness. One patient had muscle weakness for 6 months. A heat-stable toxin was suspected. Scombroid or histamine poisoning arises in certain types of fish (such as tuna, mackerel, bonito, and marlin) prone to spoilage where histidine is converted to histamine by bacterial action. A scombroid outbreak with 15 cases occurred in Spain from a fresh tuna sold at a hypermarket in June 1994. Typical symptoms of facial flushing, headache, diarrhea, nausea, and abdominal pain occurred 45 min after the tuna was eaten. Histamine was found in leftover tuna samples and in urine specimens from the cases.

1.5. EGGS AND DAIRY PRODUCTS

Salmonella in eggs in recent years has been a major problem for public health agencies. *Salmonella* Enteritidis (SE), particularly phage type (PT) 4, infects egg-laying poultry flocks and some of the eggs contain the organism. The risk of illness is more when the egg is used as an ingredient in a food that is eaten by many people rather than when it is a single egg. Many outbreaks have been reported around the world and have been associated with foods such as omelettes, quiche, meringues, desserts, and cakes with egg ingredients, egg nog, and ice cream. For instance, five persons who were suffering from SE PT 4 infection in England had attended the same gym and had consumed a protein-based beverage composed of milk powder and a raw egg as a body-building drink. The value of the drink to these persons outweighed any risk of enteric illness. In one of the largest SE outbreaks, there were an estimated 224,000 cases of SE in 1994 in several U.S. states following consumption of ice cream. The premix had been contaminated during transport in tankers that had previously been used for carrying raw liquid egg. It was determined that, for some cases, the infectious dose was no more than 28 cells. A quantitative risk assessment (QRA) for SE in eggs showed that of 46.8 billion eggs produced in the United States, 2.3 million were contaminated (1 in 20,000), and of the 10.2 million egg servings after cooking, 27% remained contaminated. Based on storage, consumption, and dose–response data, there are an estimated 661,663 SE illnesses each year from eggs, of which 621,000 are mildly ill, 36,000 visit a physician, 3300 are hospitalized, and 390 die. In Argentina, between 1986 and April 1990, 35 outbreaks of SE affected 3500 persons, largely through consumption of insufficiently cooked poultry and eggs used in mayonnaise. An example of this occurred in Brazil, where, in 1993, 280 of 400 patrons of a restaurant suffered from SE infections after eating potato salad with a raw egg dressing; 10^6 *S. aureus*/g was also found in the salad, which probably contributed to the discomfort of those ill. Even as late as 1998, the use of raw eggs was still leading to outbreaks in the United Kingdom after over a decade of public awareness; an English family was ill after a member prepared a lemon soufflé at a cookery course in a college. It was only after the outbreak that the college agreed not to use raw eggs as ingredients in prepared foods. However, since then there has been a decrease in cases attributable to vaccination of flocks, consumer awareness of the risks of undercooking, and a date stamp on each egg to help limit the home storage period.

Outbreaks from untreated milk have been well documented, with *Salmonella, E. coli* O157, and *Campylobacter* being the most frequent pathogens. A large *Campylobacter* outbreak of 72 cases occurred in 1992 at an outdoor festival at an English farm at which local unpasteurized milk was sold. Occasionally, pasteurized milk has been implicated. In 1985, more than 200,000 persons were ill from milk produced in one large Illinois dairy. The cause was never discovered, but it assumed to be a cross-connection fault that allowed contamination of the pasteurized milk. The dairy closed permanently, and there were high settlement costs. In 1994, 17 were ill after drinking a brand of pasteurized milk in Montana. The pathogen was a non-O157 vero cytotoxin-producing *E. coli* (VTEC) (*E. coli* O104). Coliform counts exceeded 10 per 100 mL of milk, and fecal coliforms were isolated from postpasteurization pipes and surfaces. A series of outbreaks have occurred since 1985, with the largest in 1997 with 100 persons affected. Contact with farm animals was the most important risk factor. Advice to farmers was to avoid visits, especially by children, not offer raw milk, and fence in cattle to reduce human contact (e.g., at beaches). *Escherichia coli* O157 has been implicated in several dairy product outbreaks, including Caerphilly-type cheese, cheese curds, and yogurt if raw milk was used in their manufacture. Risk factors for this pathogen include direct or indirect contact with farm animals or animal manure. Farming families may develop immunity to the organism. One new pathogen of concern is *Streptococcus zooepidemicus*, which causes acute pharyngitis, glomerulonephritis, meningitis, septicemia, and death. There have been outbreaks in Romania, Austria, England, and Australia from raw milk consumption. In one of these, delivered milk, meant to be pasteurized, was substituted with bottled raw milk because of supply shortages. In New Mexico in 1983, there were 16 cases and 2 deaths arising from consumption of cheese made from unpasteurized milk. In a study of 15 cases worldwide, 40% of these died. In another type of SE problem, 728 cases of phage type 8 (PT8) infection were confirmed in Canada in the spring of 1998. Most of these ate a school snack product or were secondary cases, with a median age of 9 years, and 14% were hospitalized. Shredded mild cheddar cheese from one company was found to be the contaminated vehicle, but no mishandling or processing error was found. Butter is an unusual vehicle for foodborne illness, but from December 1998 to February 1999, 18 people in Finland contracted listeriosis after eating locally produced butter. These were very vulnerable hospitalized organ transplant patients. Sixteen of these had septicemias and 4 died. The same PFGE strain of *L. monocytogenes* serotype 3 was isolated from the cases, the butter, and the producing dairy plant.

Although botulism from dairy products is rare, some serious outbreaks have been documented. In 1989, the largest outbreak in the United Kingdom took place after commercially prepared yogurt containing hazelnut puree contaminated with *C. botulinum* toxin was consumed. In Italy, in 1996, eight persons were infected, mostly children, and one death occurred after *tiramisu* made with mascarpone cheese was eaten.

1.6. VEGETABLES AND FRUITS

1.6.1 Produce

Fresh fruit and vegetable consumption has risen 50% in the United States since 1970. Premixed salad sales have risen from $83 million in 1989 to $1 billion in 1996. Processors are now venturing into new areas, sometimes with little understanding of the risks of transmitting pathogens and the requirements for producing safe food. Many products are certified as organic and chemical free, but there are no control measures regarding the presence of manure or cattle feces. The following outbreaks illustrate the extent of the problem. In Montana, in 1995, 40 were ill after eating lettuce contaminated with *E. coli* O157:H7. The source could have been cattle manure from a local dairy, irrigation water for the lettuce fields, or other animal reservoirs. In another lettuce outbreak, in 1996, 61 persons in three states were infected with *E. coli* O157:H7; 21 were hospitalized, and 3 children were severely affected. At the source California farm, the incriminated lettuce (*mesclun*) had been washed and packed less 100 ft from a cattle pen, and there were no measures to prevent contamination. Particles of cattle feces could have been blown by wind, washed in by rain, tracked in by boots, or carried by birds into the lettuce preparation area. The lettuce was carted in dirt-encrusted boxes, the wash tank swished around by workers hands, some without gloves, and there were no sanitizing facilities or chlorine used in the wash water, the source of which was 20 ft from the cattle pen. Irrigation water for the lettuce was from a cattle pasture. Environmental sources are not the only ones; food handlers are also responsible. For instance, at a community fund-raising effort in Nova Scotia, 400 locally made salad plates were sold and distributed. Thirty-nine persons eating these were infected with *E. coli* O157:H7, which was also found in potato salad. One elderly woman died and 4 children were hospitalized. The pathogen was prevalent in the community before the outbreak, and 3 of 50 persons preparing the food had gastrointestinal symptoms, one with bloody diarrhea before the event.

Radish sprouts had been implicated in a large outbreak in Japan in 1997 when nearly 8000 persons (mainly school children) were infected and 106 had hemolytic uremic syndrome (HUS). Applying a strong disinfectant to radish sprouts for 10 min did not destroy *E. coli* O157; it exists deep in plant tissues where the disinfectant cannot reach. Salmonellosis has a history of contaminated produce resulting in outbreaks. For instance, in the United States, in 1990 and 1991 cantaloupes were associated with 645 cases, in 1990 and 1993 tomatoes were associated with 258 cases, and in 1995 alfalfa sprouts were associated with 242 cases. Since 1995, health officials have attributed 13 foodborne disease outbreaks worldwide to sprouts, with 10 of these in the United States affecting at least 956 people and causing one death. The latest outbreak occurred in the spring of 1999 in Oregon and Washington from alfalfa sprouts contaminated with *Salmonella* Mbandaka. Alfalfa or bean sprouts have also been involved with outbreaks in several other countries, including Canada, the United Kingdom, and Scandinavia.

There were six outbreaks of shigellosis associated with salad bars in Denmark in 1998 with cases from all over the country. Baby maize imported from Thailand was the most likely food to be contaminated. The maize was uncooked and packaged in 125-g bags and distributed throughout the country. Also, in 1998, outbreaks with hundreds of cases involving parsley occurred in four states and Ontario. In five of the outbreaks, *Shigella sonnei* was the pathogen and enterotoxigenic *E. coli* (ETEC) in one other. The distributor for the parsley was in California and was receiving the product from a Mexican company. The same company had also sent lettuce and celery in the past. Potato salad contaminated with ETEC served at a large deli and catering facility in Illinois infected 6500 of 20,000 people who ate there on June 6–7, 1998. Inspection showed there were many errors in preparation, but once staff were trained in food safety, the operation opened up again in August. In 1993, airline passengers from North Carolina to Rhode Island ate a garden salad and guests at a mountain lodge in New Hampshire ate *tabbouleh*, with carrots being the only common element; they could not, however, be traced to a single source. In 1994, there were up to 645 cases of ETEC among 1240 attendees following a banquet in Milwaukee. Pan-fried potatoes were the only food associated with the illness. These were cooked with cheese and spices and left at room temperature for long periods of time. It is uncertain what the source of the ETEC was, but the contamination had to be extensive, and considerable growth would probably have to occur to allow so many people to become infected since volunteers require 10^6–10^{10} cells to induce diarrhea. ETEC has not been considered a frequent foodborne pathogen of temperate countries, but at least in North America there appears to be evidence for it being in the environment in recent years. Viruses and parasites have also been associated with outbreaks involving produce. In January 1996, an investigation showed that hepatitis A occurred in two workers in a Quebec restaurant kitchen and five customers. The workers had been serving hot foods and salads. Immunoglobulin was given to workers and contacts. In December 1997, 51 persons were infected with *Cryptosporidium* after eating at a dinner banquet in Spokane. The cases were epidemiogically linked to uncooked green onions. These onions had not been washed. In the same year, basil in salads and *mesclun* lettuce served on a cruise ship infected 308 and 224 cases with *Cyclospora*, respectively.

In 1996, 1400 *Cyclospora* infections in 15 U.S. states and Ontario resulted from consumption of fresh Guatemalan raspberries. Similar outbreaks in 1997 in over 500 persons were linked to fresh raspberries also from Guatemala. In 81% of the 1996 outbreaks and most of the 1997 outbreaks, the raspberries had been rinsed before they were eaten. In 1998, imports of Guatemalan raspberries were banned in the United States but not in Canada, and in Ontario in May 1998, 13 clusters of 192 *Cyclospora* cases occurred in southern Ontario. Control measures on farms either were not effective or they were not directed against the true source. The Guatemalan industry had to spend funds to use chlorinated water irrigation systems, develop HACCP plans, and improve worker hygiene, but the market for North America is tenuous at best. These episodes leave us with questions about *Cyclospora*. Why does it cause illness seasonally (spring and not fall, but there is no real season difference in Guatemala?). Where does it survive between seasons? How does it contaminate the raspberries? What are the vectors? Are there reservoir hosts? Since rinsing does not appear to be enough to remove the parasite, are there any means of inactivating it? Has *Cyclospora* now become endemic in the United States and Canada? Are there sporadic cases not associated with imported berries? *Cyclospora* is not the only agent causing illness from berries. In 1996, there were 175 cases of hepatitis A arising from frozen strawberries that originated in Mexico and were distributed in California.

Fresh produce, therefore, is becoming increasingly a vehicle for transmitting enteric diseases of many different types, and because there is no terminal heat step, consumers are vulnerable even to low doses of pathogens. Control strategies are limited because there are many stages in the production where pathogens can enter the food chain, and for imported products, there is little knowledge of the agricultural practices involved in their production.

1.6.2. Processed Products

Processed vegetables and fruits have also been implicated in foodborne disease. Botulism has been documented from several types of vegetable products. For example, in Japan, in July 1998, five persons were ill after eating bottled green olives containing type B toxin, with another six ill from the same product in August. In Italy, in August 1998, a woman drank a soup bottled locally and suffered from type A botulism. In the United Kingdom, in April 1998, two were ill and one died after eating home-bottled mushrooms from Italy containing type B toxin. Botulism from home-bottled low-acid vegetable products has been documented for decades, but it still occurs as illustrated above. Less typical is botulism from baked potatoes. In El Paso, Texas, in 1994, 30 suffered from botulism after eating a potato dip or egg plant dip at a Greek restaurant. The potatoes had been baked in oven and left at room temperature, still in the foil, for 18 h before being used in the dip. Foil allowed anaerobic conditions for surviving *C. botulinum* spores on the skin to germinate and grow into the potato. Shared utensils spread the organism to the eggplant dip. There have been five other botulism outbreaks from baked or boiled potatoes in United States since 1978. These are now considered a potentially hazardous food. In a more unusual situation, 13 members of a Native American church ingested peyote from a communal jar during a religious ceremony, and 3 of these later developed botulism. The peyote was made from a tea of the dried alkaline-ground peyote cactus that had been covered with water and stored for 2 months under refrigeration in a closed jar. The anaerobic and alkaline environment favored outgrowth of *C. botulinum* spores on the cactus. Botulism can also be caused by *Clostridium butyricum*. In 1996, 34 children in India had symptoms of botulism and 3 died. Local *sevu*, a crisp made from gram flour contained *C. butyricum* capable of producing toxin as determined by PCR studies. The *sevu* crisps had been improperly stored. Another outbreak was caused by *C. butyricum* in China in 1996.

In Australia in 1996, 57 cases of *S.* Mbandaka in three states were associated with consumption of peanut butter found to be contaminated. Settlements in 1998 ranged from $500 to $50,000, totaling $7 million for the most seriously affected. An outbreak of salmonellosis caused by *S.* Agona affected 211 persons with 47 hospitalized in 11 states from April to May 1998. Most cases were less than 10 or over 70 years old. Toasted oat cereal was the implicated food. In August, after the recall, 2000 bags were stolen from a trash bin at the company warehouse with a risk of illness to any consumers. A listeriosis outbreak in 1997 with 1594 cases arose from consumption of salad with canned corn served in Italian schools by a mass catering establishment. The same strain of *L. monocytogenes* (group 4) and deoxyribonucleic acid (DNA) profiles were found in 123 hospitalized patients, left-over corn, and floor and sink drains of the catering establishment. The pathogen grew in corn at 25°C after 6 h.

Fruit products are also being increasingly implicated in outbreaks involving different pathogens. In 1995, 62 cases of salmonellosis were associated with unpasteurized orange juice. Four *Salmonella* serovars were found in juice samples, unwashed fruit surfaces, and amphibians beside the processing facility. A case–control study showed that 22 persons were ill from May to June 1998 after eating *Salmonella* Oranienberg–contaminated cantaloupes imported into Ontario from the United States, Mexico, and Central America. Advice to consumers was to thoroughly clean melons with potable water before cutting, prepare cut melons using clean and sanitized utensils and surfaces, and hold cut melons at 7°C or colder until served or sold.

In 1993, consumers of apple juice in Massachusetts contracted cryptosporidiosis seemingly from cysts in cattle manure contaminating the apples. Apple juice outbreaks have also been documented in Canada and the United States in 1980, 1996, and 1998 where *E. coli* O157 was the confirmed or suspected agent. In the largest of these, with over 61 cases and 1 death, a widely marketed commercial product was involved. A batch of rotten apples had been included in the lot pressed on the day the juice was contaminated, despite the inspector's advice to reject it. The impact on the company was high, with a fine of $1.5 million, payments in millions of dollars to the families of the ill people, and over a $1 million in business losses. Several parents of the ill children became lobbyists for stricter food safety.

1.7. ENVIRONMENT

Pathogens can also be present in many parts of the environment that can act as a reservoir or assist in their transmission. The following limited examples show the wide variety of environmental sources that can lead to contamination. Water has long been known as a source of enteric organisms through animal or fecal contamination. The sixteenth-century Villa d'Este in Italy, with fountains, ponds, and tumbling cascades, has been a tourist attraction for centuries, but since the water source became polluted and *Salmonella* was present in all the water displays, there was a danger of aerial contamination to the visitors. These were cordoned off, with signs saying not to come in contact with the water. In one Utah community, half the town's water supply was affected because of high *E. coli* counts in October 1998. The residents had to drink boiled water or bottled water. When the 3-million-gallon reservoir was drained, a large dead raccoon was found on the bottom. In 1998 in Georgia, 24 children were infected with *E. coli* O157:H7 after being in a recreational water park. The source was probably one infected child who excreted feces accidentally into the water and the others ingested the pathogen during their play. In a Colorado zoo, 65 persons were infected with *S.* Enteritidis in 1996 after seeing a Komodo dragon exhibit. No case touched the dragons, but 83% touched the wooden barrier surrounding the dragon's pen. *Salmonella* was found in feces and on barriers. The infected dragon stood in fecally contaminated mulch and frequently placed its paws on top of these barriers. Those most likely to be ill did not wash their hands after touching the barrier and probably put fingers into their mouths or contaminated food that was eaten. Transmission from reptiles to humans is well known, but this is the first outbreak from a zoo. Birds too can be sources of pathogens. In April 1998, redpolls were dying from *S.* Typhimurium infection in the eastern part of North America. There were risks to people handling sick or dead birds and contaminated bird feed at feeders and through cats becoming infected. Goose droppings may contain *Giardia* and *Cryptosporidium* parasites. As the goose population continues to grow in urban areas, is this a risk to those who frequent parks, golf courses, soccer fields, and playgrounds? Control of burgeoning populations of geese may be controlled by a vaccine that prevents fertilization of the ova. There is evidence in the United Kingdom that plant and fruit crops may be contaminated with *E. coli* O157 through farmers spreading fields with untreated blood and guts from abattoir waste. The animals can be infected by eating the grass. However, is this any more of a problem than *E. coli* being deposited through animal dung or manure? It has also been shown that *E. coli* O157 may be spread by insects. Fruit flies infected with a strain of *E. coli* were released into a chamber with bruised apples, and the bruised areas were colonized by the *E. coli* within 48 h.

There are new pathogen risks in the oceans. *Pfeisteria* dinoflagellate killed millions of fish in North Carolina in the last few years, and apparently some fishers were adversely affected. Human viruses are transmitted from Florida's 1.6 million septic tanks into coastal waters. People are infected from swimming, windsurfing, or boating in these waters, as well as eating shellfish. New York waters have 40% of shellfish contaminated with human viruses. Waterborne viruses caused deaths and five hospitalizations in 1995 on the Mississippi coast of the Gulf of Mexico. More than one-third of water samples from Waikiki Beach, Hawaii, contained human viruses. The same scenario is taking place in a multitude of locations around the world. In addition, viruses are now suspected to be linked to heart disease and diabetes as well as meningitis and hepatitis. In Bolivia, the population of La Paz is exposed to the sewage-contaminated La Paz River, where water samples included ETEC, enteropathogenic *E. coli* (EPEC), enterinvasive *E. coli* (EIEC), and *Salmonella*. Since these were also found in children with diarrhea in the city, such a water source can cause infections through direct consumption or through food in contact with the water. Climate change is an environmental factor that we are just beginning to see as being significant. In 1997, there were 150,000 cases of cholera worldwide, with 80% in Africa. Heavy rains in east Africa and South and Central America contributed to these. An increase in foodborne disease may also be connected with higher temperatures. An argument has been made that warmer summers in England may contribute to disease. The monthly incidence of foodborne disease was found to be significantly associated with the temperature of the same and previous months. Thus, animals prior to slaughter may be more infected, and the risks of microbial contamination and growth may increase in the derived meat products with higher temperatures. It was predicted that an additional 179,000 cases may occur by 2050 as a result of climate change.

1.8. SURVEYS OF THE PUBLIC ON KNOWLEDGE OF FOODBORNE DISEASE AND ITS PREVENTION

Consumer practices in food safety and hygiene have not been effective in reducing outbreaks. Specific studies are starting to show where education may be more efficiently directed. Surveys in North America shows that only <1% of participants followed effective sanitation practices in the kitchen during meal preparation, such as practicing hand washing, time–temperature control, and avoiding cross-contamination. Some even think it is possible to avoid a potentially dangerous food by smelling or looking at it. Those who are educated, have higher incomes, and are men are more likely to eat pink hamburger and uncooked oysters than others. In a multistate survey on food safety in 1995–1996, it was found that 50% ate undercooked eggs, 24% home-canned vegetables, 20% pink hamburgers, 8% raw oysters, and 1.4% raw milk. Nineteen percent did not wash their hands or cutting boards used for preparing raw meats and chicken. Forty-five percent did not see any labels on raw meat products, and of those that did 77% read the label and 37% changed their preparation habits. In a food safety survey of students, 1.8% indicated that they had a foodborne disease attributed by a physician to a specific pathogen, mostly *Salmonella*. Female students, graduate students, and those with a course that included some food safety (dietetics, food science, nutrition, and health programs) had higher scores than other students, particularly freshmen.

1.9. HIGH-RISK POPULATIONS

Susceptible persons include children under the age of 5, pregnant women, the elderly, persons who have an impaired immune system [e.g., people taking immunosuppressive drugs, undergoing cancer therapy or organ transplants, or infected with the human immunodeficiency virus (HIV), which causes AIDS]. These should avoid raw foods of animal origin such as undercooked poultry, eggs, and meats, unpasteurized milk or cheese or yogurt made with raw milk, dishes prepared with raw or undercooked eggs, raw or undercooked mollusks and crustacea, any cooked food that has been cross-contaminated by raw food after cooking or as a result of poor personal hygiene, and any food stored above the recommended safe temperature for storage. Other populations at risk are those in refugee camps and camps for alien workers and those with specific customs that compromise food safety, such as eating feasts at funerals even during cholera epidemics, where people share plates, eat with their hands, and have poor standards of hygiene. Some high-risk populations are unpredictable, such as those exposed to bioterroism, where enteric pathogens may be put into food in random locations. Travelers to countries where food hygiene is not understood or practiced well are exposed to a higher risk of infection than at home, even if the location they are staying in seems to be high class.

1.10. POLICIES TO REDUCE FOODBORNE DISEASE

Policies around the world on how to reduce foodborne disease vary. Most traditional regulations may have specific bacterial counts that cannot be exceeded, and a few may ban products if pathogens like *L. monocytogenes* are in ready-to-eat foods or *E. coli* O157:H7 is in ground meat (zero tolerance). The emphasis today is more on preventative measures such as HACCP, either voluntary or mandatory in meat, poultry, and/or seafood operations. Warning labels for consumers are now being put on packages of raw meat and poultry (e.g., to cook well and avoid cross-contamination). For juices, these may be pasteurized or a warning label may indicate the risks of drinking unpasteurized product. Because of the concern over high numbers of foodborne illness relating to products prepared in the home from ingredient purchased in retail stores, a few supermarket chains are promoting food safety by educating consumers to keep food safe at home and working with government to prevent contaminated food from entering the distribution system. In Sweden, great emphasis is put upon eliminating the source of the pathogen (e.g., depopulating flocks with *Salmonella*, testing animals and feed, sanitary slaughter, hygienic practices, and import control). These types of controls come with a substantial cost factor. Governments are recognizing there has to be improved surveillance, and new approaches to this are being undertaken in different countries, such as measuring the burden of infectious intestinal disease or establishing sentinel sites for case–control and other epidemiological and laboratory studies, including DNA fingerprinting of strains. In the United Kingdom, the government is considering making farmers directly liable for compensation for any illness caused by the food they produce. This means that a victim of foodborne disease could sue a farmer if the source of the infection could be traced to a particular foodstuff and farm, and a plaintiff would be required merely to establish a causal link but not prove any negligence on the part of the farmer. The principle of strict liability has applied to manufacturers but up till now not to primary producers. This raises a number of questions, such who would be responsible for illnesses arising from bulked food or where raw products of animal origin were mishandled in a home or a restaurant.

The Council for Agricultural Science and Technology has issued two publications on foodborne pathogens. The first has been widely quoted and contributed to U.S. policy. The second, published in 1998, gives specific recommendations in goal setting, research needs, production control, and education. Some of these are listed below and are worth considering by all countries.

Goal Setting

1. Base food safety policy on risk assessment and include risk management and risk communication strategies.
2. Base food safety regulations on risk assessment and risk management.
3. Set federal food safety goals and priorities. Criteria include number of acute illnesses; number of chronic complications; number of deaths and disabilities; type of food products implicated; types of production, harvesting, or processing deficiencies or handling errors identified; impact on high-risk populations; and economic losses to society.

Research Recommendations

1. Improve reporting of foodborne disease by pathogen, by food, and by contributory factors.
2. Expand existing database on food animals, foods, and pathogens.
3. Conduct epidemiologic studies to establish the cause of illness.
4. Improve and regularly update foodborne disease estimates and their costs.
5. Find mechanisms of chronic illnesses and populations at elevated risk from chronic disease associated with foodborne pathogens.
6. Develop rapid, accurate detection methods for pathogen detection in foods.
7. Use dose–response modeling in the risk assessment process.
8. Identify food and pathogen/toxin associations in order to establish controls to minimize the risks.
9. Support pathogen research to understand more about the agents causing foodborne disease (e.g., biofilms, virulence factors, factors contributing to contamination, survival, and growth).

Production Control

1. Require producers to adopt effective preharvest intervention strategies to enhance public health, including foodborne pathogen control practices from food source to consumption.
2. Harmonize international food safety standards.

Education

1. Educate the general public and food handlers for safe food preparation and handling and especially for high-risk populations.
2. Use and evaluate food labeling to communicate safe food preparation.
3. Provide risk information on food choices to susceptible persons.

1.11. CONCLUSION

Foodborne disease is an increasing concern in all countries. Because of lack of foresight to invest in surveillance and research in infectious diseases over decades, we have to react to problems today rather than anticipate them. For instance, we have limited knowledge on virulence factors and their transfer between organisms, such as verotoxin production from *E. coli* to *Citrobacter* and enteroinvasive properties from *E. coli* to *Klebsiella*. Antimicrobial resistance to antibiotics is preventing adequate means of eliminating enteric pathogens from the gut. Therefore, we are seeing new varieties of pathogens, some of which become important, such as *E. coli* O157:H7, and others, such as *Aeromonas*, we are still not certain about for the normal population but can infect the immunocompromised person. Governments need to collaborate with limited resources within each country and within blocks of countries and work with interested stakeholders to develop meaningful policies. Countries need to take recommendations and research conclusions from other countries and adapt these to their own situation, through scientific experts with appropriate resources to produce the relevant policies, risk assessments, and methods to reduce foodborne disease. At present, no country can claim that the battle against the foodborne pathogen is won. Can existing long-term strategies of production control (on farm and HACCP), import inspection, trade agreements, and consumer education substantially reduce foodborne disease or will new problems continually arise to keep the numbers up? If governments are not committed to put resources into new ways to reduce or at least stabilize the impact of enteric pathogens, the population will be continually exposed to new and old hazards. Will the end of the twentieth century be looked upon as a time when foodborne disease lurks in unexpected places, and will it become in unpredictable ways a threat to the lives of our children and the increasingly aging population?

BIBLIOGRAPHY

Alterkruse, S. F., and D. L. Swerdlow. 1996. The changing epidemiology of foodborne disease. *Am. J. Med. Sci.* **311**, 23–29.

Bryan, F. L., J. J. Guzewich, and E. C. D. Todd. 1997a. Surveillance of foodborne disease. Part II. Summary and presentation of descriptive data and epidemiologic patterns; their value and limitations. *J. Food Prot.* **60**, 567–578.

Bryan, F. L., J. J. Guzewich, and E. C. D. Todd. 1997b. Surveillance of foodborne disease. Part III. Summary and presentation of data on vehicles and contributory factors. *J. Food Prot.* **60**, 701–714.

Cassin, M. H., A. M. Lammerding, E. C. D. Todd, W. Ross, and R. S. McColl. 1998. Quantitative risk assessment for *Escherichia coli* O157:H7 in ground beef hamburgers. *Int. J. Food Microbiol.* **41**, 21–44.

Faruque, S. M., M. J. Albert, and J. J. Mekalanos. 1998. Epidemiology, genetics, and ecology of toxigenic *Vibrio cholerae. Microbiol. Mol. Biol. Rev.* **62**, 1301–1314.

Foegeding, P., and T. Roberts. 1994. *Foodborne Pathogens: Risks and Consequences.* Report No. 122. Ames, IA: Council for Agricultural Science and Technology (CAST).

Foegeding, P., and T. Roberts. 1999. *Foodborne Pathogens: Review of Recommendations.* Report No. 22. Ames, IA: Council for Agricultural Science and Technology (CAST).

Glynn, M. K., C. Bopp, W. Dewitt, P. Dabney, M. Mokhtar, and F. J. Angulo. 1998. Emergence of multidrug-resistant *Salmonella Enterica* serotype Typhimurium DT104 infections in the United States. *N. Engl. J. Med.* **338**, 1333–1338.

Goldwater, P. N., and K. A. Bettelheim. 1998. New perspectives on the role of *Escherichia coli* O157:H7 and other enterohaemorrhagic *E. coli* serotypes in human disease. *J. Med. Microbiol.* **47**, 1039–1045.

Guzewich, J. J., F. L. Bryan, and E. C. D. Todd. 1997. Surveillance of foodborne disease. Part I. purpose and types of surveillance systems and networks. *J. Food Prot.* **60**, 555–566.

Johnson, R. P., R. C. Clarke, J. B. Wilson, S. C. Read, K. Rahn, S. A. Renwick, K. A. Sandhu, D. Alves, M. A. Karmali, H. Lior, S. A. McEwen, J. S. Spika, and C. L. Gyles. 1996. Growing concerns and recent outbreaks involving non-O157:H7 serotypes of verotoxigenic *Escherichia coli. J. Food Prot.* **59**, 1112–1122.

Palmer, S. R., Lord Soulsby, and D. I. H. Simpson. 1998. *Zoonoses: Biology, Clinical Practice, and Public Health Control.* New York: Oxford University Press.

Smith, J. L. 1998. Foodborne disease in the elderly. *J. Food Prot.* **61**, 1229–1239.

Smith, J. L., and P. M. Fratamico. 1995. Factors involved in the emergence and persistence of food-borne disease. *J. Food Prot.* **58**, 696–708.

Tappero, J. W., A. Schuchat, K. A. Deaver, L. Mascola, and J. D. Wenger. 1995. Reduction in the incidence of human listeriosis in the United States. *J. Am. Med. Assoc.* **273**, 118–1122.

Todd, E. C. D. 1996. Worldwide surveillance of foodborne disease: The need to improve. *J. Food Prot.* **59**, 82–92.

Todd, E. C. D. 1998. Seafood-associated diseases in Canada. *Rev. Sci. Tech. Off. Int. Epiz.* **16**, 661–672.

Todd, E. C. D., and J. Harwig. 1996. Microbial risk analysis of food in Canada. *J. Food Prot.* (Suppl.), 10–18.

Todd, E. C. D., J. J. Guzewich, and F. L. Bryan. 1997. Surveillance of foodborne disease. Part IV. Dissemination and uses of surveillance data. *J. Food Prot.* **60**, 715–723.

Unklesbay, N., J. Sneed, and R. Toma. 1998. College students'attitudes, pratices, and knowledge of food safety. *J. Food Prot.* **61**, 1175–1180.

Van Beneden, C. A., W. E. Keene, R. A. Strang, D. H. Werker, A. S. King, B. Mahon, K. Hedberg, A. Bell, M. T. Kelly, V. K. Balan, W. R. Mac Kenzie, and D. Fleming. 1999. Multinational outbreak of *Salmonella enterica* serotype Newport infections due to contaminated alfalfa sprouts. *J. Am. Med. Assoc.* **281**, 158–162.

CHAPTER 2

Arcobacter and *Helicobacter*

IRENE V. WESLEY

Campylobacter, *Helicobacter*, and *Arcobacter* are closely related microbes of the Proteobacteria ribonucleic acid (RNA) superfamily VI. These microbes are gram-negative, motile curved, or spiral rods that grow under microaerobic conditions. Their general features are summarized in Table 2.1. Several species are pathogenic for humans and animals. Herein we examine the evidence that *Arcobacter*, especially *Arcobacter butzleri*, and *Helicobacter*, in particular *Helicobacter pylori*, are potential human foodborne pathogens.

2.1. ARCOBACTER

2.1.1. Introduction

The genus *Arcobacter* was proposed in 1992, following reanalysis of the aerotolerant species *Campylobacter* (Latin, "curved rods") *cryaerophila* (Latin, "loving cold and air"). The genus *Arcobacter* includes aerotolerant *Campylobacter*-like gram-negative bacteria, which are motile by means of polar unsheathed flagella. *Arcobacter* (Latin, "arc-shaped bacterium") was first isolated from aborted bovine fetuses and was designated *C. cryaerophila. Arcobacter*, unlike other *Campylobacter* species, grows in the presence of atmospheric oxygen (aerotolerant) and at temperatures (15–25°C) that are lower than those used for incubation of *Campylobacter*.

Three species have been recovered from man or animals: *A. butzleri*, *Arcobacter cryaerophilus*, and *Arcobacter skirrowii*. With respect to public health importance, *A. butzleri* is perhaps the most likely species to cause human illness, which is manifested primarily as enteritis. *Arcobacter cryaerophilus* is recovered from livestock and infrequently from humans. *Arcobacter skirrowii* is not commonly isolated from livestock; no isolations have been reported from

Guide to Foodborne Pathogens, Edited by Ronald G. Labbé and Santos García
ISBN 0-471-35034-6

TABLE 2.1. Summary of Major Distinguishing Characteristics of Members of recombinant RNA Superfamily VI

Genus	Growth at 150°C	Oxygen Tolerance	Flagella	Urease	G + C (mol %)
Helicobacter	No	Microaerophilic	Multipolar, sheathed	Yes	33–44
Campylobacter	No	Microaerophilic	Single polar, unsheathed	No	30–46
Arcobacter	Yes	Aerotolerant	Single polar, unsheathed	Yes	35–41

humans. *Arcobacter nitrofigilis* is the type strain and is restricted to the roots of *Spartina*, a salt-marsh plant. It has not been isolated from animals. A fifth unnamed environmental species has been recovered from oil fields.

2.1.2. Nature of Illness in Animals and Humans

Arcobacter species, like *Campylobacter*, may exist as commensals in livestock, including cattle, hogs, and poultry. *Arcobacter* infections in animals are seldom reported but have been associated with abortions and enteritis.

Arcobacter is included on the list of reportable human diseases. In humans, enteritis and occasionally septicemia occur. An aerotolerant *Campylobacter*, possibly *A. butzleri*, was recovered from 2% of Thai children during a survey of pediatric gastroenteritis. This suggested that aerotolerant *Campylobacter* might be common in developing nations. In a survey of 29 patients in the United States from whom *A. butzleri* was isolated, diarrheal illness (76%), bacteremia (14%), and acute appendicitis (10%) were observed. Risk factor analysis of 19 patients with *A. butzleri*–associated illness indicated that 63% of the cases were linked to consumption or contact with potentially contaminated water associated with travel.

Arcobacter butzleri has been cultured from Italian school children with recurrent abdominal cramps. The clustering of the cases suggested person-to-person transmission. *Arcobacter butlzeri* has also been recovered from an acquired immunodeficiency syndrome (AIDS) patient with intermittent diarrhea in the absence of other enteric pathogens. Nonhuman primates are naturally infected with *Arcobacter* and may develop colitis, which may provide insight into its pathogenesis in humans.

2.1.3. Characteristics of Agent

Aerotolerance and growth at 15–30°C are the key features to distinguish *Arcobacter* from thermophilic *Campylobacter* species. However, *A. butzleri*, like *Campylobacter jejuni*, may grow at 42°C, which is a temperature that is generally regarded as characteristic of *C. jejuni*. In one study, more than 75% of *A. butzleri* field strains examined grew at 42°C. *Arcobacter* spp. are oxidase positive and hydrolyze indoxyl acetate, which are traits also exhibited by *Campylobacter*.

Arcobacter are morphologically similar to *Campylobacter* in dark-field microscopy. Rapid darting motility is observed; unusually long cells of >20 μm may be seen. Colonies are small (1 mm in diameter), generally nonpigmented, and convex with entire edges and may swarm on fresh agar.

Phenotypic traits may distinguish the three species of *Arcobacter* recovered from livestock and humans. The most reliable tests to identify *A. butzleri* are negative or weak catalase production, growth in *Campylobacter* minimal medium, abudant growth on blood agar, growth on MacConkey agar, and resistance to cadmium chloride. In contrast, *A. cryaerophilus* is strongly catalase positive, does not grow on *Campylobacter* minimal medium or on MacConkey agar, and is sensitive to cadmium chloride. *Arcobacter cryaerophilus* consists of two subgroups, which can be distinguished by either sodium dodeyl sulfate–polyacrylamide gel electrophoresis (SDS–PAGE) or by deoxyribonucleic acid (DNA) methods. Although identification of subgroups may add insight into epidemiological associations of *A. cryaerophilus*, no phenotypic tests are available to distinguish them. *Arcobacter skirrowii*, which is rarely encountered, exhibits a strong catalase activity, reduces nitrate, but fails to grow on MacConkey.

2.1.4. Epidemiology

The epidemiology of *Arcobacter* may parallel that of *Campylobacter*, to which it is closely related. *Arcobacter* species may be commensals of livestock and poultry. For cattle, *Arcobacter* (10%) and *A. butzleri* (1.5%) were detected in the feces of clinically healthy dairy cows surveyed. The presence of *Arcobacter* in live cattle predicts its presence in beef. *Arcobacter butzleri* has been cultured from 1.5% of minced beef samples ($n = 68$) examined.

Arcobacter is present in the feces of 40% of healthy pigs. In addition, *Arcobacter* was detected in 22% of ground pork samples ($n = 290$) in the United States. In contrast, *Arcobacter* was cultured from only 0.5% (1 of 194) of pork samples in the Netherlands. The difference in handling between ground pork and minimally processed pork cuts as well as isolation methods may underlie the differences between the findings of the two studies. Despite its presence in livestock and red meats, no human cases of *Arcobacter* have been linked to consumption of either beef or pork.

Arcobacter, like *Campylobacter*, has been reported more frequently from poultry products than from red meats. Thus, poultry may be a significant reservoir of *A. butzleri*. In France, *A. butzleri* was recovered from 81% of poultry carcasses examined ($n = 201$). Nearly half of the poultry isolates in that study were of serogroup 1. Because serogroups 1 and 5 are recovered from clinical cases, it has been suggested that consumption of undercooked contaminated poultry may cause human infection. No case–control studies have been reported to verify this observation. In Canada, *A. butzleri* was recovered from 97% (121 of 125) of poultry carcasses obtained from five different processing

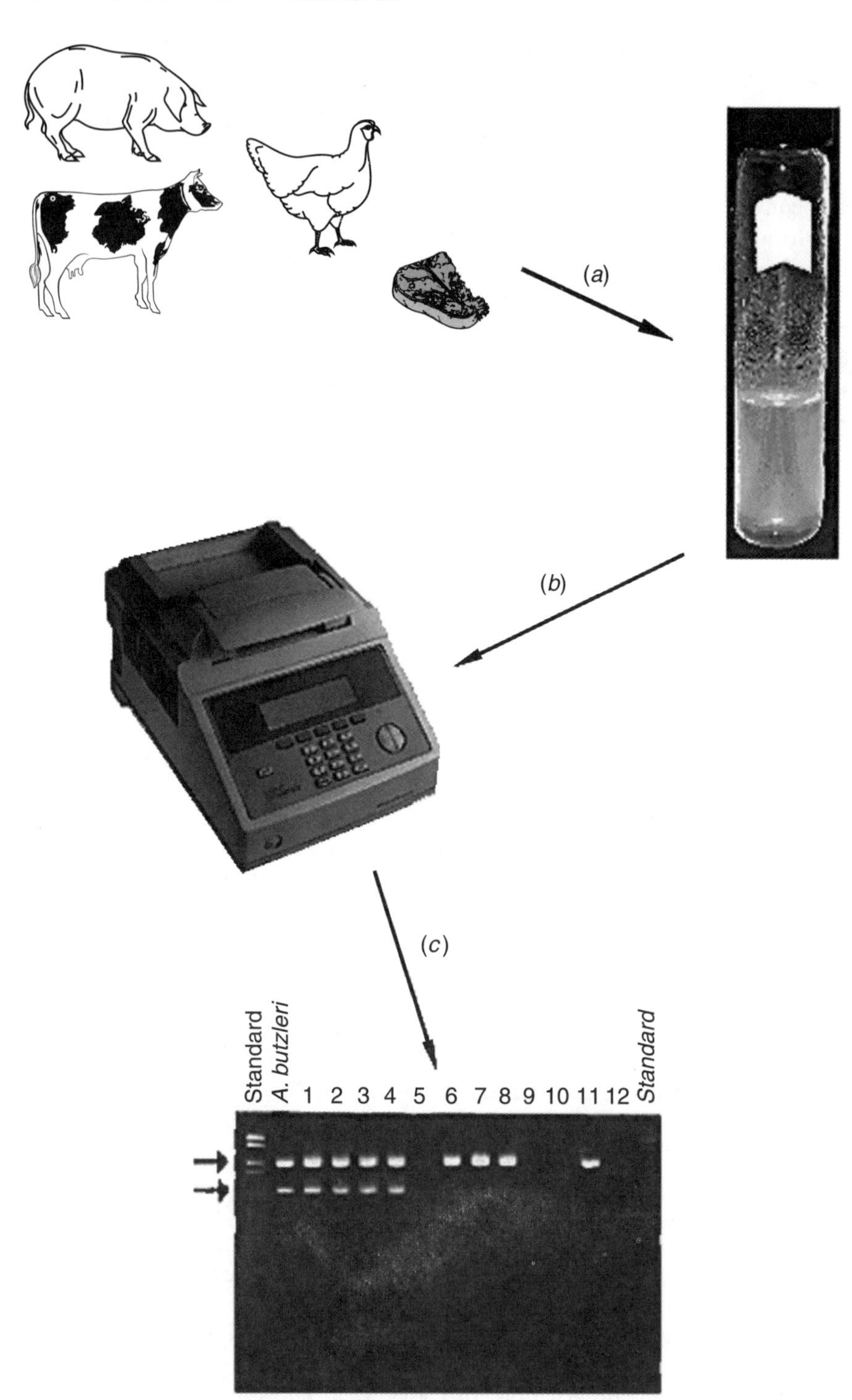

(a)
(b)
(c)
Standard
A. butzleri
1 2 3 4 5 6 7 8 9 10 11 12
Standard

plants. Serotype 1 was the predominant group isolated from Canadian poultry. In contrast, *Arcobacter* was detected in only 24% (53 of 224) of retail purchased poultry products in the Netherlands. No isolations were made from poultry in a limited study conducted in Italy.

In the United States, a pilot study was conducted to estimate the prevalence of *Campylobacter* and *Arcobacter* in whole turkey carcass rinses ($n = 300$). Overall, *Arcobacter* (88%) was recovered more frequently than *Campylobacter* (54%). More specifically, *A. butzleri* (75%) was detected more often than either *C. jejuni* (27%) or *Campylobacter coli* (27%). In a tristate study, *Arcobacter* (77% of samples) and *A. butzleri* (57% of samples) were isolated from mechanically separated turkey meat ($n = 395$ samples). This suggets that *Arcobacter*, in contrast to *C. jejuni*, survives the aeration associated with meat grinding.

Arcobacter, like *Campylobacter*, colonizes live birds. *Arcobacter* (15%), *A. butzleri* (1%), and *Campylobacter* (13%) were detected in cloacal swabs of live chickens ($n \sim 300$). Its low recovery from live birds and its high prevalence on poultry carcasses suggest that extensive contamination occurs during processing. This also may reflect the ability of *Arcobacter* to survive in cold processing waters.

The distribution of *Arcobacter* spp. in seafood, shellfish, and raw milk is unknown. Given its recovery from dairy cattle (10%) and its tolerance to cold, air, and salt, it is probable that its prevalence in these foods may exceed that of *Campylobacter*.

Transmission of *A. butzleri* to humans and livestock may occur while drinking contaminated water. *Arcobacter* spp. may be more common in developing nations with inadequate water supplies since *A. butzleri* accounted for 16% of the *Campylobacter*-like isolates made from cases of diarrhea in Thai children. *Arcobacter butzleri* has been reported in drinking water reservoirs in Germany, water treatment plants, rivers, the canal waters of Bangkok, and water samples obtained from oil fields. In the United States, *A. butzleri* was recovered from a well serving an Idaho youth camp following an outbreak of waterborne enteritis. The presence of *A. butzleri* in unchlorinated water supplies suggests this as a possible source of contamination.

Figure 2.1. Generalized scheme for PCR-based detection of *Arcobacter* in poultry, livestock feces, meats, or water. (*a*) Samples are enriched in EMJH-P80 media (1 week, 30°C). (*b*) After incubation, cells are lysed by boiling, and template DNA with PCR reagents are placed in a thermal cycler for amplification, as described. (*c*) The PCR products or amplicons are detected by gel electrophoresis. The presence of a 1223-bp PCR amplicon (upper arrow) is specific for *Arcobacter. Arcobacter butzleri* exhibits two PCR products: the 1223-bp amplicon characteristic of the genus and a smaller 686-bp (lower arrow) product unique to *A. butzleri* (lanes 1–4). The presence of the single 1223-bp amplicon indicates the presence of *Arcobacter* species other than *A. butzleri* (lanes 6–8; lane 11). The absence of a visible PCR product indicates that *Arcobacter* is not in the sample (lanes 5, 9, 10). Positive (lane *A. butzleri*) and negative (lane 12) controls are shown.

2.1.5. Detection and Differentiation of Organism

Methods for collecting and transporting foods, water and fecal samples, poultry carcass rinses, and livestock carcass swabs are conducted generally as described for *Campylobacter*. Like *Campylobacter*, samples may be transported on ice in buffered peptone water or Cary-Blair transport media.

No single method is universally accepted for the recovery of *Arcobacter* from foods, livestock, or water. Isolation protocols use formulations based on (1) Ellinghausen, McCullough, Johnson, and Harris polysorbate 80 (EMJH-P80) medium; (2) modifications of isolation media used for *Campylobacter*; and (3) formulations specific for *Arcobacter*.

Arcobacter were first described from aborted bovine and porcine fetuses cultured in EMJH-P80 medium. This media was originally formulated to recover *Leptospira* from clinical samples. The EMJH-P80 medium has also been used to recover *Arcobacter* from ground pork, mechanically separated turkey meat, water, and live birds and pigs.

Use of media designed for *Campylobacter* when incubated at 30°C should be adequate for *Arcobacter* isolation. To illustrate, *A. butzleri* was recovered from poultry using a modification of Rosef's enrichment broth described for the cultivation of thermophilic *Campylobacter*. Formulations have been developed specifically for *Arcobacter*, and at least one is commercially available.

The use of molecular methods, such as DNA probes and polymerase chain reaction (PCR) primers, has accelerated the detection and identification of *Arcobacter*. Polymerase chain reaction assays to detect all members of the genus *Arcobacter* as well as protocols specific for *A. butzleri*, *A. cryaerophilus*, and *A. skirrowii* have been described. A multiplex PCR assay to simultaneously identify *Arcobacter* and *A. butzleri* in livestock and foods has been detailed (Fig. 2.1). The multiplex PCR assay is conducted directly from enrichment, does not require pure cultures, bypasses the ambiguities of biochemical identification, and is suitable for field studies.

Field strains of *A. butzleri* may be distinguished by serotyping. At least 72 serogroups are recognized. As with *Campylobacter*, the availability of serotyping reagents limits the applicability of this method.

In contrast, the general availability of reagents for DNA-based typing methods has accelerated epidemiological studies. Molecular-based assays to characterize *A. butzleri* strains include restriction fragment length polymorphisms (RFLPs) and PCR-based DNA fingerprinting. In addition to providing species identification, the DNA profiles can be used to determine if the isolate patterns are identical, inferring an epidemiological relationship, or different, which may suggest that the isolates are not related and thus may not be associated with a common source of contamination. The PCR-derived fingerprints have suggested person-to-person transmission in a nursery school outbreak and have indicated multiple sources of contamination of mechanically separated turkey meat.

2.1.6. Physical Methods for Destruction

It is postulated that *Arcobacter* is hardier than *Campylobacter*. By comparing D10 values (the irradiation dose that reduces by 10-fold the number of viable bacteria), *A. butzleri* (0.27 kGy) was found to be more resistant to irradiation than *C. jejuni* (0.18 kGy). *Arcobacter* grows over a pH range of 5.5–7.5 with optimal growth of *A. butzleri* at pH 6 and optimal growth of *A. cryaerophilus* at pH 7.0–7.5. Thermal tolerance studies indicate the thermal death time for a single strain of *A. butzleri* to be 2.5, 5, and 15 min at 60, 55, and 50°C, respectively. Based on these data, *Arcobacter* will not survive minimum pasteurization treatment (63°C for 30 min or 71.7°C for 15 s) of milk.

Arcobacter, in contrast to *Campylobacter*, grows at 15°C, a trait reflected by the species name *cryaerophila*. This suggests that it can survive in the cold longer than typical *Campylobacter* species. For example, *A. butzleri* populations decreased by 0.5 $\log_{10}$ when held in ground water at 5°C for 14 days. In contrast, *C. jejuni* declined by six to seven $\log_{10}$ units when held for a similar interval.

Chlorine inactivates *A. butzleri* as well as *C. jejuni* and *H. pylori*. Thus, disinfection practices normally used in drinking water are adequate to control these closely related organisms. This is critical because of the epidemiological relationship noted between cases of human enteritis due to *A. butzleri* and consumption of contaminated drinking water.

2.1.8. Prevention and Control

The limited epidemiological data available for *Arcobacter* suggest that transmission to humans results from consumption of contaminated water and undercooked poultry. Thus, strategies to prevent human infection should focus on consumption of chlorinated drinking water and thorough cooking of poultry. These measures parallel those recommended for *Campylobacter*.

2.2. HELICOBACTER

2.2.1. Introduction

Members of the genus *Helicobacter* colonize humans and animals and cause enteritis, gastritis, and rarely abortions in livestock. By 1999, 21 species had been described from a variety of animal hosts, including primates, poultry, and domestic animals. Each species maintains a restricted host range. Humans are the major hosts for *H. pylori*, which is the most common human bacterial infection in the world.

2.2.2. Nature of Human Illness

Helicobacter pylori (*Campylobacter pylori*) is present in 95% of duodenal and in 70–80% of human gastric ulcer cases as well as in clinically healthy individuals, including family members of patients. It is unique among bacteria in being directly linked to human gastric carcinoma.

Antibodies to *H. pylori* are highest in individuals from rural settings and in populations of low socioeconomic status. Antibodies in >50% of the population occur in individuals living in developing countries. In contrast, antibodies are found in <50% of the population living in industrialized countries such as Australia, the United States, and France.

2.2.3. Characteristics of Agent

Helicobacter pylori is a gram-negative microbe that is motile by means of four to six unipolar sheathed flagella, microaerophilic, and spiral shaped. The sheathed flagella may be adaptations to survival in gastric juices. Growth occurs between 30 and 37°C, but not at 42 or 25°C.

2.2.4. Epidemiology

Helicobacter pylori may be transmitted to humans by (a) fecal–oral spread, (b) oral–oral spread via salivary secretions, (c) from pets (human-to-pet transmission is also possible), and (d) ingestion of contaminated foods and water (Fig. 2.2). Close contact, poor sanitary conditions, familial crowding, confinement in submarines, and clustering in institutions have also been identified as risk factors for *H. pylori* infection.

Fecally contaminated foods and water may serve as potential vehicles of transmission. *Helicobacter pylori* has been detected in human feces either by culture or by PCR.

Fecally contaminated fruits, vegetables, and shellfish may transmit *H. pylori*. In a study of 1815 Chileans under the age of 35, *H. pylori* antibodies were detected in >60% of lower socioeconomic groups. Seropositivity correlated with age, low socioeconomic status, and consumption of uncooked vegetables. Another risk factor that reached marginal significance was consumption of uncooked shellfish. In contrast, no significant difference in *H. pylori* seroprevalence was found between vegetarians and meat eaters in the United States.

No isolations of *H. pylori* have been reported from vegetables, fruits, shellfish, or seafoods.

Water has been proposed as a vehicle of transmission. *Helicobacter pylori* has been detected via PCR in water, including rivers and sewage-contaminated water. In communities of Lima, Peru, the water source may be a more important risk factor than socioeconomic status in acquiring *H. pylori* infection. One study evaluated 407 Peruvian children (aged 2 months to 12 years) from families of low and high socioeconomic status. Children from high-income families whose homes received municipal drinking water were 12 times more likely to be infected than those from high-income families whose water supply came from community wells. Thus, the municipal water supply may present a greater risk of infection than the socioeconomic status.

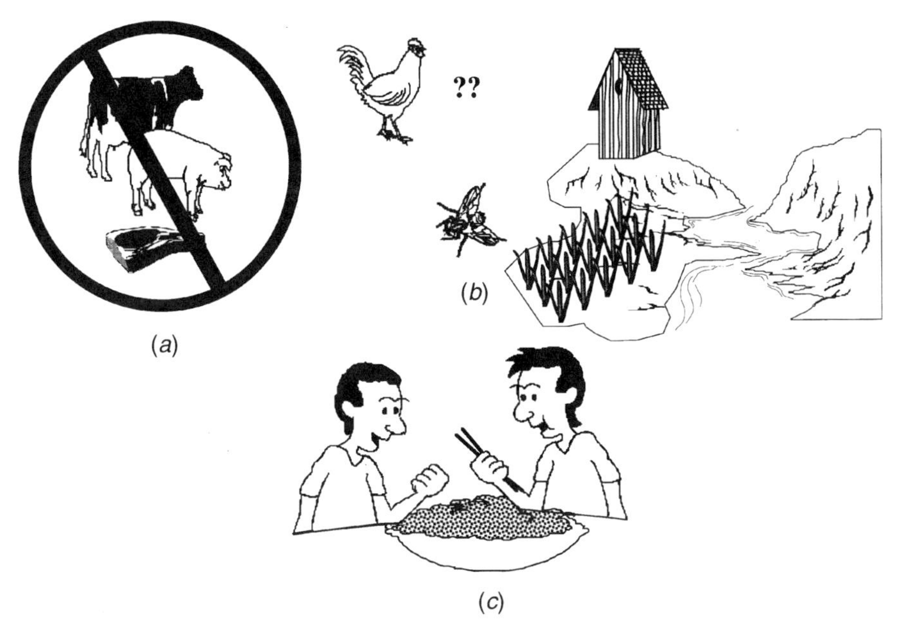

Figure 2.2. Possible food-related routes of transmission of *Helicobacter pylori.* (*a*) *Helicobacter pylori* has not been detected in cattle or swine. There is no evidence in support of transmission of *Helicobacter* species to humans via consumption of beef and pork. However, *H. pullorum* may be transmitted to humans by consumption of contaminated poultry products. (*b*) *Helicobacter pylori* has been cultured from human feces. Contaminated feces may pollute water that is then used to irrigate vegetables. Recently, *H. pylori* has been shown to survive in flies. Thus, flies may transmit *H. pylori* and thus may disseminate it from feces by routes similar to other enteric pathogens. (*c*) Since *H. pylori* is present in saliva, transmission by chopsticks has been suggested.

Results of serosurveys of veterinarians and slaughterhouse workers have suggested that *H. pylori* is a zoonosis and is therefore transmitted from livestock to humans. Antibodies to *H. pylori* were detected more frequently in slaughterhouse workers exposed to animal carcasses than in clerks employed at the same abattoir. However, no preemployment serological titers were included in that study to evaluate preexisting infection status. Also, comparisons between the clerical and nonclerical staff with respect to age, socioeconomic status, and country of origin were not provided. As previously noted, infection status increases with age and reflects social economic status as well as country of birth. That antibodies to *H. pylori* are higher in persons in contact with sheep is provided as evidence of zoonotic transmission.

Evidence of infection by *Helicobacter*-like organisms in livestock, including catttle and hogs, is based in part on serosurveys. Use of nonspecific test antigens, however, may lead to false-positive reactions. Of all of the *Helicobacter* species, only *Helicobacter pullorum* may be transmitted from poultry to humans via consumption of undercooked chicken.

2.2.5. Detection of the Organism

Methods to detect *Helicobacter* in foods are limited. Isolation protocols that have been used to culture the microbe from clinical specimens may be adapted to its recovery from foods. In general, a suspect food product, for example, poultry samples contaminated with *H. pullorum*, are homogenized and plated onto nonselective and selective media. Nonselective media include chocolate agar or brain heart infusion agar supplemented with 5–10% defibrinated blood. Selective agars are commercially available. Freshly poured plates are inoculated, incubated (35°C, 3–5 days) microaerobically (5–7% O_2, 5–10% CO_2), with the critical addition of hydrogen (8% H_2), in high humidity. Plates are examined for the presence of small pinpoint translucent nonhemolytic colonies. The presence of gram-negative, curved, or spirally shaped bacilli that grow at 35°C but not at 25°C and are strongly urease positive is indicative of *Helicobacter*.

Polymerase chain reaction assays have accelerated the identification of the diverse species of *Helicobacter*, especially *H. pylori* in fecal, gastric biopsy, and oral cavity samples. The PCR assays specifically designed for *H. pylori* target the *vacA* and *cagA* genes and the genes encoding the urease enzyme.

2.2.6. Physical Methods for Destruction

Thermal and cold tolerance data are limited for *H. pylori*. However, because of their phylogenetic relationship, it is probable that *H. pylori* is as sensitive to heat, cold, and irradiation as *C. jejuni* and *A. butzleri*. *Helicobacter pylori* survived longer in experimentally inoculated milk that was refrigerated (6 days) than in milk held at room temperature (3 days).

With respect to water treatment, chlorination studies completed in the United States by the Environmental Protection Agency indicated that *H. pylori* is as sensitive to standard chlorination regimens as *C. jejuni*, *A. butzleri*, and *Escherichia coli*.

2.2.7. Prevention and Control

To summarize, although other *Helicobacter* species have been recovered in livestock, *H. pylori* does not colonize pigs, cattle, or poultry, thus eliminating these meat animals as a source of infection. Because *H. pullorum* has caused human enteritis, thorough cooking of poultry, as recommended for other foodborne agents, is indicated.

Risk analyses indicate that *H. pylori* may be transmitted to humans via consumption of fecally contaminated water. However, routine chlorination inactivates *Helicobacter* as well as *Campylobacter* and *Arcobacter*. Flies may transport *H. pylori* from feces to foods. Thus, intervention strategies that have effectively halted fecal spread by insects of other foodborne agents should be effective.

Epidemiological evidence suggests that in developing countries consumption of sewage-contaminated vegetables may be a risk factor for *H. pylori* infection. Therefore, cleaning of raw vegetables with chlorinated water would most likely reduce the risk of infection.

BIBLIOGRAPHY

Arcobacter

Bastyns, K., D. Cartuyvels, S. Chapelle, P. Vandamme, H. Goossens, and R. de Wachter. 1995. A variable 23S rDNA region is a useful discriminating target for genus-specific and species-specific PCR amplification in *Arcobacter* species. *Syst. Appl. Microbiol.* **28**, 353–356.

Johnson, L. G., and E. A. Murano. 1999. Comparison of three protocols for the isolation of *Arcobacter* from poultry. *J. Food Prot.* **62**, 610–614.

Kiehlbauch, J. A., D. J. Brenner, M. A. Nicholson, C. N. Baker, C. M. Patton, A. G. Steigerwalt, and I. K. Wachsmuth. 1991. *Campylobacter butzleri* sp. nov. isolated from humans and animals with diarrheal illness. *J. Clin. Microbiol.* **29**, 376–385.

Manke, T. R., I. V. Wesley, J. S. Dickson, and K. M. Harmon. 1998. Prevalence and genetic variability of *Arcobacter* species in mechanically separated turkey. *J. Food Prot.* **61**, 1623–1628.

Neill, S. D., J. N. Campbell, J. J. O'Brien, S. T. C. Weatherup, and W. A. Ellis. 1985. Taxonomic position of *Campylobacter cryaerophila* sp. nov. *Int. J. Syst. Bacteriol.* **35**, 342–356.

On, S. L. W., B. Holmes, and M. J. Sackin. 1996. A probability matrix for the identification of campylobacter, helicobacters and allied taxa. *Appl. Bacteriol.* **81**, 425–432.

Rice, W. E., M. R. Rodgers, I. V. Wesley, C. H. Johnson, and A. S. Tanner. 1998. Isolation of *Arcobacter* spp from ground water. *Lett. Appl. Microbiol.* **28**, 31–35.

Vandamme, P., B. A. J. Giesendorf, A. van Belkum, D. Pierard, S. Lauwers, K. Kersters, J. P. Butzler, H. Goossens, and W. G. V. Quint. 1993. Discrimination of epidemic and sporadic isolates of *Arcobacter butzleri* by polymerase chain reaction-mediated DNA fingerprintin. *Clin. Microbiol.* **31**, 3317–3319.

Vandamme, P., M. Vancanneyt, B. Pot, L. Mels, B. Hoste, D. Dewerttinck, L. Vlaes, C. Van den Borre, E. Higgins, J. Hommez, K. Kersters, J. P. Butzler, and H. Goossens. 1992. Polyphasic taxonomic study of the emended genus *Arcobacter* with *Arcobacter butzleri* comb. nov. and *Arcobacter skirrowii* sp. nov., an aerotolerant bacterium isolated from veterinary specimens. *Int. J. System. Bacteriol.* **42**, 344–356.

Wesley, I. V. 1994. *Arcobacter*. Infections. In: G. W. Beran and J. H. Steele (Eds.), *Handbook of Zoonoses*. Boca Raton, FL: CRC Press, pp. 181–190.

Wesley, I. V. 1997. *Helicobacter* and *Arcobacter*: Potential human foodborne pathogens? *Trends Food Sci. Technol.* **89**, 293–299.

Helicobacter

Blaser, M. J. 1997. Ecology of *Helicobacter pylori* in the human stomach. *J. Clin. Invest.* **100**, 759–762.

Dunn, B. E., H. Cohen, and M. J. Blaser. 1997. *Helicobacter pylori. Clin. Microbiol. Rev.* **10**, 720–741.

Fox, J. G., and A. Lee. 1997. The role of *Helicobacter* species in newly recognized gastrointestinal tract diseases of animals. *Lab. Anim. Sci.* **47**, 222–255.

Goodman, K. H. J., and P. Correa. 1995. The transmission of *Helicobacter pylori*: A critical review of the evidence. *Int. J. Epidemiol.* **224**, 875–887.

Goodman, K. H. J., P. Correa, H. J. K. J. Tengana Aux, H. Ramirez, J. P. DeLany, M. Pepinosa, O. G. Lopez Quinones, and T. C. Parra. 1996. *Helicobacter pylori* infection in the Colombian Andes: A population based study of transmission pathways. *Am. J. Epidemiol.* **1244**, 290–299.

Goodwin, C. S., J. A. Armstrong, T. Chilvers, M. Peters, M. D. Collins, L. Sly, W. McConnel, and W. E. S. Harper. 1989. Transfer of *Campylobacter pylori* and *Campylobacter* mustelae to *Helicobacter* gen. nov. as *Helicobacter pylori* comb. nov. and *Helicobacter mustelae* comb. nov., respectively. *Int. J. Syst. Bacteriol.* **39**, 397–405.

Goodwin, G. S., and B. W. Worsley. 1993. *Helicobacter pylori: Biology and Clinical Practice*. Boca Raton, FL: CRC Press.

Johnson, C. H., E. W. Rice, and D. J. Reasoner. 1997. Inactivation of *Helicobacter pylori* by chorination. *Appl. Environ. Microbiol.* **12**, 4969–4970.

Klein, P. D., D. Y. Graham, A. Gaillour, A. R. Opekun, and E. O'Brian Smith. 1991. Water source as risk factor for *Helicobacter pylori* infection in Peruvian children. *Lancet* **337**, 1503–1506.

Stanley, J., D. Linton, A. P. Burnens, G. F. E. Dewhirst, S. L. W. On, A. Porter, R. J. Owen, and M. Costas. 1994. *Helicobacter pullorum* sp nov. genotype and phenotype of a new species isolated from poultry and from human patients with gastroenteritis. *Microbiology* **140**, 3441–3449.

Taylor, D. N., and H. J. Parsonnet. 1995. Epidemiology and natural history of *Helicobacter pylori* infection. In: M. J. Blaser (Ed.), *Infection of the Gastrointestinal Tract*. New York: Raven, 1995, pp. 1–13.

Vandamme, P., E. Falsen, R. Rossau, B. Hoste, P. Segers, R. Tytgat, and J. de Ley. 1991. Revision of *Campylobacter*, *Helicobacter*, and *Wolinella* taxonomy: Emendation of generic descriptions and proposals of *Arcobacter* gen. Nov. *Int. J. Syst. Bacteriol.* **41**, 88–103.

Wesley, I. V. 1997. *Helicobacter* and *Arcobacter*: Potential human foodborne pathogens? *Trends Food Sci. Technol.* **89**, 293–299.

CHAPTER 3

Aspergillus

DEEPAK BHATNAGAR and SANTOS GARCÍA

3.1. INTRODUCTION

Aspergillus is a genus name for a group of fungi. Members of this particular group of molds are among the most widely distributed and abundant of living things. *Aspergillus* finds high temperatures in tropical and subtropical areas particularly favorable conditions for growth. These conditions also favor the production of mycotoxins by these fungi. Several species of *Aspergillus* are ubiquitous, live as saprotrophs or parasites, and play an important role in the environment in the decomposition and decay of organic materials. Aspergilli are metabolically diverse and have been used in the fermentation industry for the production of organic acids, enzymes, and certain food products. This genus provides the major source of enzymes for the food industry. Aspergilli also produce gluconic acid and kojic acid (which derive from sugars) and citric, itaconic, malic, oxalic, and epoxy-succinic acids (which are related to tricarboxylic acid intermediates). As a result, these fungi are involved in a wide range of food processing in the dairy, wine and juice, distilling, brewing, baking, textile, animal feed, leather, and tea industries. Aspergilli have been known to be contaminants of food and feeds and some are etiologic agents of human diseases.

Many species of *Aspergillus*, including the common species *A. flavus* and *A. niger*, lack a sexual stage in their life cycle. They make asexual or mitotic spores (called conidia) but lack sexual or meiotic spores (called ascospores). In some cases these fungi may have lost their sexual phase thousands or millions of years ago. This is an evolutionary paradox: Without sex they should not be able to adapt to changing environments and should eventually die out, yet they are among the most successful organisms on the earth. Many theories have been proposed to explain this, but it remains an open question.

The asexual lives of the aspergilli also create a taxonomic nightmare. Fungi are classified by their sexual stages, and the aspergilli clearly belong to the class

Guide to Foodborne Pathogens, Edited by Ronald G. Labbé and Santos García
ISBN 0-471-35034-6 John Wiley & Sons, Inc. 2001

Ascomycetes. But fungi without a sexual stage are put in the Deuteromycetes, or *Fungi imperfecti*. So some species that make both conidia and ascospores are classified in both an ascomycete genus (e.g., *Emericella nidulans*) and a deuteromycete genus (*Aspergillus nidulans*). How can one organism belong to two genera in two different classes? At first it sounds ludicrous. But the fungi are difficult organisms in many respects and defy biological concepts developed for animals and plants.

Aspergilli cause mycotoxicoses and mycoses. A diverse group of secondary metabolites are produced by these fungi, including mycotoxins. The mycotoxins produced by these fungi result in a host of adverse health effects (mycotoxicoses). Aflatoxicosis is the most common illness caused by fungal products, aflatoxins, produced by *A. flavus* and *A. parasiticus*. Aspergilli are allergens, and human infections (mycoses) caused by aspergilli exhibit a broad range of consequences. Most acute cases are originated by *A. fumigatus*, followed by *A. flavus* and *A. niger*. Different risk factors are associated with each of these diseases, particularly alterations in the immune system.

3.2. ECOLOGY OF *Aspergillus*

Species of *Aspergillus* are ubiquitous and grow on a wide variety of substrates and under varied environmental conditions. These fungi are saprotrophs found in soils, stored grain, food products, and decaying vegetation. Several species of this genus parasitize insects, plants, and animals, including human beings. Species of *Aspergillus* have been most extensively studied from soils. Aspergilli have been reported from diverse kinds of soils in various regions of the world. During warm, dry periods, several aspergilli, such as those from the *A. flavus* group, can increase rapidly on crops or crop debris. In soil, *Aspergillus* can exist as fungal hyphae, spores (conidia or ascospores), or sclerotia; those with a sexual stage produce cleistothecia. In stored cereal grains, temperature, moisture content, and grain quality determine the type and density of populations of *Aspergillus* species. As food contaminants, *Aspergillus* species can survive or even multiply under conditions of high temperature and very low water activity. Some aspergilli thrive at temperatures of 40°C or above (even up to 65°C) and water potentials of −40 MPa or less. Spores (conidia) of aspergilli germinate over a broad range of temperatures (12–37°C). Germination can occur under very low water potentials (−27 MPa), but optimal germination occurs at about −3 MPa. Within 3 days of germination, a conidium can develop into a colony capable of producing millions of conidia. Under optimal conditions, the aspergilli can colonize substrates very, very quickly.

3.3. ISOLATION OF *Aspergillus*

For preliminary isolation of *Aspergillus*, dichloran–glycerol (DG-18) medium has been used routinely. DG-18 medium, which is also available commercially, consists of the following ingredients dissolved in 1 L of water: peptone 5.0 g,

Figure 3.1. Magnified view of *Aspergillus* conidiophore (from Bhatnagar et al., 1999).

glucose 10.0 g, KH_2PO_4 1.0 g, $MgSO_4 \cdot 7H_2O$ 0.5 g, dichloran (botran) 0.002 g, agar 15.0 g, chloramphenicol 0.1 g, and glycerol 220.0 g.

When viewed visibly or at higher magnifications under a stereomicroscope, colonies of *Aspergillus* display heads of different colors. These colonies have been traditionally isolated further by subculturing separately on oatmeal agar (OMA). Several additional, optimal media have been defined more recently. Color is an important diagnostic feature but often varies with, for example, age of the colony and growth medium. Aspergilli, even when they exhibit the same color, can be distinguished by the structure of conidial heads when viewed under a compound microscope (Fig. 3.1). Cultures should be made from all colonies showing such differences.

Isolation and enumeration of *Aspergillus* species can also be carried out in potato dextrose agar or malt extract agar. Bacterial growth can be controlled on these media by the addition of rose bengal or by antibiotics such as chlortetracycline, chloramphenicol, gentamycin, or streptomycin. To control other fungi, such as *Rhizopus* and *Mucor*, rose bengal, dichloran, or sodium chloride are incorporated into the media.

3.4. IDENTIFICATION OF *Aspergillus* SPECIES

Cultures grown on OMA, routinely used in the identification of *Aspergillus* spp., are incubated at 25°C for 7 days in complete darkness. Oatmeal agar is made up of 60.0 g oatmeal and 12.5 g agar per 1000 mL of water.

Laboratory guides for identification of common aspergilli are mostly based on morphological and colony characteristics when cultivated in Czapek agar (CZA), Czapek yeast agar (CYA), and malt extract agar (MEA). For teleomorphic species, cornmeal agar, OMA, and CBS malt agar are recommended. For xerophilic species, low-water-activity media such as CZA or CYA with 20% sucrose or malt yeast extract agar with 40% sucrose are commonly used. *Aspergillus* cultures should be incubated at 25°C for 5–7 days in the dark for anamorphic species and 10–14 days for most teleomorphic species.

Singh et al. (1991) have suggested that CZA dishes should be inoculated at three points with pure cultures already grown on OMA. The ingredients of CZA per liter of distilled water are sucrose 30.0 g, $NaNO_3$ 3.0 g, K_2HPO_4 1.0 g, KCl 0.5 g, $MgSO_4 \cdot 7H_2O$ 0.5 g, $FeSO_4 \cdot 7H_2O$ 0.01 g, agar 15.0 g, and trace metal solution 1.0 mL.

The method of Singh et al. for a *three-point* inoculation requires taking approximately 0.5 mL of molten agar (0.2%) and detergent (0.05%) in a small vial with screw cap (sterilized). A needle carrying conidia or other propagules is dipped in the medium and mixed. Petri dishes are inoculated with flamed needle dipped in the spore suspension. During inoculation, holding the petri dishes containing medium upside down is recommended. This helps in restricting the colonies to the three points, and any fine droplets of the suspension, if formed, fall down by shaking the needle. Using a separate vial for each isolate is essential.

The colony characters are recorded by making slides in lactic acid for observing morphological characters under a compound microscope.

The primary morphological criterion for delineation of this fungal genus is the conidiophore (Fig. 3.2). This conidiophore and the conidia that it bears reminded early mycologists of an aspergillum, the tool used by priests to sprinkle holy water, hence the name *Aspergillus*. Other important characteristics useful in identification are the color and shape of conidial heads, Hülle cells, and presence of sclerotia. The former has a long aseptate stipe terminating in a swollen vesicle on which one or two layers of specialized cells are borne. The cells bearing the conidia are called phialides. A second layer of specialized cells (metulae) could be present between the phialides and the vesicle. Conidia are produced during the asexual reproduction (anamorphic state). In addition, some species also reproduce sexually (ascosporic or teleomorphic state). Hülle cells are thick walled and are frequently associated with cleistothecia, or sexual fruiting bodies containing the meiotically derived ascospores. Sclerotia are hard, compact masses of hyphae with a darkened rind capable of surviving unfavorable environmental conditions.

New approaches have been developed to distinguish or relate species. These techniques include isozyme polymorphisms, distribution of ubiquinone systems, restriction fragment polymorphism (RFLP), profiles of secondary metabolites, enzyme-linked immunosorbent assay (ELISA), latex agglutination, polymerase chain reaction (PCR), and karyotyping. However, the most informative technique has been deoxyribonucleic acid (DNA) sequencing combined with phy-

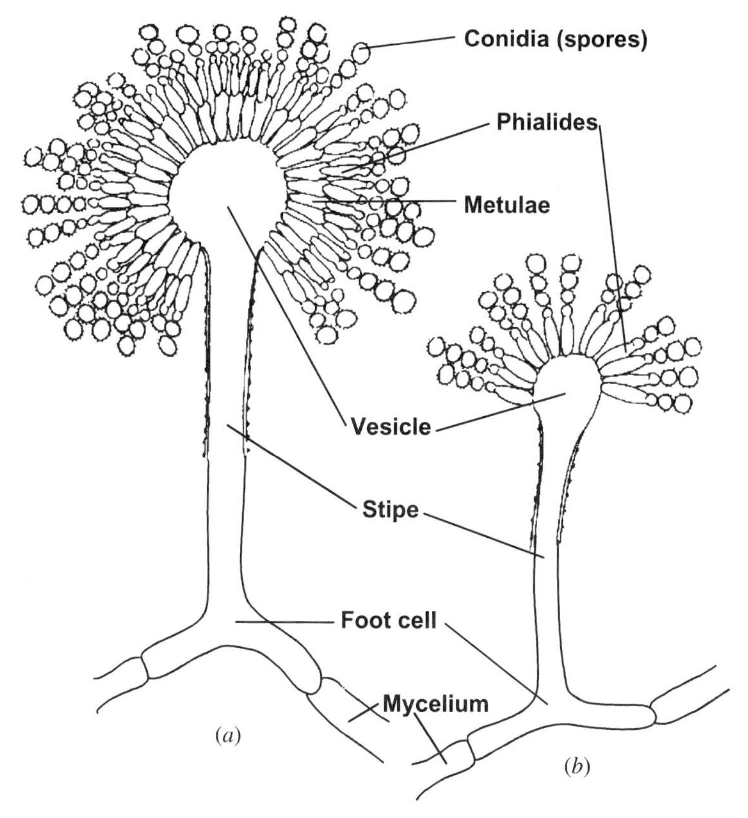

Figure 3.2. Basic morphological structure of *Aspergillus* showing (*a*) biseriate and (*b*) uniseriate conidial heads (from Bhatnagar et al., 1999).

logenetic analysis. Thus, more than 180 species with 72 teleomorphs have been described in the *Aspergillus* genus.

3.5. DETECTION OF MYCOTOXINS

There are several methods used for isolation and detection of mycotoxins, particularly from toxin-contaminated commodities. Various reviews of these methods show that there is no "best" technique, but methods are available that can be modified to fit a specific need.

The selection of an analytical method will depend on personal preference, mycotoxin of interest, availability of equipment and supplies, economics of assay procedure, time, and location and number of samples to be analyzed. For rapid, qualitative analysis, chemical and immunochemical methods are commercially available for specific toxins. Immunological methods, although useful for screening or quantitation of mycotoxins within a limited concentration

TABLE 3.1. TLC Separation of Sterigmatocyctin and Aflatoxins B_1 and G_1

		R_f		
Solvent System	Ratio (vol/vol)	Sterigmatocystin	Aflatoxin B_1	Aflatoxin G_1
Ether–methanol–water	96:3:1	0.97	0.37	0.28
Chloroform–acetone	10:0.5	0.74	0.22	0.11
Chloroform–methanol	10:0.5	0.93	0.63	0.55

Note: The TLC plates (20 × 20 cm) are spotted with approximately 50 ng of metabolites and developed for a distance of 15–17 cm, with a developing time of approximately 45 min.

range, are most useful for systems for evaluation of acute or chronic exposure to mycotoxins. However, thin-layer chromatography (TLC) has been the separation technique of choice for almost all mycotoxins. Analytical separations carried out by high-performance liquid chromatography and gas chromatography are the most accurate methods for determination of *Aspergillus* toxins such as aflatoxins, ochratoxins, patulin, and citrinin.

The agar plug method suggested by Singh et al. is a relatively rapid, accurate method for isolating mycotoxins. Seven-day-old cultures of *Aspergillus* spp. grown on CZA or yeast extract sucrose (YES) agar are particularly useful for mycotoxin production. The YES medium consists of yeast extract 20.0 g, $MgSO_4 \cdot 7H_2O$ 0.5 g, sucrose 150.0 g, agar 20.0 g, water 1000.0 mL, and trace metal solution 1.0 mL.

According to the described method, an agar plug approximately 0.5 cm across is removed and a drop of chloroform–methanol (2:1) or dichloromethane–acetone (2:1) is added to the mycelial side of the plug and placed directly onto a silica gel TLC plate.

The spot is dried and the TLC plate developed in a developing solvent of choice, for example, toluene–ethylacetate–90% formic acid (TEF, 5:4:1 v/v/v) for most *Aspergillus* mycotoxins, such as aflatoxins. Chloroform–acetone (10:0.5 v/v) and ether–methanol–water (96:3:1 v/v/v) are also recommended for aflatoxins (Table 3.1).

The dry, developed TLC plates can be viewed under light (daylight, short-wave UV 254 nm or long-wave UV 366 nm), and the unknown mycotoxin is compared to standards or published results on colors and R_f values (Table 3.1). Under UV light aflatoxins B_1 and B_2 fluoresce blue and G_1 and G_2 fluoresce green. Other mycotoxins such as sterigmatocystin will appear brick red on the TLC plates under UV light but can be verified after treatment with different spray reagents ($AlCl_3$ in the case of sterigmatocystin).

3.6. NATURE OF ILLNESS

The growth and profuse sporulation of several *Aspergillus* species on a variety of substrates ensure frequent exposure of animals and humans to these impor-

TABLE 3.2. Mycotoxins Produced by *Aspergillus* spp. on Various Agricultural Substrates

Mycotoxin	Producing Fungus	Substrate	Occurrence
1. Aflatoxins	*A. flavus* *A. parasiticus* *A. nomius* *A. tamarii*	Corn, peanuts, cottonseed, most treenuts, figs, cereals, milk, sorghum	Field/storage
2. Cyclopiazonic acid	*A. flavus* *A. oryzae* *A. versicolor*	Peanut, corn, cheese	Field
3. Sterigmatocystin	*A. versicolor* *A. nidulans*	Corn, grains, cheese, coffee beans	Field/storage
4. Penicillic acid	*A. ochraceus*	Barley, corn, meat	Field/storage
5. Ochratoxin A	*A. ochraceus* *A. sulphurous* *A. sclerotiorum* *A. melleus* *A. awamori* *A. citricus* *A. fonsecaeus*	Barley, beans, cereals, coffee, feeds, corn, oats, rice, rye, wheat	Storage
6. Patulin	*A. clavatus* *A. giganteus* *A. terreus*	Apples, apple juice, apple cider and some other fruit juices	Storage
7. Citreoviridin	*A. terreus*	Corn, other feeds	Storage
8. Xanthomegnin, Viomellein	*A. ochraceus*	Cereals, feeds	Storage

tant agents of disease. These fungi can cause disease by either infection (mycoses) or intoxication (mycotoxicosis). Most of the diseases caused by *Aspergillus* species from food and feed contamination result from mycotoxins in a variety of foods (Table 3.2). In addition to causing mycotoxicoses, several *Aspergilllus* species cause mycoses (aspergillosis) or allergic reactions in animals and humans.

3.6.1. Mycotoxicoses

Aspergillus mycotoxins can cause a broad range of adverse health effects such as suppression of the immune system, teratogenesis effects, and even carcinogenesis. These mycotoxins generally affect specific target organs such as the liver, kidney, thymus, skin, enteric mucosa, bone marrow, and lymph tissue. When high levels of toxin are consumed, acute clinical episodes occur, usually with specific symptoms. But when low levels of toxin are ingested, the clinical symptoms are less obvious and may include such effects as reduced rate of growth, reduced production of milk and eggs, and increased susceptibility to infectious diseases.

In many cases, it is not clear what function the mycotoxins have for the fungus that makes them. Aflatoxins are the classic example. It is tempting to speculate that they defend the fungus from microbial competitors or animal predators, but there is very little evidence that they are effective. As with many secondary metabolites, their adaptive function is a mystery.

Aflatoxins are the most important and widespread toxins that contaminate foods. Most other *Aspergillus* mycotoxins are rarely detected or occur at low levels. Briefly, some of the most significant molecules and their biological effects are outlined here.

Aflatoxins. Many strains of *A. flavus* and *A. parasiticus* and only a few of *A. tamarii* and *A. nomius* produce aflatoxins ("A" from *Aspergillus*, "fla" from the species *flavus*, and "toxin" for poison). Aflatoxins have been found in many foods of animal and plant origin, including corn meal, peanut, cottonseed, spices, cassava, pistachio nuts, rice, cocoa, bread, macaroni, copra, Brazil nuts, oilseeds, pumpkin seeds, meat pies, milk, cheese, sausage, and cooked meat. Among these products, frequent preharvest contamination of corn, cotton, peanuts, and treenuts is of the most concern because of their level of contamination.

The toxicity of these metabolites is widely recognized. Since their discovery in 1960, they have been implicated in carcinogenicity, mutagenicity, teratogenicity, hepatotoxicity, and aflatoxicosis. Currently 18 different aflatoxins are known. The most important are B_1, B_2, G_1, G_2 (Fig. 3.3), M_1, and M_2. Of these, aflatoxin B_1 (AFB_1) is the most common and carcinogenic. Aflatoxin M_1 (AFM_1) is a contaminant in the milk of cows fed with AFB_1-contaminated feed.

Aflatoxicosis is the major syndrome associated with aflatoxins. Signs and symptoms depend on the intake and the frequency. Acute aflatoxicosis, which is a relatively infrequent occurrence, results after high to moderate consumption; it provokes fatty, pale, and decolorized livers, derangement of normal blood-clotting mechanisms resulting in hemorrhages, reduction in total serum proteins of the liver, accumulation of blood in the gastrointestinal canal, glomerular nephritis, and lung congestion. Chronic aflatoxicosis appears when moderate to low concentrations are consumed for a long duration. In this case symptoms include congested liver with hemorrhagic and necrotic zones, proliferation of the hepatic parenchyma and epithelial cells of the bile duct, congested kidneys, and occasional hemorrhagic enteritis. When low concentrations are consumed, impairment of native resistance and immune systems could occur.

Hepatotoxicity results when aflatoxin is absorbed by hepatic tissue, causing damage to cells. Aflatoxin B_1 is degraded by transformation to its reactive form and then to a less toxic and easily excretable metabolite. Reactive intermediates such as 2,3-epoxy-aflatoxin B_1 react with liver macromolecules, resulting in fatty and pale livers, moderate to extensive necrosis, and hemorrhage.

Epidemiological studies provide evidence of the carcinogenicity of aflatoxins to man. Aflatoxin B_1 is also a potent mutagen. The liver is the primary target organ in many animal species. However, tumors in other organs have also been

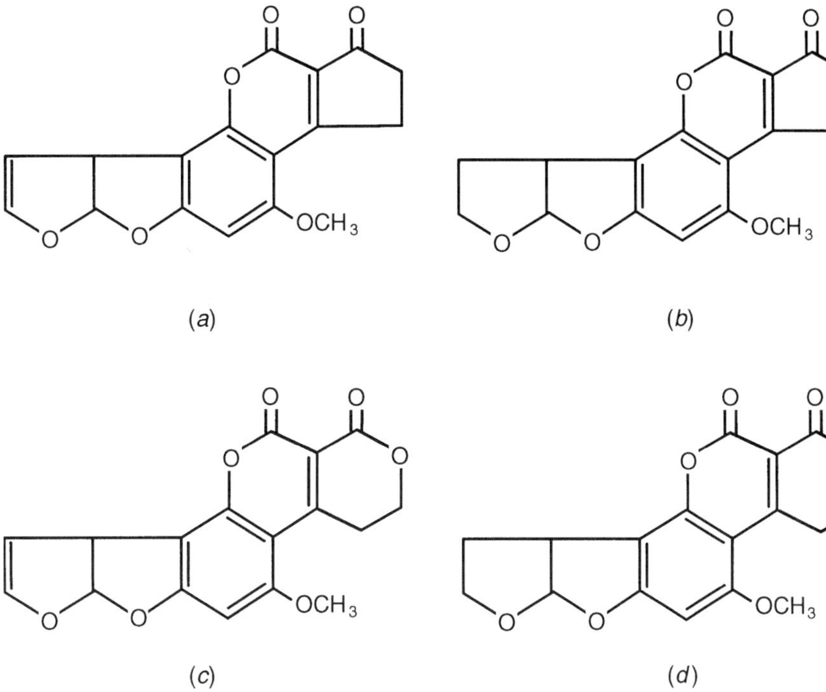

(*a*) (*b*)

(*c*) (*d*)

Figure 3.3. Chemical structures of the most significant *Aspergillus* toxins; (*a*) aflatoxin B_1; (*b*) aflatoxin B_2; (*c*) aflatoxin G_1; (*d*) aflatoxin G_2.

observed in aflatoxin-treated species. This effect arises as a result of the metabolism of the aflatoxin and detoxification outlined above. Highly reactive aflatoxin derivatives (2,3-epoxy-aflatoxin B_1) can interculate DNA, resulting in cellular repair of these adducts often leading to G-to-T transversions and subsequent mutations in the coding region of genes, particulary the oncogene *p53*. It has also been demonstrated that AFB_1 binds to DNA leading to the formation of single-stranded gaps as a result of inhibition of DNA polymerase activity at DNA binding sites. This stimulates an error-prone repair system that may induce mutation. Furthermore, it has been suggested that AFB_1 is teratogenic due to its prenatal effects on certain animals. The inhibitory effect on protein synthesis of eukaryotic cells can impair differentiation in sensitive primordial cells.

Although aflatoxins are among the most mutagenic compounds known for animal systems, it is not clear just how mutagenic they are in humans. A very high incidence (67%) of liver carcinomas in Senegal, China, Swaziland, and Mozambique bear the characteristics of aflatoxin-induced mutation of the *p53* tumor suppressor gene. This mutation has also been associated with liver cancer in Mexico. According to epidemiological studies, raising the permissible limits of aflatoxins in foods in the United States would not greatly increase the

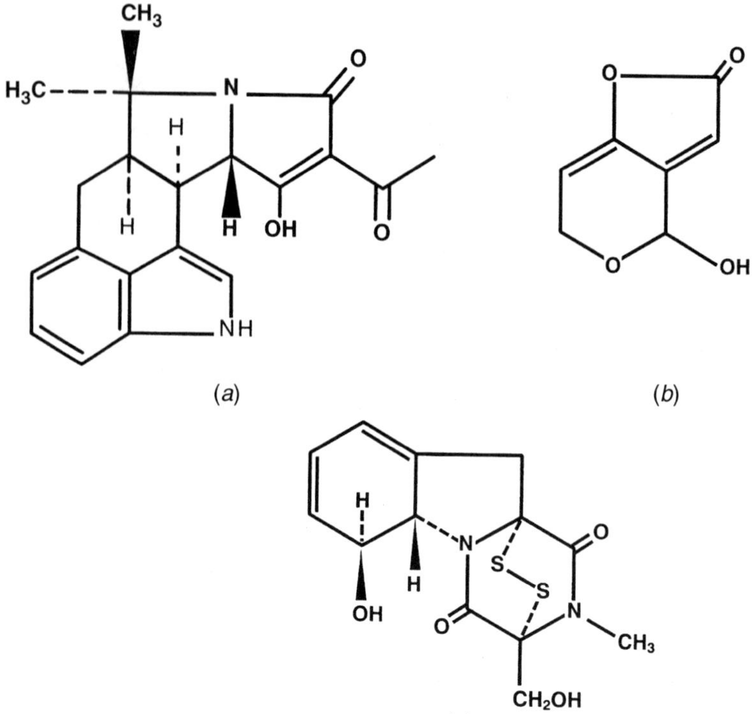

Figure 3.4. Selected toxic secondary metabolites (other than aflatoxins) produced by *Aspergillus* spp.: (*a*) cyclopiazonic acid; (*b*) patulin; (*c*) gliotoxin.

incidence of liver cancer. However, there is a worldwide epidemic of hepatitis C, and hepatitis C patients have impaired liver function and may be much more susceptible to aflatoxins than healthy people.

Cyclopiazonic Acid. Cyclopiazonic acid (Fig. 3.4) is produced primarily by species of *Penicillium* as well as by *A. flavus*, *A. oryzae*, *A. tamarii*, and *A. versicolor* (but not by *A. parasiticus*). This secondary metabolite is toxic to many species of animals, for example, pigs, chickens, rabbits, rats, and dogs. Symptoms include weight loss, diarrhea, depression, convulsions, and death. This compound has been found as a natural contaminant of agricultural raw materials and compounded feeds, particularly in corn and peanuts, and as a residue in poultry meat after its administration to the animal.

Sterigmatocystin. Sterigmatocystin is produced by *A. versicolor* and *A. nidulans* as well as by other fungal genera in corn, barley, moldy rice, coffee beans, and cheese. It is a precursor in the biosynthesis of AFB_1. Epidemiological studies on the occurrence of sterigmatocystin in foods suggest a role for this toxin in carcinogenesis. Evidence of contamination of foodstuffs is scarce;

however, contaminated samples are likely to occur in moldy products that may also contain other mycotoxins or in food items suspected of being responsible for food-related problems or in low-quality foods. Although the carcinogenicity of this molecule in test animals is 10–100 times less than AFB_1, it has also been found to be mutagenic in in vitro systems.

Aspertoxin. A molecule closely related to sterigmatocystin, aspertoxin has been isolated from *A. flavus* and shown to be embryotoxic in the chicken. It is not considered of relevance in animal feedstuffs.

Ochratoxin. Ochratoxins are nephrotoxic metabolites produced by several species of *Aspergillus* and *Penicillium* primarily in stored cereal grains such as barley, rye, wheat, corn, and oats. Natural occurrence in beans, coffee, nuts, and olives has also been reported. Processed and nonprocessed animal products such as dairy products, fish, pork, and sausage have been shown to contain this toxin. Ochratoxin A, the most toxic member, is also a hepatotoxin and causes enteritis in intoxicated animals. Furthermore, its immunosupressive, teratogenic, and genotoxic activities have been demonstrated. *Aspergillus* species are associated with ochratoxin production in tropical areas, whereas production in *Penicillium* species occur in colder climates.

Patulin. This mycotoxin has frequently been found in damaged apples, apple cider, apple and pear juices, and other foods. Although the occurrence of patulin in these products is primarily due to *Penicillium* species, *A. clavatus* has the ability to produce it and could also account for its occurrence. The compound is toxic and produces tumors in rats when injected subcutaneously, but there are no published toxicological or epidemiological data to indicate whether consumption of patulin is harmful for humans.

Xanthomegnin and Viomellein. Although contamination of food and feeds by these mycotoxins has been attributed to *Penicillium* species, *A. ochraceus* has the ability to produce them in many items. Poorly stored cereals and animal feeds have been contaminated by these toxins. Xanthomegnim and viomellein are considered to be nephrotoxic and hepatotoxic.

Penicillic Acid. This hepatotoxin produced primarily by *Penicillium* sp. is also produced by a number of aspergilli in the *A. ochraceus* group. It causes necrosis of the liver, kidney, and thyroid tissue. Penicillic acid can be carcinogenic as well.

Aspergillic Acid. This mycotoxins does not produce tremoring in mice on ingestion but instead produces severe convulsions followed by death. Aspergillic acid is produced by *A. flavus* and *A. sojae*.

Citrinin. This mycotoxin exhibits nephrotoxic properties. Although the presence of this compound in food is more likely due to *Penicillium*, it is also produced by species of *Aspergillus* such as *A. ochraceus* and *A. terreus*.

Gliotoxin. This compound is a metabolite of *A. fumigatus*, *A. chevalieri*, and *A. terreus*. It is considered as a tremorgenic agent and is known to interfere with macrophage function and acts as a hemotopoetic inhibitor. Gliotoxin can be produced in animals and humans and is believed to be a virulence factor for *A. fumigatus*.

Citreoviridin. A potential mycotoxin of domestic animals, it is produced by *Penicillium* species but can also be produced by *A. terreus* on corn and other feeds. This toxin is a neurotoxin causing paralysis and muscular atrophy.

β-Nitropropionic Acid. This is a mycotoxin that causes apnea, convulsions, congestion in lungs, and liver damage. It is produced by *A. flavus*, *A. wentii*, and *A. oryzae*.

Territrem, Aflatrem, and Fumitremorgen. These mycotoxins produced by *A. terreus* as well as *A. flavus* and *A. fumigatus* are tremorgenic toxins that act by blocking acetylcholinesterase activity, causing mild tremoring to convulsive muscle spasm, depending on concentration ingested.

3.6.2. Mycoses

True infections caused by fungi are termed mycoses. Aspergillosis is a general term used for infections in animals and humans caused by *Aspergillus*. *Aspergillus fumigatus* is the most common species causing mycoses, followed by *A. flavus* and *A. terreus* and occasionally *A. niger* and *A. clavatus*. The respiratory tract is the most frequent site of infection, and other tissues may also be involved, such as the sinuses, bones, skin, and meninges. A clinical classification of aspergillosis includes allergic, saprophytic (colonizing), and invasive mycoses. In allergic aspergillosis, the disease is preceded by inhalation of conidia, sensitizing the internal tissue and eliciting allergic responses. Saprophytic aspergillosis results from conidia germinating in the lungs and establishing mycelial colonies, especially in lung cavities previously affected by tuberculosis or other diseases, whereas invasive aspergillosis is a true infection in which the normal animal host barrier is invaded by the fungus to establish disease.

Five lung diseases are commonly caused by *Aspergillus*:

1. *Disseminated aspergillosis* is associated with invasion of the bronchial wall resulting in a definite bronchitis and is often accompanied by pneumonia, mycotic abscesses, and systemic spread in individuals with an immune response that has been compromised naturally or by other primary diseases.

2. *Aspergilloma* or *mycetoma* is the most common form of aspergillosis and consists of superficial invasion of an abnormality in the bronchial tube.
3. *Bronchial asthma* is a general allergic response to molds or pollen.
4. *Extrinsic allergic alveolitis* (Malt worker's lung) is caused by sensitivity to *A. clavatus* growing in the moldy barley on the floors of breweries.
5. *Allergic bronchopulmonary aspergillosis* is the most common form of *Aspergillus* mycoses. It is a hypersensitivity response in the lung to several species of fungal genus *Aspergillus*.

3.7. PREVENTION AND CONTROL

The ubiquitous nature of aspergilli, primarily *A. fumigatus* and *A. flavus* as opportunistic pathogens, ensures wide involvement of these fungi in animal and human disease as well as contamination of food and feed with toxins. These fungi also survive and multiply under a wide range of environmental conditions and are therefore very difficult to eliminate from the surroundings. Mycotoxins are generally heat resistant so they are not destroyed under many food and feed processing procedures. These compounds are also nonantigenic so they do not promote acquired immunity in animals or humans exposed to the toxins.

It is nearly impossible to eliminate fungal populations, particularly aspergilli, from the environment. However, attempts have been made to reduce the fungal populations in areas such as hospitals or nursing homes, where the impact of mycoses can be higher. Removal of spores by including specific filters in air ducts has been very useful. Using sterilized soils and frequent changes of soil for potted plants is another way of reducing *Aspergillus* populations. However, these approaches are not feasible in many developing countries, and a high probability of occurrence of mycoses persists in these regions of the world.

Nearly one-quarter of the world's food crops are affected by mycotoxins each year. This contamination results in enormous negative economic impact arising out of direct losses of crops and livestock as well as from health-related costs associated with mycotoxicosis in the human population. Several countries have instituted regulatory guidelines for permissible levels of toxins in food and feed. Amongst all the *Aspergillus* toxins, only aflatoxins and patulin are currently considered as fungal metabolites of significant concern.

Patulin is generally not considered a serious threat to human and animal health as revealed by subacute and chronic toxicity studies. Therefore, regulatory action based on safety evaluation would not be required. However, several countries consider patulin contamination of food as a problem and have regulatory limits around 50 μg/kg in various foods and juices. Presence of patulin is an indicator of bad manufacturing practices, such as use of moldy raw material. Prevention of the presence of this toxin, therefore, requires careful examination of raw materials prior to food processing. Patulin can also be degraded or

TABLE 3.3. Aflatoxins in Foodstuffs: Maximum Tolerated Levels in Some Representative Countries

	Tolerance (μg/kg)	
Country	Aflatoxin B_1 in Food	Aflatoxin M_1 in Milk/Milk Products
Austria	1	0.01–0.05
Belgium	5	0.01
Brazil	3–15	0.1–0.5
China	50	
France	5–10	0.2
Germany	5	0.01–0.05
India	30	
Japan	10	
Kenya	20	
Mexico	20	
The Netherlands	5	0.02–0.2
Nigeria	20	1
Sweden	5	0.05
Switzerland	1	0.05
United States	20	0.5

removed from solutions by actively fermenting yeasts. During commercial cider making in England, two strains of yeast were able to remove patulin from the medium. The observed decrease in patulin was attributed to the fermentation process including the yeasts rather than to the growth of the yeast itself.

Statutory limits for aflatoxins are imposed by several countries on particular foods (Table 3.3). Generally the limits imposed by many countries range between 5 and 20 ppb (parts of aflatoxin per billion parts of food or feed substrate, i.e., micrograms per kilogram). Consequently, contaminated commodities have to be either discarded or destroyed, putting an enormous burden on the food-related industries.

Economic loss due to aflatoxin contamination is an issue that drives the need for prevention and control of aflatoxins. This issue is primarily relevant to developed countries where food is in ample supply. In developing countries, where sometimes food is in short supply, long-term health implications of aflatoxin contamination are generally overlooked. However, economies of developing countries could be seriously affected when the presence of even the smallest amount of toxin in export commodities is not accepted by developed countries. Elimination of aflatoxins from the food and feed supply would sharply reduce this economic cost. Therefore, research in many laboratories throughout the world is focused on pre- and postharvest control of these toxins.

Several physical and chemical methods have been tested for postharvest management of aflatoxins. Physical methods have included sorting of contaminated seed by, for example, density gradients and discoloration. Chemical treatments of contaminated commodities include use of oxidizing agents (e.g., ozone and hydrogen peroxide). The most promising chemical treatment for aflatoxin destruction involves the use of anhydrous ammonia gas at elevated temperatures and pressure, with greater than 95% reduction in total aflatoxin in commodities such as peanut meal or corn. This technique is used commercially for detoxification of animal feeds in Senegal, France, and the United States.

Preharvest contamination of crops with aflatoxigenic fungi is considered to be the primary source of aflatoxins in food and feed. Therefore, the most adequate method for controlling aflatoxin contamination is to prevent the crop plants from becoming infected with aflatoxigenic strains of *A. flavus* and *A. parasiticus*.

Several agronomic practices can reduce preharvest aflatoxin contamination of certain crops. These include the use of pesticides (fungicides and insecticides), altered cultural practices (such as irrigation), and the use of resistant varieties. However, such procedures have demonstrated only a limited potential for reducing aflatoxin levels in the field, especially in years of drought when environmental conditions favor contamination.

Recent research has therefore focused on acquiring knowledge of (1) the molecular regulation of aflatoxin formation within the fungus, (2) environmental factors and biocompetitive microbes influencing growth of *A. flavus* and aflatoxin synthesis in crops, and (3) host plant resistance factors to aflatoxin accumulation. This research has led to the development of biotechnological strategies and, in the long term, is expected to lead to the development of elite crop lines "immune" to aflatoxin-producing fungi. The genetic regulation of aflatoxin synthesis is now well understood. Genes required for toxin synthesis have been used as expression "probes" in fungal "tester" strains to identify plant factors that may naturally inhibit toxin synthesis. Information gained from these studies has established the identity of plant mechanisms that inhibit aflatoxin formation in the plants and has aided ongoing plant breeding and genetic engineering strategies to suppress the aflatoxin biosynthetic process in crops. Early infection of plants with nontoxigenic strains of *A. flavus* to prevent the subsequent entry of toxigenic strains has been extremely effective in reducing preharvest aflatoxin contamination in cottonseed and peanuts. Identification of critical genes governing aflatoxin formation has also led to the rational design of nonaflatoxigenic biocompetitive strains of toxigenic fungi through use of gene disruption techniques.

REFERENCES

Bhatnagar, D., T. E. Cleveland, and G. A. Payne. 1999. *Aspergillus flavus*. In: *Encyclopedia of Food Microbiology*. New York: Academic, pp. 72–79.

Singh, K., J. C. Frisvad, U. Thrane, and S. B. Mathur. 1991. *An Illustrated Manual on Identification of Some Seed-Borne Aspergilli, Fusaria, Penicillia and Their Mycotoxins*. Hellerup, Denmark: Danish Government Institute of Seed Pathology for Developing Countries.

BIBLIOGRAPHY

Bennett, J. W., and M. A. Klich (Eds.). 1992. *Aspergillus—Biology and Industrial Applications*. Stoneham, MA: Butterworth-Heinemann.

Bhatnagar, D., E. B. Lillehoj, and D. K. Arora. (Eds.). 1992. *Handbook of Applied Mycology*, Vol. 5: *Mycotoxins in Ecological Systems*. New York: Marcel Dekker.

Chang, P.-K., D. Bhatnagar, and T. E. Cleveland. 1999. *Aspergillus*: Introduction. In: *Encyclopedia of Food Microbiology*. New York: Academic, pp. 62–66.

Chu, F. S. 1998. Mycotoxins—occurrence and toxic effects. In: *Encyclopedia of Food and Nutrition*. New York: Academic, pp. 858–869.

Cole, R. J., and R. H. Cox. 1981. *Handbook of Toxic Fungal Metabolites*. New York: Academic.

Cotty, P. J. 1994. Comparison of four media for the isolation of *Aspergillus flavus* group fungi. *Mycopathologia* **125**, 157–162.

Cotty, P. J., P. Bayman, D. S. Egel, and K. S. Elias. 1994. *Agriculture, Aflatoxins and Aspergillus*. In: K. A. Powell et al. (Eds.), *The Genus Aspergillus*. New York: Plenum, pp. 1–27.

Klich, M. A., and J. Pitt. 1992. *Laboratory Guides to Common Aspergillus Species and Their Teleomorphs*. North Ryd NSW, Australia: CSIRO Division of Food Processing.

Powell, K. E., J. F. Renwick, and J. F. Peberdy (Eds.). 1994. *The Genus Aspergillus—From Taxonomy and Genetics to Industrial Application*. New York: Plenum.

Sinha, K. K., and D. Bhatnagar (Eds.). 1998. *Mycotoxins in Agriculture and Food Safety*. New York: Marcel Dekker.

CHAPTER 4

Bacillus Cereus

R. W. BENNETT

4.1. INTRODUCTION

Bacillus cereus has long been considered to be a harmless saphophyte; its role as a food-poisoning microbe was only established in the 1950s. Among the first to report the *B. cereus* gastroenteritis syndrome with precision was U. Plazikowski. A number of documented outbreaks have been presented in relative detail by Schultz and Smith (1994).

Historical accounts have been presented by a number of investigators to establish *B. cereus* as a foodborne pathogen. The ubiquitous distribution of bacilli in the environment makes this organism conducive for the natural contamination of foods through such sources as soil, dust, and materials of plant and animal origin. Although various *Bacillus* have been isolated from a wide variety of foods, *B. cereus* is the species most commonly associated with food-poisoning outbreaks. In a survey of Nigerian spices, 60% were found to contain *B. subtilis* as well as other bacilli. Skim milk powder has been shown to contain *B. subtilis*, *B. polymyxa*, and *B. coagulans*. Psychrotrophic *B. circulans*, *B. mycoides*, and other *Bacillus species* have been isolated from raw and pasteurized milk samples. Although finding *Bacillus* as frequent contaminants in foods does not mean that they can cause foodborne illnesses, documented evidence shows that food-poisoning episodes have occurred due to *B. licheniformis*, *B. subtilis*, *B. pumilus*, and *B. brevis*. Both diarrhea and vomiting were present in these outbreaks, but the nature of the toxins was not determined.

4.2. NATURE OF ILLNESS

Food-poisoning strains of *B. cereus* and other species of this genus produce a number of toxins and other defined extracellular products. The metabolic components most commonly associated with such strains include lecithinase,

Guide to Foodborne Pathogens, Edited by Ronald G. Labbé and Santos García
ISBN 0-471-35034-6

TABLE 4.1. Toxins Found in *B. cereus* Group

Toxin	Gene(s)	Comments
Haemolysin		A tripartite protein that contains a haemolysin and two binding proteins. It also has enterotoxin activity as demonstrated in a rabbit ileal loop assay
BL	*hblA*	
B	*hblC*	
L_1	*hblD*	
L_2		
Enterotoxin	*bceT*	A single protein whose activity has been established on the basis of a mouse ileal loop assay
Cereolysin		Two genes encoded in a single operon
A	*cerA*	Phospholipase C
B	*cerB*	Sphingomyelinase
Cereulide	Not identified	Small dodecadepsipeptide produces vacuole response in Hep-2 cells.

Source: Batt (1997).

proteases, hemolysin, β-lactamase, mouse lethal toxin, cereloysins, and food-poisoning toxins. The enterotoxins are responsible for two food-poisoning syndromes caused by this organism. A number of these metabolic components including the toxins have been identified genetically and are documented by Batt (1997) and listed in Table 4.1.

4.2.1. Diarrheal Syndrome

This syndrome was the first recognized as the illness caused by the consumption of foods contaminated with large numbers of enterotoxigenic *B. cereus*. This syndrome is primarily characterized by abdominal cramps with profuse watery diarrhea, rectal tenesmus, and occasionally nausea that seldom results in vomiting. It has an incubation period within the range of 8–16 h. The symptoms generally resolve in 12–24 h.

4.2.2. Emetic Syndrome

The second type of illness described as the "emetic type" of intoxication caused by *B. cereus* is characterized by an acute attack of vomiting that occurs 1–5 h after consumption of contaminated food.

4.3. CHARACTERISTICS OF AGENT

4.3.1. Organism

This aerobic, endospore-forming bacterium have been suspect as an agent of foodborne illness since the early days of microbiology. While some skepticism has surrounded the association with gastroenteritis, only recently has this organism gained full acceptance as a potential foodborne pathogen of significant consequence.

The cells of *B. cereus* are observed microscopically as large, gram-positive rods that are motile by means of peritrichous flagella. The cells are typically 1.0–1.2 μm in diameter by 3.0–5.0 μm in length. Their physiological properties have been documented by Kramer and Gilbert (1989). The metabolic features of *B. cereus* and other culturally related *Bacillus* spp. are discussed later in this chapter. This organism has a minimum growth range of around 10–12°C and a maximum of 48–50°C. The organism grows rapidly in foods held in the 30–40°C range. Growth has been demonstrated over the pH range 4.9–9.3. The spores possess a resistance to heat that mimics other mesophilic spore formers.

4.3.2. Toxins

The diarrheal strains of *B. cereus* produce an extracellular mature protein that elicits diarrhea in monkeys and can be identified in culture fluids by serological assay. This toxin has been reported to be a protein with a molecular weight of approximately 55–60 kDa. The toxin is antigenic, and specific antibodies to this protein have been produced for its serological identification. Less is known about the emetic component than the diarrheal factor, but apparently it is not produced by all strains of *B. cereus*. When fed to monkeys with rice, a large number of strains implicated in outbreaks with symptoms of vomiting elicited a vomiting response. The emetic activity is associated with a peptide. Recent evidence shows this component to be a dodecadepsipeptide that correlates to vacuole formation in Hep-2 cells. This entity is of low molecular weight, extremely resistant to heat (withstanding 120°C for more than 1 h), and produced as a cell-free component in broth cultures. In addition to the production of emetic responses in monkeys, heat-treated culture fluids from emetic strains injected intravenously induced vomiting in cats. A similar response in cats is also produced by the injection of heated (100°C for 90 min) culture fluids of *B. subtilis* and *B. licheniformis* culture that have been implicated epidemiologically in food poisoning. It is not known whether the toxin (or toxins) produced by these species are the same as those produced by emetic *B. cereus* strains.

4.4. EPIDEMIOLOGY

Each year many thousands of food-poisoning outbreaks in all countries of the world go unreported to health agencies responsible for the collection of epidemiological data for tracking cases and outbreaks of foodborne illnesses. Proper documentation of specific outbreaks is imperative to construct epidemiological profiles. Such documentation should include brief histories of the illness (symptoms, incubation, and duration), details on those affected (type of population involved and attack rate), and information concerning the incriminated food (nature, where obtained, preparation, storage, and handling). This documentation is helpful in the laboratory investigation of such illnesses.

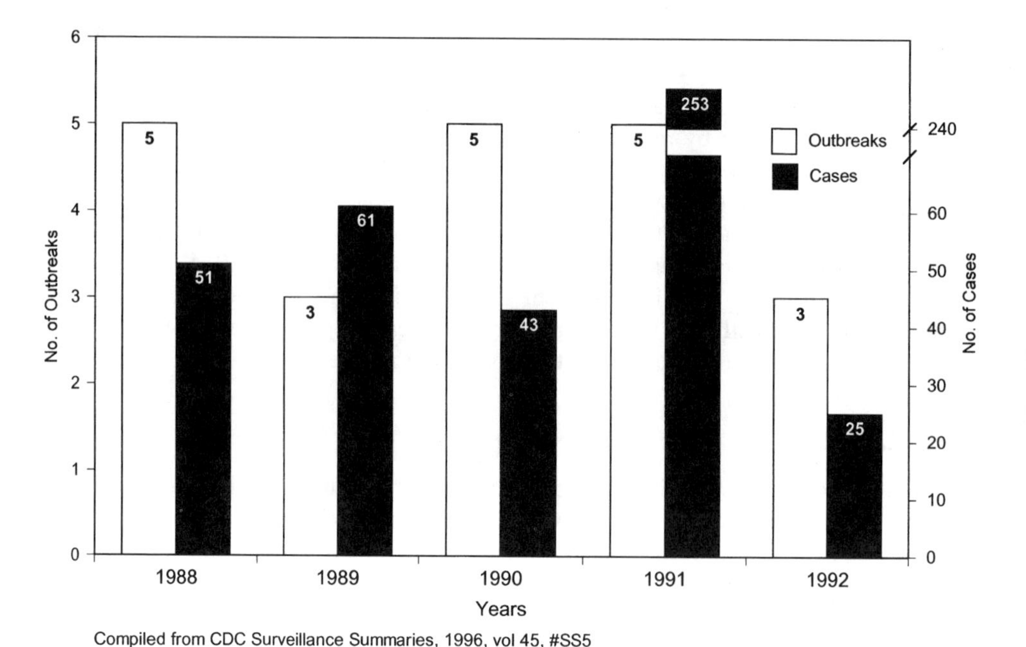

Figure 4.1. Comparative numbers of reported outbreaks caused by *B. cereus* and other species in the United States, 1988–1992.

In spite of improved harvesting techniques, processing approaches, packaging, and storage improvements, outbreaks caused by *Bacillus* continue to occur. The number of documented outbreaks caused by *B. cereus* and related species occurring in the United States from 1988 to 1992 is given in Figure 4.1.

One of the difficulties in the investigation of foodborne illnesses is that symptoms of those afflicted might be similar to symptoms caused by other etiological agents. The *B. cereus* diarrheal syndrome symptoms mimic the clinical features of *Clostridium perfringens*, while symptoms elicted by the emetic toxin are not unlike those of staphylococcal intoxication. *Bacillus cereus*–related illnesses and the clinical features of *C. perfringens* and *Staphylococcus aureus* food poisonings are given in Table 4.2.

As a consequence of these symptomatic parallels, it is imperative that laboratory investigations be carried out completely. Laboratory analysis should include enumeration, identification of the organism, determination of enterotoxigenicity of the recovered organism, and demonstration of preformed toxin in the incriminated food. The presence of the diarrheal factor is usually associated with proteinaceous foods, vegetables, sauces, and puddings. In contrast, the emetic form of the illness is associated with farinaceous foods, particularly cooked rice and other starchy foods. A vast number of foods previously involved in food poisoning outbreaks have been documented by other investigations.

TABLE 4.2. Comparative Clinical and Epidemiological Similarities in *B. cereus* Foodborne Illness and Those Caused by *C. perfringens* and *S. aureus*

		B. cereus Toxins		
	C. perfringens Enterotoxin	Diarrheal Syndrome	Emetic Syndrome	*S. aureus* Enterotoxin
Onset of symptoms, h	8–22	8–16	1–5	2–6
Duration of illness, h	12–24	12–14	6–24	6–24
Diarrhea, abdominal cramps	Predominant	Predominant	Fairly common	Common
Nausea, vomiting	Rare	Occasional	Predominant	Predominant
Pathogenesis	Toxin mediated[a]	Toxin mediated[b]	Toxin mediated[c]	Toxin mediated[c]
Principal food vehicles	Cooked meat and poultry	Meat products, soups, vegetables, puddings, sources	Cooked rice and pasta	Cold cooked meat and poultry, dairy products

[a]Sporulation-associated toxin released in the small intestine.

[b]Toxin may be preformed in food or produced in the small intestine.

[c]Toxin preformed in food.

Source: Modified from Kramer et al. (1982).

4.5. DETECTION OF ORGANISM AND TOXINS

Although *B. cereus* has a ubiquitous distribution in the environment and can be isolated from a variety of processed foods, its presence in foods is not a significant hazard to consumer health unless it is able to grow and produce toxins. Consumption of food containing 10^5 viable enterotoxigenic *B. cereus* cells/g of food has resulted in outbreaks of food poisoning.

4.5.1. Metabolic Indicators

The most important features of the four biotypes of the *B. cereus* group are summarized in Table 4.3. However, even within a species, variability and strain heterogeneity are relatively common. One of the basic metabolic characteristics of *B. cereus* is that it produces what is termed as an "egg yolk reaction," which is used to describe the turbidity developed in agar that contains egg yolk. The responsible agent is an extracellular substance termed egg yolk turbidity factor, lecithinase, or phospholipase. While it has been established that *B. cereus* produces phospholipase C, there is evidence that the turbidity developed in egg yolk may be due to a more complex series of events than the action of a single enzyme. Quantitation of *B. cereus* in a food is, generally, performed by a simple surface-plating procedure. In many laboratories, use is made of (1) the ability of *B. cereus* to produce turbidity surrounding colonies growing on agar containing

TABLE 4.3. Characteristics of *B. cereus* and Culturally Similar Species

Feature	*B. cereus*	*B. cereus* var. *mycoides*	*B. thuringiensis*	*B. anthracis*
Egg yolk reaction	+	+	+	(+)
Acid from mannitol	–	–	–	–
Catalase	+++	+	+	+
Gram reaction	+	+	+	+
Motility	±	–	±	–
Hemolysis (sheep red blood cells)	+	(+)	(+)	–
Rhizoid growth	–	+	–	–
Toxin crystals produced	–	–	+	–
Anaerobic utilization of glucose	+	+	+	+
Reduction of nitrate	+	+	+	+
VP reaction	+	+	+	+
Tyrosine decomposition	+	(+)	+	(+)
Resistance to lysozyme	+	+	+	+

Note: + = 90–100% of strains are positive; (+) = usually weakly positive; ± = 50–90% of strains are positive; – = less than 10% of strains are positive.

Source: From *Compendium of Methods for the Microbiological Examination of Foods*, APHA, Washington, DC, 2000, p. Reprinted with permission of the American Public Health Association.

egg yolk and (2) the resistance of *B. cereus* to the antibiotic polymyxin B to create a selective and differential component in plating medium.

4.5.2. Enumeration

Four such methods have been described for the enumeration of *B. cereus*. Of these, three methods have incorporated mannitol to enhance differentiation. The use of blood agar has been used by other investigators. Mannitol egg yolk polymyxin (MYP) medium is specified for use as an AOAC International surface plating technique; however, other equally efficient methods may be used. In instances where *B. cereus* counts are expected to be $<1000\ g^{-1}$, the alternative AOAC International most probable number (MPN) method using tryptic–soy polymyxin broth is recommended. Colonies of appropriate morphological appearance that show the characteristic turbidity surrounding colonies on the plating medium are considered to be presumptive *B. cereus*.

4.5.3. Isolation and Identification

Presumptive colonies from plating media should be transferred to nutrient agar slants and tested to confirm their identify, as shown in Table 4.3. The differential and confirmatory tests as specified in official AOAC methods are preferred; however, other methods could be used. Lecithinase-positive, mannitol-negative colonies on MYP agar are highly indicative of *B. cereus*; however, because of the similarity of reactions by other members of the *Bacillus* group, isolates must be further tested to differentiate them from other *Bacillus* species before definite identification can be made. The procedures for the enumeration, isolation, differentiation, and subsequent identification have been thoroughly detailed in AOAC International's *Official Methods of Analysis* (1995) and by other associations dedicated to the validation of microbiological methods.

4.5.4. Toxins and Toxin Assays

A number of investigators have studied the production of toxins in laboratory media as well as in a variety of foods and have developed assays for the routine identification of toxins and their pathological effects. These studies have included the diarrheal and emetic enterotoxins, hemolysins, phospholipase C, and lethal toxins (mouse assay); however, the toxins (diarrheal and emetic) responsible for foodborne illnesses appear to be unrelated to hemolysins and phospholipase C.

4.5.5. Diarrheal Toxin

The heat-labile purified diarrheal enterotoxin has been reported to cause marked increased vascular permeability in the skin of rabbits and a positive fluid accumulation response in rabbit ligated ileal loop test. The diarrheal factor is antigenic and can be used to produce specific antibodies in rabbits and other animals for use in serological diagnostic assays. A microslide gel double-diffusion assay for the identification and semiquantitation of the diarrheal factor has been developed. Commercial immunoassays for the identification of the diarrheal toxin are currently available, including a reversed passive latex agglutination (RPLA, Oxoid) assay and an enzyme-linked immunosorbent assay (ELISA, TECRA International).

4.5.6. Emetic Toxic

The emetic toxin recently isolated and purified on the basis of its capability to induce vacuolation in Hep-2 cells is a 1.2-kDa cyclic peptide that is a highly stable to heat, acidic pH and proteolytic activity entity. This entity in strains of *B. cereus* incriminated in emetic syndrome outbreaks has been demonstrated by monkey feeding and in kittens by intravenous injection of heated-cell-free culture fluids. The tissue culture assay employing Hep-2 cells and a recently described spermatozoa toxicity bioassay may be useful for the identification and quantitation of this toxin. An assay using microbial cells that is as sensitive as the Hep-2 cell assay has been recently developed (N. Belay, unpublished observations). This simple assay is based on the potassium ionophore activity of the emetic toxin cyclic peptide.

4.6. PHYSICAL METHODS FOR DESTRUCTION

Low temperature is often used as a means of limiting or preventing proliferation of *B. cereus*, although some studies have shown that some strains can grow and produce toxins at reasonably low temperature. Studies on growth temperature requirements with 50 strains showed some strain variation. In this study, all the strains grew at temperature ranges from 14 to 40°C, although only half of the strains grew at 45°C and three strains grew at 49°C. More than half the strains grew at 10°C, six strains grew at 8°C, and one strain grew at 6°C. Other studies have shown that strains of *B. cereus* isolated from foods incriminated in food-poisoning outbreaks were able to grow and produce toxins at 4–7°C. Alkaline conditions lead to a growth lag while pH 5.0 conditions have been shown to be more inhibitory to growth with clinical and veterinary strains of *B. cereus* than to food-poisoning strains. Methods involving the inhibition and inactivation of *B. cereus* spores include the determination of *D* values. A study of *B. cereus* spores in heated custard shows *D* values of 3.6, 2.8, and 2.2 min at 90, 95, and 100°C, respectively; however, when the product was adjusted to pH 6.2, the *D* values decreased. Conversely, increasing the pH (7.6) also increased the *D* value. Studies have shown that *B. cereus* spores inoculated into uncooked rice were not inactivated completely during the cooking process (100°C for 30 min), although viable cells were destroyed in fried rice when subjected to 180–190°C for 5–7 min. Studies on the effect of various solutes on spore germination and growth of *B. cereus* show that spore germination is less sensitive to reduced A_w than outgrowth or growth of vegetative cells of *B. cereus*.

Other methods investigating the inhibition and inactivation of *B. cereus* include microwave, treatment by irradiation, and chemicals. Studies have been conducted on the effect of NaCl, pH, and temperature combinations on the growth of *B. cereus*. At 0.5% NaCl, growth occurred over a range of 14–41°C with a pH range of 4.7–6.8. Increasing the NaCl level to 3.0% restricted the growth to temperatures ranging from 15 to 40°C and to pH values of 4.9–6.8; however, no growth was observed at 7.0% NaCl. Some studies have shown that a few strains of *B. cereus* can grow in the presence of 9.5% NaCl. Nitrite acts as a membrane-directed sulfhydryl agent against *B. cereus*, although nitrite inhibition is reversed when the exposure to nitrite is removed.

Nicotinamide-treated spores germinate poorly and lose their capability to germinate over extended storage. It has been suggested that the addition of nicotinamide to foods could be useful as an anti–*B. cereus* agent.

4.7. PREVENTION AND CONTROL

The ubiquitity of *B. cereus* in the general environment, the stability and resistance of their spores, and their presence on raw agricultural products provide justifiable concern for *Bacillus* spp. as actual or opportunistic foodborne pathogens. Their presence on raw agricultural products ensures possible contamination of the food-processing environment and equipment. As a consequence, effective prevention and control measures would include (a) the control of *Bacillus* spore germination and (b) prevention of proliferation of the vegetative cells in foods. Effective heat or irradiation treatment may be necessary where complete destruction of the organism is desired. The creation of unfavorable conditions such as low temperatures, low A_w, or low pH in foods may greatly reduce the spore germination of enterotoxigenic *Bacillus* spp., thus preventing toxin formation in foods. A number of cooking methods such as steaming under pressure, thorough roasting, frying, and grilling are likely to destroy both vegetative cells and spores, although cooking at temperatures below 100°C might allow survival of *Bacillus* spores. Of major concern to the consumer is the multiplication of the organism during inadequate cooling or the holding of moist foods in a nonrefrigerated state over periods that would allow cell proliferation. Favorable conditions for enterotoxigenic *Bacillus* are sometimes provided by cooking procedures that activate spores and then by slow cooling and mass storage of foods at temperatures above 10°C and below 60°C. If food storage is necessary, it should be cooled rapidly to a temperature (8°C) that prevents growth of the organism. If a food must be held in a warm state, such as might be necessary in institutional settings, the temperature should be maintained above 60°C. Of major concern to the food processor and retailer in the prevention of food-poisoning outbreaks should be the effective utilization of hazard analysis critical control point (HACCP) systems by all who are involved in the harvest, manufacture, distribution, storage, and serving of food.

REFERENCES

Batt, C. 1997. *Bacillus cereus*. In: *Encyclopedia of Food Microbiology*. London: Academic, pp. 1–6.

Kramer, J. M., and R. J. Gilbert. 1989. *Bacillus cereus* and other *Bacillus* species. In: M. P. Doyle (Ed.), *Foodborne Bacterial Pathogens*. New York: Marcel Dekker, pp. 21–70.

Kramer, J. M. et al. 1982. Identification and characterization of *Bacillus cereus and* other *Bacillus* species associated with foods and food poisoning. In: J. E. L. Corry, E. Roberts, and F. A. Skinner (Eds.), *Isolation and Identification Methods for Food Poisoning Organisms*. Society of Applied Bacteriology, Technical Series 17. London: Academic, pp. 261–286.

Official Methods of Analysis, 15th ed. 1995. Gaithersburg, MD: AOAC International, Method 980.31.

Schultz, F. J., and J. L. Smith. 1994. *Bacillus:* Recent advances in *Bacillus cereus* food poisoning research. In: Y. H. Hui, J. R. Gorham, K. D. Murrell, and D. O. Cliver (Eds.), *Foodborne Disease Handbook*, Vol. 1: *Diseases Caused by Bacteria*. New York: Marcel Dekker, pp. 29–62.

BIBLIOGRAPHY

Bennett, R. W., and S. M. Harmon. 1988. *Bacillus cereus* food poisoning. In: A. Balows, W. J. Hausler, Jr., M. Ohashi, and A. Turano. (Eds.), *Laboratory Diagnosis of Infectious Diseases, Principles and Practice*. New York: Springer-Verlag, pp. 83–93.

Bennett, R. W., and N. Belay. 2000. *Bacillus cereus*. In: K. Ito (Ed.), *Compendium of Methods for the Microbiological Examination of Foods*. Washington, DC: American Public Health Association.

Bennett, R. W. 1995. *Bacillus cereus, Diarrheal Enterotoxin*. In: *FDA Bacteriological Manual*, 8th ed. Gaithersburg, MD: AOAC International, pp. 15.01–15.09.

Centers for Disease Control and Prevention. 1996. *Surveillance Summaries* (1988–1992). Atlanta, GA: Centers for Disease Control and Prevention, 45, #SS 5.

Gilbert, R. J., et al. 1981. *Bacillus cereus* and other *Bacillus* species: Their part in food poisoning and other clinical infections. In: R. C. Berkely and M. Goodfellow (Eds.), *The Aerobic Endospore-Forming Bacteria: Classification and Identification*. Society of General Microbiology Publications No 4. London: Academic, pp. 298–314.

Gilbert, R. J. 1979. *Bacillus cereus* gastroenteritis. In: H. Riemann and F. L. Bryan (Eds.), *Foodborne Infections and Intoxications*, 2nd ed. New York: Academic, pp. 495–518.

Goepfert, J. M., et al. 1972. *Bacillus cereus J. Milk Food Technol.* **35**, 213–227.

Jay, J. M. 1992. Food poisoning caused by gram-positive sponeforming bacteria. In: J. M. Jay (Ed.), *Modern Food Microbiology*, 4th ed. New York: Van Nostrand Reinhold, pp. 479–509.

Johnson, E. A. 1990. *Bacillus cereus* food poisoning. In: D. O. Cliver (Ed.), *Foodborne Diseases*. San Diego, CA: Academic, pp. 127–135.

CHAPTER 5

Campylobacter

NORMAN J. STERN

5.1. INTRODUCTION

Campylobacter spp. are important bacterial pathogenic agents, can cause human gastroenteritis, and are transmitted primarily through foods of animal origin (zoonotic). The most significant of the human enteropathogens in the genus consist mainly of the species *C. jejuni* and to a far more limited degree also include *C. coli*, *C. lari*, and *C. upsaliensis*. Subsequently, these species will be referred to as *Campylobacter*. These human pathogens have previously been labeled as "newly emerging" or "newly recognized," but this is certainly inaccurate. In 1957, E. O. King was first to suggest that these could be an important zoonotic cause of human disease. Forty-three years later, the word "emerging" no longer seems appropriate. Corroboration of King's hypothesis was provided in the 1970s, and it is only a reflection of our limited response times to suggest that these organisms are newly recognized.

Campylobacter spp. are fragile by nature and only pose a health problem if foods are prepared using poor hygiene or if food is consumed before being thoroughly cooked. Because the organism is so fragile, the food handler should be able to prevent transmission through food products, although as indicated by the frequency of human disease, foodborne transmission is widespread.

The primary ecological niche for *Campylobacter* is the intestinal tract of warm-blooded animals. The minimum growth temperature for the genus is about 30°C, and for *C. jejuni* about 37°C; therefore there are rare situations in which they are likely to multiply in foods. Measures to reduce contamination by *Campylobacter* during animal production and during processing of raw products should be used to reduce the incidence and levels of these organisms.

Guide to Foodborne Pathogens, Edited by Ronald G. Labbé and Santos García
ISBN 0-471-35034-6

5.2. NATURE OF DISEASE

Symptoms of infection range from mild to severe: Watery and/or bloody diarrhea, abdominal pain, malaise, fever, and headaches may occur. The onset of disease follows ingestion by 2–5 days. Symptoms usually diminish within a week and are most unpleasant 2–5 days after onset. Reappearance of the disease symptoms can recur. In feeding studies the infectious dose for humans has been reported as low as 500 cells. Variation in infectious dose likely exists because of differences in virulence of *Campylobacter* strains and susceptibility in the host. Campylobacteriosis is typically a self-limiting disease, although erythromycin is a highly effective treatment. Hospitalization or death due to *Campylobacter* is reported less often than for salmonellosis or listeriosis but can occur.

The mechanisms by which *C. jejuni* produces disease are not agreed upon. Factors such as toxins, motility, and adherence have been described as possible virulence factors. *Campylobacter* may colonize the human in a manner similar to other spiral bacteria normally found in the gut. These bacteria are also corkscrew-like and microaerophilic and invade deeply into intestinal crypts. The adaptation to the intestinal mucous niche may be an important factor in expressing virulence for *Campylobacter*. The capability of the organism to move may be an important virulence factor for *C. jejuni*. The profuse watery diarrhea observed in some *Campylobacter* infections may indicate that these strains could be enterotoxigenic. *Campylobacter* cytotoxic and cytotonic factors have been described. The lack of an easy-to-use, inexpensive animal model that depicts human infections has seriously hampered the identification of virulence factors in *C. jejuni*. Although many animal models have been proposed, none has gained wide acceptance.

Guillain-Barré syndrome (GBS) is a human neurological pathology that can result in acute neuromuscular paralysis. Approximately 40% of patients with GBS have evidence of recent *Campylobacter* infection. Perhaps 20% of GBS patients have long-term disability and approximately 5% die. Another sequel to *Campylobacter* infection is Reiter syndrome, a reactive arthropathy. There appears to be an autoimmune component to both GBS and Reiter syndrome in its involvement with *Campylobacter* infection.

5.3. CHARACTERISTICS OF AGENT

The taxonomic status of the genus is in flux and is likely to continue changing for some time to come. *Campylobacter* grows between 25 and 43°C, are gram-negative, motile, curved or spiral rods (Fig. 5.1), are oxidase positive, and do not ferment or oxidize carbohydrates. Catalase-positive *Campylobacter* species are most frequently associated with human disease but catalase-negative species can cause disease.

Campylobacter is generally inactive in many conventional biochemical tests, and identification is based on only a few morphological and biochemical fea-

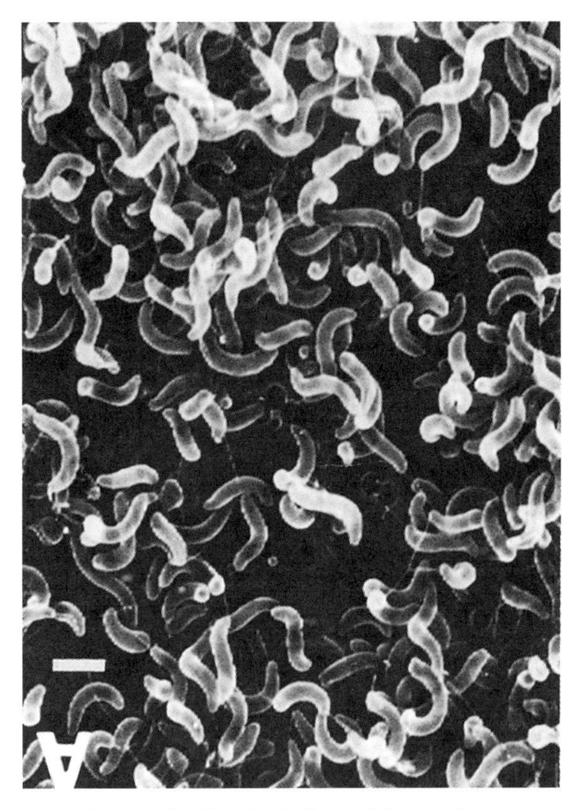

Figure 5.1. Electron micrograph of typical *Campylobacter* during exponential growth. From D. Rollins and R. Colwell, 1986, *Applied and Environmental Microbiology*, vol. 52, pp. 531–538. Reprinted with permission of American Society for Microbiology.

tures. With certain strains, species identification is based on the results of only one test, and in some cases definitive identification is not possible with routinely available laboratory tests. Confirmation of *Campylobacter* (*jejuni*, *coli*, and *lari*) isolation may be achieved through application of immunologically based latex agglutination assays or by specific gene-based tests. The need for species differentiation appears less important than is the genetic characterization of isolates.

5.4. EPIDEMIOLOGY

Rather than manifesting disease in humans through large outbreaks, the epidemiology of human *Campylobacter* infection reflects a pattern of numerous sporadic cases. Harris and colleagues estimated that 48% of *Campylobacter* spp. cases could be traced back to the mishandling or consumption of contaminated poultry product. These were accounted for primarily in sporadic cases occurring among members of a health maintenance organization in

Seattle. Among university students in Georgia 70% of cases were accounted for by eating chicken, often undercooked or raw. In another study handling raw chicken, as opposed to eating it, emerged as a risk factor. Poor kitchen hygiene associated with the frequency of using soap to clean the kitchen cutting board. Work has shown that *Campylobacter* serogroups isolated from poultry sources are the same as some of the predominant serogroups isolated from humans. Improperly treated milk has been a source of outbreaks for the organism.

In the 1970s isolation rates of 7.1 and 5.1% in the United Kingdom and the United States, respectively, of *Campylobacter* were reported from the stools of patients with gastroenteritis. Prior to these early reports, clinical laboratories responsible for isolation and identification of bacterial agents did not culture stools at 42°C under microaerobic atmospheres. Even after numerous reports on the significance of *Campylobacter*, many clinical laboratories still fail to routinely culture under these specified conditions.

Campylobacter is now reported as the most frequently identified causative bacterial agent of acute infectious diarrhea in developed countries, including the Netherlands, the United Kingdom, France, Italy, Sweden, the United States, and Australia. *Campylobacter* infections are likely to be more common than salmonella infections. In the United States in 1996, FoodNet began collecting population-based active surveillance on culture-confirmed cases of seven potentially foodborne diseases, including *Campylobacter*. Incidence rates in stool specimens or specimens from normally sterile sites were highest for campylobacteriosis at 25 per 100,000 population. Rates for this disease ranged from 14 (Georgia) to 58 (California). In 1996, the Public Health Laboratory Service (United Kingdom) Communicable Disease Surveillance Centre reported a total of just over 43,000 laboratory isolations of campylobacteriosis. Numbers of reports received in the first 20 weeks of 1997 were 7% higher than in the same period of 1996. A study conducted in the Netherlands in 1992 and 1993 yielded an annual incidence of *Campylobacter* infection of 69 cases per 100,00 persons. However, the true incidence of campylobacteriosis in the general population is likely to be much higher, since only a small proportion of persons with gastrointestinal complaints consult a physician. In developed countries campylobacteriosis usually occurs most frequently in children under the age of 4 years and in young adults between 20 and 30 years of age. In developing countries *Campylobacter* infection occurs most frequently in children under 5 years of age, with a prevalence of infection much greater than in developed countries.

Campylobacter isolates from swine (typically *C. coli*) do not typically belong to serotypes commonly found in humans. Therefore, pigs do not appear to be a major vector for human campylobacteriosis. Although cattle are an important source in milk-borne transmission of *Campylobacter*, cattle are not strongly implicated in transmission through meat products.

Serological surveys demonstrated persons employed in poultry and red meat plants or having exposure to these animals have a significantly higher rate of seropositivity to *Campylobacter* antigens than rural field laborers. Contact with household pets such as cats and dogs may also be responsible for transmission of *Campylobacter* infection. Similar *Campylobacter* biotypes or serotypes are frequently isolated from humans and animals. Newly acquired puppies and kittens are generally a greater risk factor than are older adult animals, especially if the pet develops diarrhea. Studies conducted in England reported three cases of *Campylobacter* infection associated with cats and 97 cases associated with dogs.

5.5. DETECTION OF ORGANISM

Because *Campylobacter* is so fragile and cannot grow outside of its host, they frequently die under normal atmospheric concentrations of oxygen and normal ambient temperatures. Therefore, low concentrations or injured *Campylobacter* may be present in foods. Selective enrichment may be needed to detect the few culturable cells of *C. jejuni* that may be present. In food products such as freshly processed poultry, where large numbers may be expected, direct plating is appropriate. Large sample sizes, selective enrichment broth, suitable microaerobic conditions, and selective isolation media or filtration techniques are important for isolating *Campylobacter* from nonpoultry foods. Various approaches for selection and methods to produce a microaerobic atmosphere have been developed for isolation and are described in the *Compendium of Methods for the Microbiological Examination of Foods* as well as in standard methods procedures of various countries.

Food samples must be processed rapidly to ensure optimum isolation of *Campylobacter*, which is sensitive to many environmental conditions. Improper handling or storage of samples before testing may negate the value of a sensitive isolation procedure. As indicated above, the organism is sensitive to drying, oxygen, and storage at room temperature. Therefore, if the samples are improperly handled before testing, the value of any isolation procedure is compromised. Analysis of samples must address the fragile nature of the organism and should take the following facts into consideration: *Campylobacter* is not likely to grow in foods held under typical storage conditions. The organism dies more quickly at 25°C than at either 4°C or 30°C. Studies indicate that *C. jejuni* survives best in foods held at refrigeration temperature but is very susceptible to freezing conditions. Within 5 days of refrigerated storage, *Campylobacter* on processed chicken carcasses becomes injured and requires enrichment for recovery. The optimum temperature for growth for *C. jejuni* is 42–43°C, and thermal inactivation begins at 48°C. Heat injury and repair of *C. jejuni* begins at 46°C, and the existence of an injured, chick-colonizing, but nonculturable *Campylobacter* has been documented. *Campylobacter* does not survive pasteurization treatment for milk or meat products.

Campylobacter jejuni is very sensitive to sodium chloride. Campylobacters are quite sensitive to drying conditions, although it survives for up to 6 weeks at 4°C in an environment of 14% or less relative humidity. In laboratory media, *C. jejuni* grows well at pH 5.5–8.0. Optimal growth is in the pH range of 6.5–7.5, while no growth occurs at pH 4.9 or lower.

The surface rinse procedure is useful for sampling poultry carcasses and moderately large pieces of foods. Briefly, the undivided sample (about 1–2 kg) is placed in a sterile, "zipping" plastic bag with 100 to 400 mL of buffered peptone water (BPW) or other appropriate liquid. The surface is rinsed by shaking and massaging for approximately 1 min as desired. For enrichment culture, a 25-mL portion of the rinse may then be added to 25 mL of double-strength enrichment media. The Hunt enrichment technique and the Bolton enrichment broth have met with favorable results.

Selective agar media can be used for isolating *C. jejuni* from foods. Most prominent among these selective media are the modified CCDA-Preston blood-free medium and Campy-Cefex. The Preston blood-free and the Campy-Cefex media provide reasonable selectivity and allow for good quantitative recovery from poultry carcasses. A recently developed medium, Campy-Line agar, has been proposed for use in enumerating campylobacters from poultry carcass rinses: This plating medium simplifies enumeration by increasing color contrast between growing colonies and the agar.

Comparison of most probable number (MPN) enrichment techniques and direct Campy-Cefex plating indicated no statistical advantage gained in enumerating *Campylobacter* levels from freshly processed carcasses by either direct plating or MPN techniques. For freshly processed chicken carcass samples we estimate that direct plating enumeration procedures cost is approximately 5 times less than comparable MPN enrichment procedures. The time required for direct plating for enumeration is greatly diminished when compared to the MPN technique.

In this case two loops from the above enrichment broths should be transferred to the surface of each selective plate and streaked for isolated colonies. Swab samples should be directly inoculated to the selective plates. Twenty-five grams of foods may be added directly to enrichment broth.

Inoculated plates are held in a microaerobic atmosphere (5% O_2, 10% CO_2, 85% N_2) at 42°C for 18–48 h. The plates can be inspected for characteristic colonies after 18–24 h but must be reincubated under the same conditions for an additional day if typical colonies are not observed. Typically on moist media, *Campylobacter* may swarm, which is a useful diagnostic growth characteristic; however, this type of confluent growth makes it difficult to enumerate or to obtain isolated colonies. Individual colonies can be produced on dry plates.

In a recent comparison, the Hunt enrichment procedures were compared to methods using Bolton enrichment broth (BEB). When incubated under microaerobic conditions, the simpler BEB procedure provided similar recovery to the Hunt method. However, these enrichment methods are qualitative and often do not provide adequate quantitative information. A thorough comparison between numerous enrichment procedures still is required before one method can be considered superior to another.

The optimum atmospheric composition for growth of *Campylobacter* is 5% oxygen, 10% carbon dioxide, and 85% nitrogen. The most effective means to create this required atmosphere is by introducing a mixture of 5% O_2, 10% CO_2, and 85% N_2 from a gas cylinder into an anaerobic jar or sealed plastic bag. The container, with the petri plates, is then placed into a 42°C incubator.

Campylobacter cells are curved, S-shaped, gull-winged or spiral rods, 0.2–1 μm wide, and 0.5–5 μm long with darting or corkscrew-like motility. Cells in older cultures may be coccoid and nonmotile. Colonies of *Campylobacter* spp. can be identified by latex agglutination assays, which are commercially available (Integrated Diagnostics, Baltimore, MD; Mercia Diagnostics, Shalford, UK; Becton Dickinson Microbiology Systems, Cincinnati, OH). Enzyme-linked fluorescent assay (ELFA) systems such as the VIDAS system (bioMerieux Vitek, Hazelwood, MO) offer fully automated qualitative detection of *Campylobacter* from a 24- to 48-h enriched sample in about 70 min. Immunomagnetic separation of campylobacters coupled with polymerase chain reaction (PCR) based detection systems are being developed for detection of *Campylobacter* from food samples.

Subtyping of *Campylobacter* strains is needed for epidemiological purposes. By 1982, the most widely used serological schemes were published and in use.

5.6. PREVENTION AND CONTROL

Despite substantial effort by international researchers, contamination of domestic meat animals, especially poultry, continues to be a major problem. This strong association has led to various studies designed to diminish the flow of the organism from the poultry reservoir. When contaminated animals are presented for slaughter, the opportunity for spread of foodborne bacterial pathogens during processing is substantial, even under the best of conditions. Means required to remove this pathogen once it has become attached to carcasses and fresh meat products would likely decrease the organoleptic quality of the meat. However, if means were developed to reduce the incidence *Campylobacter* in the chicken gastrointestinal tract, then processed carcasses with reduced levels of the pathogen could be achieved. Concomitantly, the reduced incidence and level of *Campylobacter*-contaminated poultry products would reduce consumer exposure and should result in diminished frequency of the human disease.

Campylobacter is a fragile organism inactivated by all commonly applied approaches used to rid foods of pathogens. However, the organism is also the leading causative agents of bacterial induced human gastroenteritis. It is incumbent upon the industry and researchers to develop strategies that will control or reduce levels of consumer exposure. As this organism is incapable of multiplying on the raw product, controlling the colonization of the intestinal tract in farm animals or finding ways to limit carcass contamination could accomplish this ambitious goal.

BIBLIOGRAPHY

Blaser, M. J., and L. B. Reller. 1981. Campylobacter enteritis. *New Engl J. Med.* **305**, 1444.

Bolton, F. J. VIDAS Campylobacter product insert, bioMerieux Vitek, Hazelwood, MO.

Bolton, F. J., and L. Robertson. 1982. A selective medium for isolating *Campylobacter jejuni/coli. J. Clin. Pathol.* **35**, 462.

Doyle, M. P., and D. J. Roman. 1982. Sensitivity of *Campylobacter jejuni* to drying. *J. Food Prot.* **45**, 507.

Hams, N. V., D. Thompson, D. C. Martin, and C. M. Nolan. 1986. A survey of *Campylobacter* and other bacterial contaminants of pre-market chicken and retail poultry and meats. King County, Washington. *Am. J. Publ. Health* **76**, 401.

Hutchinson, D. N., and F. J. Bolton. 1984. An improved blood-free selective medium for isolation of *Campylobacter jejuni* from fecal specimens. *J. Clin. Pathol.* **37**, 956.

Kazmi, S. U., B. S. Roberson, and N. J. Stern. 1984. Animal-passed, virulence-enhanced *Campylobacter jejuni* causes enteritis in neonatal mice. *Curr. Microbiol.* **11**, 159.

Line, J. E. 1999. Development of a selective differential agar for isolation and enumeration of *Campylobacter* spp. Abstract. Presented at the Tenth International Workshop on *Campylobacters, Helicobacters* and Related Organisms, Baltimore, MD.

Mishu-Allos, B. 1997. Association between *Campylobacter* infection and Guillian-Barré syndrome. *J. Infect. Dis.* **176**, 125.

Palumbo, S. A. 1984. Heat injury and repair in *Campylobacter jejuni. Appl. Environ. Microbiol.* **48**, 477.

Robinson, D. A. 1981. Infective dose of *Campylobacter jejuni* in milk. *Brit. Med. J.* **282**, 1584.

Rollins, D., and R. Colwell. 1986. Viable but non-culturable stage of *Campylobacter jejuni* and its role in survival in the natural aquatic environment. *Appl. Environ. Microbiol.* **52**, 531–538.

Stern, N. J. 1992. Reservoirs for *Campylobacter jejuni* and approaches for intervention in poultry. In: I. Nachamkin (Ed.), *Campylobacter jejuni: Current Status and Future Trends*. Washington, DC: American Society for Microbiology, pp. 49–60.

Stern, N. J. 1995. Influence of season and refrigerated storage on *Campylobacter* spp. Contamination of broiler carcasses. *J. Appl. Poultry Res.* **4**, 235.

Stern, N. J., S. S. Green, N. Thaker, D. J. Krout, and J. Chiu. 1984. Recovery of *Campylobacter jejuni* from fresh and frozen meat and poultry collected at slaughter. *J. Food Prot.* **47**, 372.

Stern, N. J., D. Jones, I. Wesley, and D. Rollins. 1994. Colonization of chicks by non-culturable *Campylobacter* spp. *Lett. Appl. Microbiol.* **18**, 336.

Stern, N. J., B. Wojton, and K. Kwiatek. 1992. A differential-selective medium and dry ice-generated atmosphere for recovery of *Campylobacter jejuni. J. Food Prot.* **55**, 514.

CHAPTER 6

Clostridium botulinum

BARBARA M. LUND and MICHAEL W. PECK

6.1. INTRODUCTION

Foodborne botulism in adults is caused by eating food in which *Clostridium botulinum* has multiplied and formed a powerful neurotoxin. The disease was described as long ago as 1817 by a German physician, Kerner. *Clostridium botulinum* is an anaerobic, gram-positive, spore-forming bacterium; it was isolated first in 1897 from raw salted ham that caused an outbreak of disease in Belgium. Since that first isolation, botulism has been shown to occur in many animals and birds; different isolates of *C. botulinum* have been shown to produce toxins that differed antigenically, and seven toxin types (A–G) have been differentiated. Botulism in humans is almost always caused by toxin type A, B, or E and occasionally by toxin type F. On the basis of their physiological properties, strains of *C. botulinum* have been placed in different groups that cut across the differences in the type of toxin formed (Table 6.1). The majority of strains of *C. botulinum* appear to produce toxin of a single antigenic type, but some strains form two types of toxin, usually a major amount of one toxin and a minor amount of another. Such strains have been designated Af (indicating the major and minor toxin, respectively), Bf, Ab, and Ba. On rare occasions isolates of *C. butyricum* and of *C. baratii* have been isolated that form botulinum neurotoxin.

Clostridium botulinum occurs widely in the environment, and many types of food have caused outbreaks of botulism. As a result of studies of the properties of the bacteria and of factors affecting their survival and growth, the procedures and controls necessary in the production of foods to minimize the risk of foodborne botulism have been described clearly; many of the outbreaks of botulism that occur are due to a failure to implement these controls, often due to ignorance of them.

Guide to Foodborne Pathogens, Edited by Ronald G. Labbé and Santos García
ISBN 0-471-35034-6

TABLE 6.1. *Clostridium botulinum* Groups and Neurotoxins Formed

Group	Neurotoxin Formed	Main Species Affected
I	A	Humans
	B	Humans, cattle, horses
	F	Humans
II	B	Humans
	E	Humans, fish, some water birds
	F	Humans
III	C	Birds, particularly water birds, farmed chicken and pheasants
	D	Cattle, sheep
IV	G	Not known

In addition to botulism caused by consumption of food containing botulinum neurotoxin, *C. botulinum* causes botulism in infants due to ingestion of the bacterium and its establishment in the intestine, where toxin is formed. Infant botulism was described first in California in 1976; in the United States the number of cases of infant botulism reported annually is now greater than that of foodborne botulism in adults.

There is evidence indicating that a few cases of botulism in adults have been caused, probably, by colonization of the intestinal tract by *C. botulinum* or *C. baratii*, with in vivo production of toxin. In some suspected cases the patients previously had undergone gastrointestinal surgery or had suffered from illness such as inflammatory bowel disease, which may have resulted in susceptibility to intestinal colonization by these bacteria.

Wound botulism results when the bacterium becomes established in a wound and forms toxin that circulates to the rest of the body. In the United States the number of reported cases of wound botulism is comparable with that of foodborne botulism, and the condition is often associated with drug abuse.

The structure and mode of action of *C. botulinum* neurotoxins have been elucidated, and minute doses of type A toxin are used therapeutically to treat certain diseases that are due to muscle spasm. A range of techniques are available to detect the bacteria in the environment and in foods; isolation of the organisms is important also in order to determine the properties of the toxin-forming bacteria.

6.2. BOTULISM

The major symptoms of foodborne botulism are caused by the neurotoxins, which are formed by growth of the bacterium in the food before consumption. Symptoms may occur within 12–36 h or after as long as 8 days. The inital symptoms may be nausea and vomiting, which are caused, probably, not by the neurotoxin but by other products of metabolism of *C. botulinum*. Initially the toxin affects neuromuscular junctions in the head and neck and causes symptoms such as double vision, inability to focus, drooping eylids (ptosis), dry mouth, difficulty in speaking clearly (dysphonia), and inability to swallow (dysphagia). Increasing failure of muscles occurs, and death can result from failure of the muscles needed for breathing or of the heart muscles. Sometimes botulism has been diagnosed mistakenly as another disease such as Guillain-Barré syndrome or myasthenia gravis.

Infant botulism occurs usually in infants less than 35 weeks old. It is caused by ingestion of spores of the toxinogenic organism; as few as 10–100 spores can cause infection. The infant suffers from constipation, generalized weakness and progressive paralysis, and other neurological symptoms. The organism is able to become established in the intestine of young infants probably because the normal intestinal flora has not yet been established sufficiently to prevent this colonization. As many as 6×10^8 colony-forming units (CFU) of *C. botulinum* per gram can be present in the feces of infants with infant botulism.

Treatment of botulism very often requires respiratory support. Intravenous injection of trivalent antitoxin to types A, B, and E toxin is recommended as soon as botulism is diagnosed. The antitoxin will only neutralize toxin that is circulating in the body; if this treatment is delayed, toxin has a chance to bind irreversibly to nerve cells, and it will then not be neutralized by the antitoxin. The antitoxin is prepared from horse serum, and there is a risk of hypersensitivity reactions. Treatment of infant botulism relies largely on respiratory support; in the United States the use of human botulinum immune globulin for treatment of infant botulism is being developed.

In the United States the Centers for Disease Control and Prevention (CDC) releases trivalent antitoxin for treatment of botulism through an emergency distribution system and has an agreement with the Pan American Health Organization to supply the antitoxin to other countries in the Western hemisphere.

6.3. PROPERTIES OF *C. botulinum* AND BOTULINUM NEUROTOXINS

6.3.1. *Clostridium botulinum*

On the basis of differences in their physiology and genetic relatedness, the four groups of *C. botulinum* are sufficiently different to be classified as different species. Renaming has not been introduced because of the confusion that is liable to be caused in medical microbiology, veterinary microbiology, and food microbiology. The grouping of *C. botulinum* strains cuts across the toxin types; this means that some proteolytic strains belonging to group I and some non-proteolytic strains belonging to group II produce type B or type F toxin. Very rarely, strains of *C. butyricum* have been isolated that form type E toxin and strains of *C. baratii* that form type F toxin. Such strains were isolated originally from cases of infant botulism; apart from their ability to form botulinum toxin, these isolates appear to be very similar to nontoxigenic strains of the species. Properties of bacteria that form botulinum toxin are summarised in Table 6.2.

Foodborne botulism, infant botulism, and wound botulism are caused pre-

TABLE 6.2. Properties of Bacteria That Form Botulinum Neurotoxin

	C. botulinum Group I	*C. botulinum* Group II	*C. botulinum* Group III	*C. botulinum* Group IV (*C. argentinense*)	*C. butyricum*	*C. baratii*
Toxins formed	A, B, F	B, E, F	C, D	G	E[a]	F[a]
Proteolysis[b]	+	−	−	+	−	−
Liquefaction of gelatin	+	+	+	+	−	−
Lipase	+	+	+	−	−	−
Optimum growth temperature, °C	30–40	25–37	40	37	30–37	30–45
Minimum temperature for growth, °C	10–12	3	15	NR	~10	NR
Spore heat resistance	High	Low	Moderate	Moderate	NR	NR
Similar nontoxic organisms	*C. sporogenes*	No species name assigned	*C. novyi*	*C. subterminale*	Typical *C. butyricum*	Typical *C. baratii*

[a]Strains that form *C. botulinum* toxin have been reported very rarely.

[b]Proteolysis denotes the ability to degrade native proteins such as coagulated egg white, cooked meat particles and casein; groups I–IV of *C. botulinum* can degrade the derived protein, gelatin.

NR = not reported.

Source: Modified from B. M. Lund and M. W. Peck, 2000, *Clostridium botulinum*, in B. M. Lund, A. C. Baird-Parker, and G. W. Gould (Eds.), *The Microbiological Safety and Quality of Food*, Aspen Publishers, Gaithersburg, MD, pp. 1057–1109.

TABLE 6.3. Properties of Group I and Group II *C. botulinum*

	Group I	Group II
Toxin types formed	A, B, F	B, E, F
Proteolysis	+	−
Optimum temperature for growth	30–40°C	25–37°C
Minimum temperature for growth	10–12°C	3°C
Minimum pH for growth	4.6	5.0
Minimum water activity for growth		
NaCl as humectant	0.96	0.97
Glycerol as humectant	0.93	0.94
NaCl concentration preventing growth	10%	5%
Heat resistance of spores in 0.1 *M* phosphate buffer, pH 7.0	$D_{121°C}$ maximum 0.21 min	$D_{82.2°C}$ in general up to 2.4 min[a]
Radiation resistance of spores at −50 to −10°C	$D = 2.0$–4.5 kGy	$D = 1.0$–2.0 kGy

[a]The *D* value for spores in the absence of lysozyme. For spores recovered in the presence of lysozyme $D_{82.2°C}$ can be up to 231 min.

Source: Modified from B. M. Lund and M. W. Peck, 2000, *Clostridium botulinum*, in B. M. Lund, A. C. Baird-Parker, and G. W. Gould (Eds.), *The Microbiological Safety and Quality of Food*, Aspen Publishers, Gaithersburg, MD, pp. 1057–1109.

dominently by *C. botulinum* groups I and II; this chapter will concentrate on these organisms.

Spores of *C. botulinum* can survive for long periods in air and can germinate in the presence of oxygen. The vegetative cells, however, are sensitive to oxygen and gradually become nonviable. The concentration of oxygen present and the resulting oxidation–reduction potential can, therefore, have a controlling effect on survival and growth of the bacterium in the environment, in foods, during attempted isolation, and during experiments to investigate the properties of the organism. In foods and other environments that appear to be aerobic, the presence and metabolism of aerobic and facultative microorganisms can create microenvironments that are anaerobic and allow growth of *C. botulinum*.

Table 6.3 shows the contrast between the properties of group I, proteolytic and group II, nonproteolytic *C. botulinum*. Spores of group I *C. botulinum* have a much greater heat resistance than that of group II strains, which makes these group I strains of major concern in determination of the heat processing given to canned, low-acid foods. These are foods with a pH higher than 4.5; the canning process for these foods should, in general, include a minimum heat treatment equivalent to maintaining the food at 121°C for 3 min (an $F_0 3$ process), which is sufficient to give a reduction of 10^{12} in the number of viable spores of group I *C. botulinum*. Heating spores of group II strains at about 80°C has been reported frequently to give rapid inactivation. There is evidence, however, that this heat treatment results in sublethal damage to the spores by inactivating enzymes involved in germination. Lysozyme can enable germination of these heat-damaged spores and subsequent growth of the organism; the presence of lysozyme in the recovery medium or in a food can result in growth from spores of group II *C. botulinum* that have been sublethally damaged by heat treatment. The presence of lysozyme in the recovery medium does not result in a marked increase in the estimated number of spores of group I strains that survive heat treatment.

Group I *C. botulinum* are able to grow at a lower water activity and a lower pH than group II strains. Group II strains, however, can multiply at temperatures as low as about 3°C, making these bacteria of concern in foods that are produced using a mild heat treatment and are designed to have an extended shelf-life at refrigeration temperatures.

6.3.2. Botulinum Neurotoxins

The botulinum neurotoxins are formed initially as single-chain polypeptides with a molecular weight of about 150 kDa; the single-chain form has a relatively low toxicity. The single-chain form is changed to a dichain form by a process termed "nicking," brought about by an extracellular bacterial protease or by an added enzyme such as trypsin. The molecule formed consists of a light chain (L) (molecular weight approximately 50 kDa) and a heavy chain (H) (molecular weight approximately 100 kDa) linked by a disulfide bond. Usually group I, proteolytic *C. botulinum* form the dichain molecule in media and in foods, but group II, nonproteolytic *C. botulinum* form the single-chain form and it is necessary to treat with trypsin in order to demonstrate the activity of the toxin. The neurotoxins act primarily at peripheral, cholinergic synapses, where they block the release of the neurotransmitter acetylcholine. The heavy chain of the toxin is involved in entry of the toxin into the motor neuron, and the light chain acts as an endopeptidase enzyme, degrading neuron proteins involved in the release of acetylcholine. Botulism symptoms can be caused by the consumption of food containing as little as 30–100 ng of toxin.

In group I and II *C. botulinum* the gene for toxin production appears to be on the chromosome; many strains that form type A toxin contain an unexpressed gene for type B toxin. The genes for botulinum type C and D neurotoxins are present in pseudolysogenic phages, and the gene for type G toxin is reported to be on a plasmid.

In foods and in culture media the toxin is present usually as a complex (M or T complex) that is much more stable than the derivative toxin (Table 6.4). The stability of the toxin complex may be sufficient to prevent or reduce breakdown of the toxin in the stomach. The composition of the food or culture medium in which growth of the bacterium occurs can affect the size of toxin complex that is formed.

The primary concern in preventing foodborne botulism is to control food production so that growth of *C. botulinum* and formation of toxin are prevented. Toxin in food can be inactivated fairly readily by heating; the temperature and time of heating needed will depend on the composition of the food.

TABLE 6.4. *Clostridium botulinum* Neurotoxins and Toxin Complexes

Complex	Approximate Size	Components	Toxin Types
S "derivative toxin"	150 kDa (7 S)[a]	Single chain	A, B, C, D, E, F
M "progenitor toxin"	300 kDa (12 S)	S + nontoxic, nonhemagglutinin	
L "progenitor toxin"	450–500 kDa (16 S)	M + hemagglutinin	A, B, C, D, G
LL	900 kDa (19 S)	Probably a dimer of L	A

[a] Sedimentation coefficient, S = Svedberg unit.

Source: From B. M. Lund and M. W. Peck, 2000, *Clostridium botulinum*, in B. M. Lund, A. C. Baird-Parker, and G. W. Gould (Eds.), *The Microbiological Safety and Quality of Food*, Aspen Publishers, Gaithersburg, MD, pp. 1057–1109.

Heating food at 79°C for 20 min or at 85°C for 5 min has been suggested as a guideline, and the latter heat treatment was reported to inactivate about 10^5 mouse intraperitoneal (IP) median lethal dose (LD_{50}) per gram. Toxin in food is usually very stable during freezing and frozen storage.

6.4. EPIDEMIOLOGY

6.4.1. Incidence of *C. botulinum* in the Environment and in Foods

Clostridium botulinum can be found usually in a proportion of soil samples. The numbers present are generally fewer than 500 kg^{-1}, whereas other clostridia are much more numerous; for example, *C. pasteurianum* and *C. perfringens* are often present at 10^3–10^4 g^{-1}. In some samples of sediments from lakes and coastal waters the incidence and the numbers of *C. botulinum* can be much greater than in soil.

The majority of types of food are liable to contain spores of *C. botulinum*; for example, the bacterium has been isolated from fish, meat, vegetables, fruits, honey, mushrooms, cheese, nuts, as well as less well-known foods such as bamboo shoots. Of the various types of food the highest incidence of *C. botulinum* is found probably in fish from certain parts of the world; the strains present in fish are usually nonproteolytic and produce type E toxin. Fish from certain trout farms with earth bottoms have been reported to contain between 340 and 5300 spores of *C. botulinum* type E per kilogram, and *C. botulinum* was detected in 63% of samples of salted carp from the Caspian Sea at a most probable number (MPN) of up to 490 kg^{-1}. In meats and in vegetables and fruits the reported incidence is usually lower and the major toxin types are A and B; the relative incidence of group I and group II strains producing type B toxin is not clear.

Some samples of honey have been reported to contain relatively high numbers of spores of *C. botulinum* (>1000 spores kg^{-1}). Several cases of infant botulism have resulted from the use of honey on pacifiers for infants. *Clostridium botulinum* cannot multiply in honey because of the high sugar content.

There is evidence that if bees are diseased, for example affected by foulbrood, *C. botulinum* can multiply in the dead bees and this can result in a high number of *C. botulinum* spores in the honey.

6.4.2. Outbreaks of Foodborne Botulism

The reported incidence of botulism depends on the quality of the local reporting and surveillance system. In some countries lack of botulinum antitoxin for treatment and of facilities to detect the toxin may prevent a conclusive diagnosis. Thus, the true incidence of botulism in many parts of the world is probably not known.

In European countries for which information is available, the numbers of reported cases of foodborne botulism in 1997 were as follows: Poland, 81; Italy, 32; France, 17; Germany, 9; Spain, 9; and Belgium, 3. No cases were reported in 1997 in Denmark, England and Wales, Finland, Greece, Sweden, Scotland, and the Netherlands. In the United States there were 31 reported cases of foodborne botulism in 1997 and in Argentina 23 cases of suspected foodborne botulism were reported. A high incidence of botulism was reported in China for the period 1958–1983. Large outbreaks can result in a considerable increase in the number of reported cases in a year. For example, an outbreak in the United Kingdom in 1989 associated with hazelnut yoghurt resulted in 27 cases, an outbreak in Egypt in 1991 due to uneviscerated, salted fish affected more than 91 people, an outbreak in the United States in 1994 associated with foil-baked potatoes affected 30 people, and one in Iran in 1997 due to a cheese preserved in oil affected 27 people. Important factors that influence the toxin type associated with foodborne botulism are the local preference for particular types of food and local methods of food preservation.

Because of the presence of a relatively well-developed reporting system in the United States, detailed reports of the incidence of botulism for many years are available (Fig. 6.1). Since 1983 the main form of botulism reported in the United States has been infant botulism; in 1996 the number of reported cases of infant botulism was almost four times that of cases of foodborne botulism, and cases of wound botulism were almost as frequent as those of foodborne botulism.

Some examples of relatively recent major outbreaks of foodborne botulism are presented in Table 6.5. These examples have been selected to illustrate the large number of people affected in some outbreaks, the varied types of food implicated, and the different factors that have led to outbreaks. Some of these outbreaks were caused by failure to give an $F_0 3$ heat treatment during heat processing of low-acid, canned foods; others were caused by failure to control inhibitory conditions properly.

One of the largest reported outbreaks of botulism in the United States occurred in 1977. The outbreak was associated with the consumption of a hot sauce made with home-canned, Jalapeno peppers and served in a Mexican restaurant. The canning process consisted of boiling the peppers in jars, a pro-

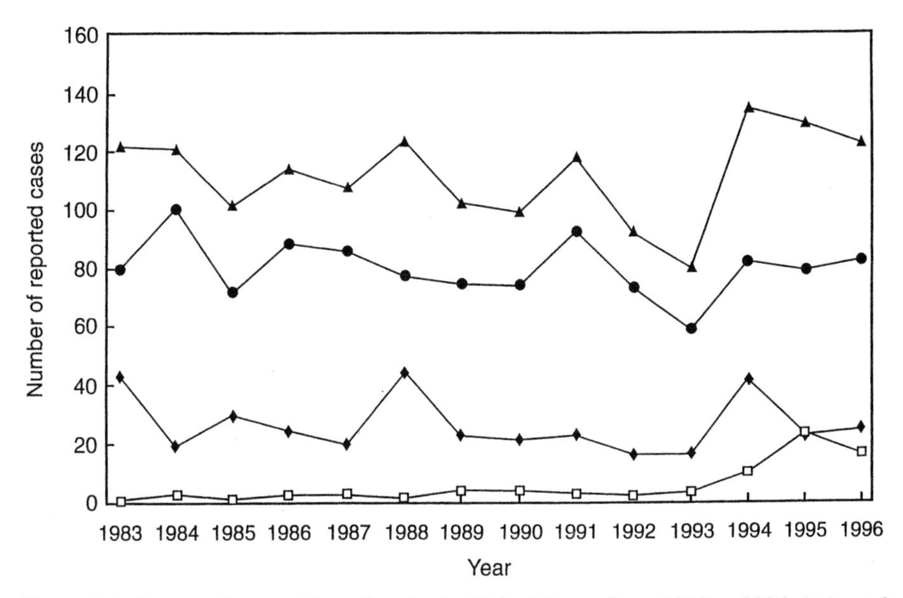

Figure 6.1. Reported cases of botulism in the United States from 1983 to 1996: (▲) total number of cases; (●) infant botulism; (◆) foodborne botulism; (□) wound botulism.

Reprinted with permission from B. M. Lund and M. W. Peck, 2000, *Clostridium botulinum*, in B. M. Lund, A. C. Baird-Parker, and G. W. Gould (Eds.), *The Microbiological Safety and Quality of Food*. Aspen Publishers, Gaithersburg, MD, pp. 1057–1109.

cess that would be less than an $F_0 3$ treatment. It was reported that after a number of days some of the jars began to explode on their shelves at the restaurant; the jars that did not explode were kept.

An outbreak in Taiwan in 1986 was associated with jars of unsalted peanuts in water. These had been produced by a small company that was not licensed to produce canned foods and did not possess the appropriate equipment. The product was a low-acid food but had been processed only by boiling. Of 104 jars of a batch of the product, 34 contained type A *C. botulinum* toxin. At the time of the outbreak local physicians were said to be largely unaware of the existence of botulism in Taiwan, but this outbreak was followed by two further reports of outbreaks associated with home-preserved meats.

In the United Kingdom foodborne botulism is very rare, but the largest outbreak ever reported occurred in 1989 and was caused by the consumption of commercially produced hazelnut yogurt. The canned hazelnut conserve that was an ingredient of the yogurt was routinely given a mild heat treatment, less than an $F_0 3$ process, during the canning; normally it contained a high concentration of sucrose that would have prevented growth from any *C. botulinum* spores that were present and survived the heat treatment. A change in the formulation of the product, involving substitution of sucrose by aspartame, resulted in a conserve that no longer inhibited growth of *C. botulinum*. Spores

TABLE 6.5. Examples of Outbreaks of Foodborne Botulism

Outbreak	Food	Number of Cases	Number of Deaths	Toxin Type
	Caused by Insufficient Heat Processing			
1977, U.S.	Restaurant-canned peppers	59	0	B[a]
1986, Taiwan	Commercially produced, heat-processed peanuts in water	9	2	A
1989, U.K.	Commercially prepared hazelnut yoghurt	27	1	Bp[b]
1998, Thailand	Home-canned bamboo shoots	13	2	A
1998, Argentina	Home-prepared *matambre* (Argentine meat roll)	9	0	A
	Caused by Insufficient Inhibition			
1985, Canada	Commercially bottled, chopped garlic in soybean oil	36	0	Bp
1989, U.S.	Commercially bottled, chopped garlic in olive oil	3	0	A
1991, Egypt	Commercially prepared, uneviscerated, salt-cured fish	>91	18	E
1993–1994, Switzerland	Commercially prepared, dry-cured ham	12	0	B
1994, China	Home-prepared, salted, fermented paste made of soybeans and wax gourds	6	NR	E[c]
1994, U.S.	Restaurant-prepared potato dip and egg-plant dip	30	0	A
1997, Iran	Locally made cheese preserved in oil	27	1	A
1997, Germany	Commercially smoked, vacuum-packed fish, imported	2	0	E
1997, Germany	Home-smoked, vacuum-packed fish	4	0	E

[a] B, type B toxin, not reported whether associated with a group I or a group II strain of *C. botulinum*.

[b] Bp, type B toxin produced by a group I strain of *C. botulinum*.

[c] Type E toxin produced by a strain of *C. butyricum*. In all other examples type E toxin was produced by strains of *C. botulinum*.

NR = not reported.

of the organism survived the heat treatment and germinated in the canned conserve, in which growth and toxin formation occurred before the conserve was added to the yogurt. The yogurt was too acidic to allow further growth of the bacterium, but the amount of toxin in the added conserve was sufficient to cause a severe outbreak of botulism.

In Thailand in 1998 an outbreak of botulism occurred that was due to underprocessing of home-canned bamboo shoots. This was the first laboratory-confirmed outbreak of botulism in Thailand.

An outbreak in Argentina in 1998 was associated with the consumption of *matambre*, Argentine meat roll. It was reported that usually this traditional food was consumed fresh. Commercial producers were licensed and used nitrites, acidifiers, or other preservatives. The food implicated in this outbreak was prepared domestically but sold commercially. The product was cooked at 78–80°C for about 4 h, then packed warm in plastic wrap from which the the air was expelled, cooled, and stored in a refrigerator for up to 2 weeks. The temperature inside the refrigerators was reported to be 9–10°C, but the food must have been at temperatures higher than 12°C for several days in order for growth of group I *C. botulinum*-producing type A toxin to occur.

An outbreak in Canada in 1985 was associated with commercially bottled, chopped garlic in soybean oil. This product was not heat processed; therefore a range of bacteria would survive in the food. The chopped, dehydrated garlic had been rehydrated, then covered with oil, which must have created anaerobic conditions and allowed germination of *C. botulinum* spores followed by growth and toxin formation. The bottled product was labeled with instructions to keep in refrigerated conditions, but it had been kept at room temperature for a considerable time. In 1989 another small outbreak due to this type of product occurred in the United States. Following this, the U.S. Food and Drug Administration (FDA) ordered companies to stop making products of this type that are only protected against *C. botulinum* by refrigeration; according to the FDA, such products must contain specified levels of antimicrobials or must be acidified, for example with phosphoric or citric acid.

The largest reported outbreak of foodborne botulism due to type E toxin occurred in Egypt in 1991. This outbreak was caused by the consumption of a traditional food prepared by salting uneviscerated mullet fish. Traditionally the fish were obtained from the Red Sea or from salt lakes but more recently had been obtained from fish farms. The fish were stored for between several hours and 1 day to allow swelling and putrefaction before they were salted and stored for between 2 weeks and 1 year. Some time would elapse before sufficient salt penetrated into the fish to inhibit *C. botulinum*, and maintenance of the fish at temperatures above 3–5°C would be liable to allow any *C. botulinum* spores in the gut of the fish to germinate, resulting in growth of vegetative bacteria and toxin formation before the salt concentration became inhibitory.

An outbreak in Switzerland in 1993–1994 due to type B toxin was caused by a commercially prepared, dry-cured ham. This type of food has been implicated in many botulism outbreaks in Europe, particularly in France, associated with type B toxin. In most cases, whether group I or II strains of *C. botulinum* were implicated was not determined, but workers in Germany reported that the majority of isolates of *C. botulinum* from raw hams were group II strains.

An outbreak in the United States in 1994 was associated with a potato dip. This followed several outbreaks with a similar cause. The potatoes had been baked in a restaurant, but the process would not be the equivalent of an $F_0 3$

heat treatment and spores of *C. botulinum* survived. The baked potatoes were then wrapped in foil and stored for several days (probably 4 days) at ambient temperature. During this time conditions at the surface of the potato must have become sufficiently anaerobic to allow growth of *C. botulinum* and formation of toxin.

Two outbreaks in Germany in 1997 were associated with vacuum-packed, smoked fish; in one outbreak the food was prepared commercially and in the other outbreak it was prepared domestically. The safety of these foods depends on a combination of inhibitory factors, including maintenance at low temperature; failure to maintain at a sufficiently low temperature was likely to have been a contributory factor to these outbreaks.

6.5. DETECTION AND ISOLATION

It may be necessary to test samples directly for *C. botulinum* toxin, for example testing food involved in an outbreak or samples of serum or stools from a patient. In other types of work, for example when testing for the presence of *C. botulinum* in an environmental sample, first an enrichment will be made in culture medium and then the enrichment is tested for toxin.

Since the earliest work on *C. botulinum*, tests involving intraperitoneal injection into mice have been used to detect *C. botulinum* neurotoxin. The test involves observation of typical symptoms of botulism in the mice and demonstration that the mice can be protected by a preparation containing antitoxin to all the neurotoxins or by antitoxin to a specific neurotoxin. The mouse test will detect 5–10 pg *C. botulinum* toxin ml^{-1}. International standards for *C. botulinum* antitoxins A, B, C, D, E, and F have been distributed through the World Health Organization (WHO), and reference diagnostic antitoxins calibrated in comparison with the WHO standards have been made available through the CDC.

In order to avoid the use of animals and to provide a more rapid test, many alternative methods of testing have been investigated. At present there is still a need for the mouse test, particularly in situations where an outbreak has occurred.

Many enzyme-linked immunosorbent assay (ELISA) tests have been developed for detection of *C. botulinum* toxins. Some tests are specific for one toxin type; others can detect types A, B, and E toxin. Many of these tests react with biologically inactive toxin; some antibodies react differently with toxin of the same antigenic type produced by different strains of *C. botulinum*, and some tests use antibodies that are not specific for *C. botulinum* toxins but react with other metabolic products of *C. botulinum*. Nevertheless these tests can be very valuable; they can be used in experiments to investigate factors that prevent growth of *C. botulinum* in foods and in culture media and were used for rapid detection of toxin in hazelnut puree and yogurt associated with the botulism outbreak in the United Kingdom in 1989. To give a high sensitivity, amplification is required. An efficient form of amplification involves use of an alkaline phosphatase conjugate. A system is available from Gibco (Gaithersburg, MD) in which nicotinamide adenine dinucleotide phosphate, reduced from (NADPH) is transformed to NADH, and in a secondary reaction this is used to reduce a tetrazolium salt to a formazan dye that is intensely colored and is measured spectrophotometrically.

Enzyme-linked clotting assay (ELCA) involves an amplification that appears to give the sensitivity of the mouse test. This test uses the addition of coagulation factors and the production of enzyme-labeled fibrin as a measure of the antibody conjugate formed. The amplification system is available commercially but is relatively expensive.

In principle, endopeptidase activity could be used to detect *C. botulinum* neurotoxin. Tests based on a combination of immunological reaction and endopeptidase activity are being developed and can be as sensitive as the mouse test. These in vitro tests rely on detection of the biological activity of the light chain of the *C. botulinum* neurotoxin but are independent of the role of the heavy chain in cell binding and internalization of the toxin; this activity of the heavy chain is essential for the in vivo effect of the toxin.

Probes have been constructed that enable the polymerase chain reaction (PCR) to be used either for the nonspecific detection of genes for all types of *C. botulinum* toxin or for the specific detection of genes for each of the toxins. An initial cultural enrichment of the organism is valuable because it increases the number of target bacteria, reduces the effect of inhibitors of the PCR that are present in many food samples, and provides evidence of detection of genetic material from living bacteria rather than from dead organisms. The inclusion of an MPN series of dilutions in the cultural enrichment has been used for the quantitative detection of *C. botulinum* and enabled the detection of about 210 spores kg^{-1} inoculated into fish.This method has been used very successfully recently in Finland to investigate the incidence of *C. botulinum* type E in the environment and in fish.

The equipment and reagents that are available for in vitro testing for *C. botulinum* toxin tend to be expensive; less expensive reagents can be developed in individual laboratories, but this does not enable the development of a common protocol that can give comparable results in different laboratories. For the mouse bioassay diagnostic reagents are available and have been calibrated against international standards. For this reason the bioassay in mice remains, at present, the most widely available, reliable test for *C. botulinum* neurotoxin for use in suspected outbreak situations. The FDA in collaboration with the CDC is working to develop reagents that can be used with a common protocol for in vitro testing.

Isolation of *C. botulinum* follows, usually, detection of the toxin in a food or clinical sample or in a culture that has been made to allow growth of the organism. Enrichment in selective conditions is needed, and because of the physiological differences between the groups of *C. botulinum*, different selective conditions are needed for their isolation. Heat treatment at 80°C for 10 min is used to inactivate competing bacteria and allow growth from spores of group I strains. This would probably inactivate spores of group II strains; heating at about 60°C may be useful in the isolation of these strains, and the addition of lysozyme to the recovery medium can increase the recovery of heated spores. Treatment with ethanol is used sometimes as an alternative to heat treatment in the isolation of these strains. Culture at 35°C is suitable for isolation of group I strains, whereas culture at about 26°C is recommended for group II strains. In addition to competing bacteria, particularly clostridia, some samples of soil and sediment contain inhibitors of *C. botulinum*, including bacteriophages and bacteriocins. The addition of filter-sterilized trypsin to the culture medium may be necessary to inactivate bacteriocins. In attempting to isolate *C. botulinum* from the environment or from food, it is important, therefore, to assess the adequacy of the isolation method by demonstrating that strains of the relevant groups of *C. botulinum* can be reisolated following the addition of low numbers of the organisms to the sample.

Detailed information on isolation of *C. botulinum* is given in the *Bacteriological Analytical Manual* published by the FDA. It is important that media used for isolation should be anaerobic. Suitable enrichment media are cooked meat medium (CMM) incubated at 35°C for group I strains and CMM glucose medium, chopped meat glucose broth medium, or trypticase–peptone–glucose–yeast extract broth containing trypsin (TPGYT) incubated at 26°C for isolation of group II strains. Cultures should be incubated for sufficient time (e.g, 7 days) to allow growth, sporulation, and toxin formation. A sample of the culture can then be used to test for toxin. A suitable nonselective medium for plating is egg-yolk agar, on which colonies of *C. botulinum* (except group IV) show lipase activity. Selective plating media for isolation of group 1 strains are *C. botulinum* isolation agar (CBI agar) and *C. botulinum* selective medium (BSM), but the trimethoprin in these media inhibits group II strains.

6.6. PREVENTION AND CONTROL

Spores of *C. botulinum* are liable to be present in many types of raw food, and it would be unwise to rely for safety on the absence of the bacterium. The ability of spores of *C. botulinum* to survive heat treatment and the effect of pH, temperature, salt concentration, and other factors on growth of the bacteria (Table 6.3) determine the methods that are used to prevent survival and growth of *C. botulinum* in foods and to assure safety with respect to *C. botulinum*. Examples of factors that are used to control *C. botulinum* in foods are given in Table 6.6.

The term *low-acid foods* is applied in the United Kingdom to foods with a pH higher than 4.5 and in the United States to foods with a pH higher than 4.6. These are considered to be foods that could allow growth of *C. botulinum*. If canned, these foods must, in general, receive a heat treatment that is sufficient to reduce the number of viable spores of group I *C. botulinum* by a factor of 10^{12} (an $F_0 3$ process). In the case of most canned, low-acid foods a heat treat-

TABLE 6.6. Examples of Main Factors Used to Control *C. botulinum* in Foods

Category	Main Controlling Factors	Examples of Foods
Shelf-stable foods		
1	Heat treatment > $F_0 3$ (>12D process)[a]	Low-acid canned foods, pH > 4.5 (U.K.) or pH > 4.6 (U.S.)
2	Heat process < $F_0 3$, NaCl, nitrite	Shelf-stable, canned cured meats
3	Heat process < $F_0 3$, pH	Acid, canned foods, pH ≤ 4.5 (U.K.) or pH ≤ 4.6 (U.S.), e.g., many types of canned fruit and vegetables
4	a_w[b] (NaCl + drying), ±nitrite; refrigeration below 5°C during salting	Raw salted and salt-cured meats, e.g., salt pork, salt bacon, salt hams, dry-cured hams, and bacon
5	a_w (NaCl + drying) nitrite, pH	Fermented sausages, e.g., summer sausage, pepperoni, acidulated sausages
Perishable foods		
6	Refrigeration alone	Raw meat, fish, and vegetables
7	NaCl; refrigeration; nitrite in many products; in some cases sorbate, benzoate, nitrate, or smoke	Salted raw fish; lightly preserved fish; semipreserved fish; hot-smoked fish; perishable cooked, cured meats
8	a_w (NaCl and other components), pH, refrigeration of some products, mild heat treatment	Process cheese and process cheese spreads
9	Mild heat treatment, refrigeration, limited shelf-life	Ready meals, "sous-vide" foods

[a] 12D process = reduction of number of viable spores of *C. botulinum* by a factor of 10^{12}.

[b] a_w = water activity.

Source: From B. M. Lund and M. W. Peck, 2000, *Clostridium botulinum*, in B. M. Lund, A. C. Baird-Parker, and G. W. Gould (Eds.), *The Microbiological Safety and Quality of Food*, Aspen Publishers, Gaithersburg, MD, pp. 1057–1109.

ment greater than this is necessary to inactivate spores of other species of *Clostridium* and *Bacillus*. Numerous cases of botulism have resulted, however, from failure to give an adequate heat treatment during home canning or bottling of low-acid foods, and some outbreaks have been caused by failures in commercial processes. In addition to the use of an adequate heat treatment, precautions are required to prevent the postprocessing contamination of canned foods by access of bacteria through the seams of cans during cooling.

Shelf-stable, canned cured meats are given a heat treatment less than an $F_0 3$ process. This is because preservation of these foods relies on the combined effect of the heat treatment and inhibition of the heated spores by sufficient concentrations of sodium chloride and nitrite in the product.

The term *acid foods* is applied to foods with a pH ≤ 4.5 (U.K.) or ≤4.6 (U.S.). The great majority of evidence shows that *C. botulinum* does not multiply in foods below pH 4.6. Canned acid foods can be given a heat treatment much lower than an $F_0 3$ process, but the spores are liable to survive in these foods. A few cases of botulism have resulted after survival of another microorganism in an acid canned food led to growth of the organism and a rise in pH that allowed growth and toxin formation by *C. botulinum*.

Many types of food rely on a combination of factors to control *C. botulinum*. In the case of foods that rely in part on salting for their preservation, there is a need to ensure that growth and toxin formation by *C. botulinum* do not occur before an inhibitory concentration of salt has penetrated throughout the food. Refrigeration below 5°C is recommended during salting and, in the case of fish, evisceration before preserving removes the probable main source of *C. botulinum* and the part of the fish that would be most inaccessible to the salt.

The safety of certain foods, including "sous vide" and other vacuum-packed foods, relies mainly on a mild heat treatment combined with refrigeration, although in some foods other preservative factors may be present. Some of these foods may be expected to have a shelf-life of up to 42 days. Spores of group I *C. botulinum* would survive the heat treatment, but provided the temperature is maintained reliably below 10°C, this organism will not multiply. However, group II *C. botulinum* can multiply at temperatures as low as 3°C. For such foods with an extended shelf-life and that do not contain sufficient preservative factors, it has been concluded that the combination of heat treatment and preservative factors together with storage at $<10°C$ should be sufficient to reduce consistently the probability of survival and growth of group II *C. botulinum* by a factor of 10^6.

6.7. CONCLUSION

In many parts of the world diagnosis and reporting of botulism are limited by a lack of awareness, a lack of antitoxin for treatment, and limited availability of reagents for diagnosis. Information about botulism needs to be disseminated more widely. Antitoxin for treatment and standard reagents for testing for *C. botulinum* neurotoxin need to be available worldwide.

Many outbreaks and cases of botulism result from failure to implement controls that are known and used widely to prevent foodborne botulism. Dissemination of information about control methods is needed in sections of the food industry and to the general public in countries where home preservation of food is practiced widely.

In some parts of the world the risk of botulism posed by traditional foods appears not to have been assessed. Such risks need to be considered, in particular because changes in the source of ingredients or in methods of production could increase the risk in those foods.

As new types of foods are introduced, often intended to have extended shelf-lives, and new processing methods, such as high-pressure treatment and ohmic heating, are used more widely, controls are needed to ensure that appropriate levels of safety, such as those that are presently built into foods, are applied in these new situations.

BIBLIOGRAPHY

Centers for Disease Control and Prevention. 1998. *Botulism in the United States, 1899–1996. Handbook for Epidemiologists, Clinicians and Laboratory Workers*, Atlanta, GA: Centers for Disease Control and Prevention.

Centers for Disease Control and Prevention. 1999. Foodborne botulism associated with home-canned bamboo shoots—Thailand, 1998. *MMWR* **48**(21), 437–439.

Dodds, K. I., and J. W. Austin. 1997. *Clostridium botulinum*. In: M. P. Doyle, L. R. Beuchat, and T. J. Montville (Eds.), *Food microbiology. Fundamentals and frontiers*. Washington, DC: American Society for Microbiology, pp. 288–304.

Hatheway, C. L., and J. L. Ferreira. 1996. Detection and identification of *Clostridium botulinum* neurotoxins. *Adv. Exper. Med.* **391**, 481–498.

Hauschild, A. H. W. 1989. *Clostridium botulinum*. In: M. P. Doyle (Ed.), *Foodborne Bacterial Pathogens*. New York: Marcel Dekker, pp. 111–189.

Hauschild, A. H. W., and K. L. Dodds. (eds.) 1993. *Clostridium botulinum: Ecology and Control in Foods*. New York: Marcel Dekker.

Lund, B. M., and M. W. Peck. 2000. *Clostridium botulinum*. In: B. M. Lund, A. C. Baird-Parker, and G. W. Gould (Eds.), *The Microbiological Safety and Quality of Food*. Gaithersburg, MD: Aspen, pp. 1057–1109.

Shapiro, R. L., C. Hatheway, J. Becher, and D. L. Swerdlow. 1997. Botulism surveillance and emergency response. A public health strategy for a global challenge. *J. Am. Med. Assoc.* **278**, 433–435.

Smith, L. D. S., and H. Sugiyama. 1988. *Botulism. The Organism, Its Toxins, the Disease*, 2nd ed. Springfield, IL: Charles C. Thomas.

Solomon, H. M., E. J. Rhodehamel, and D. A. Kautter. 1995. *Clostridium botulinum*. In: *Bacteriological Analytical Manual*, 8th ed. Washington, DC: Food and Drug Administration, pp. 17.01–17.10.

Therre, H. 1999. Botulism in the European Union. *Eurosurveillance* **4**(1), 1–7.

Villar, R. G., R. L. Shapiro, S. Busto, et al. 1999. Outbreak of type A botulism and development of a botulism surveillance and antitoxin release system in Argentina. *J. Am. Med. Assoc.* **281**, 1334–1340.

CHAPTER 7

Fusarium

LLOYD B. BULLERMAN

7.1. INTRODUCTION

Toxigenic *Fusarium* species are most often found as contaminants of plant-derived foods, especially cereal grains. As such, these molds and their metabolites (mycotoxins) find their way into animal feeds and human foods. Animals, both food-producing animals and pets, are more often affected by the toxins of these molds than are humans because of the nature of animal feed and the way it is stored and handled. Human food is generally higher in quality and more carefully protected, particularly in countries in temperate climates. However, in other areas with more tropical and subtropical climates this may not be the case, and there is evidence that grains and processed human foods may be contaminated with mycotoxins produced by *Fusarium* species. Also, exposure of food-producing animals to mycotoxigenic molds may have an impact on the human food supply by causing death of animals, reducing their rate of growth, or depositing toxins in meats, milk, and eggs. Diseases in animals caused by mycotoxins may also suggest that similar conditions may occur in humans.

Many *Fusarium* species are plant pathogens, while others are saprophytic, and most can be found in the soil. *Fusarium* species are most often encountered as contaminants of cereal grains, oil seeds, and beans. Corn, wheat, and products made from these grains are most commonly contaminated. However, barley, rye, triticale, millet, and oats can also be contaminated.

7.2. NATURE OF ILLNESS IN ANIMALS AND HUMANS

7.2.1. Alimentary Toxic Aleukia

During World War II a very severe human disease occurred in the former Soviet Union, particularly in the Orenburg area of Russia. The disease, known as alimentary toxic aleukia (ATA), is believed to be caused by T-2 and HT-2

Guide to Foodborne Pathogens, Edited by Ronald G. Labbé and Santos García
ISBN 0-471-35034-6

toxins produced by *F. sporotrichioides* and *F. poae* of the Sporotrichiella section of *Fusarium*. Because of the war, there was such a shortage of farm workers that much grain was not harvested in the fall and overwintered in the field. By spring, near-famine conditions existed and people were forced to eat the overwintered cereal grains that were milled into flour and made into bread. This resulted in a severe toxicosis manifested as ATA. The disease developed over several weeks with increasingly severe symptoms with continued consumption of the toxic grain.

There are actually four stages to the disease. The first stage is characterized by symptoms that appear a short time after ingestion of the toxic grain. These symptoms include a burning sensation in the mouth, tongue, esophagus, and stomach (Table 7.1). The stomach and intestinal mucosa become inflamed resulting in vomiting, diarrhea, and abdominal pain. This first stage may appear and disappear rather quickly, within 3–9 days. During the second stage the individual experiences no outward signs of the disease and feels well. However, during this stage the hematopoietic system is being damaged or destroyed by progressive leukopenia, granulopenia, and lymphocytosis. The blood-making capacity of the bone marrow is being destroyed, platelet count decreases, and anemia develops. The leukocyte count decreases and secondary bacterial infections occur. There are also disturbances in the central and autonomic nervous system. The third stage develops suddenly and is marked by petechial hemorrhage on the skin and mucous membranes. The hemorrhaging becomes more severe with bleeding from the nose and gums and hemorrhaging in the stomach and intestines. Necrotic lesions also develop in the mouth, gums, mucosa, larynx, and vocal cords. At this stage the disease is highly fatal with up to 60% mortality. If death does not occur, the fourth stage of the disease is recovery or convalescence. It takes 3–4 weeks of treatment for the necrotic lesions, hemorrhaging, and bacterial infections to clear up and 2 months or more for the blood-making capacity of the bone marrow to return to normal. Alimentary toxic aleukia has been reproduced in cats and monkeys with T-2 toxin isolated from the strain of *F. sporotrichioides* involved in the fatal human outbreaks of the disease in Russia. The related compound HT-2 toxin may also contribute to the disease.

7.2.2. Urov or Kashin–Beck Disease

Urov or Kashin–Beck disease has been observed among the Cossack people of eastern Russia for well over 100 years. The disease has been endemic in areas along the Urov River in Siberia and was studied extensively by two Russian scientists, Kashin and Beck, thus the name Urov or Kashin–Beck disease. The disease is a bone-joint deforming osteoarthrosis that is manifested by a shortening of long bones and thickening deformation of joints plus muscular weakness and atrophy. The disease most commonly occurs in preschool and young school-age children. The disease has been reproduced in puppies and rats by feeding by *F. poae* isolates obtained only from endemic regions of the disease. No specific toxins have yet been isolated and proven to be the definite cause, so

TABLE 7.1. Human Diseases Associated with *Fusarium* Species and Toxins

Disease	Food(s)	Organism(s)	Toxin(s)	Symptoms/Effects
ATA	Cereal grains; wheat, rye, bread	*F. sporotrichioides, F. poae*	T-2 toxin	Burning sensation in mouth and throat, vomiting, diarrhea, abdominal pain, bone marrow destruction, hemorrhaging, death
Urov/Kashin–Beck disease	Cereal grains	*F. poae*	Unknown	Osteoarthritis, shortened long bones, deformed joints, muscular weakness
Drunken bread	Cereal grains, wheat, bread	*F. graminearum*	Unknown	Headache, dizziness, tinnitus, trembling, unsteady gait, abdominal pain, nausea, diarrhea
Akakabi-byo scabby grain intoxication	Cereal grains; wheat, barley, noodles	*F. graminearum*	Unknown; possibly deoxynivalenol	Anorexia, nausea, vomiting, headache, abdominal pain, diarrhea, chills, giddiness, convulsions
Foodborne illness outbreaks	Cereal grains; wheat, barley, corn, bread	*F. graminearum*	Deoxynivalenol, acetyl-deoxynivalenol, nivalenol, T-2 toxin	Irritation of throat, nausea, headaches, vomiting, abdominal pain, diarrhea
Esophageal cancer	Corn	Possibly *F. moniliforme*	Unknown; possibly fumonisins and other toxins	Precancerous and cancerous lesions in the esophagus

the actual etiology has not been completely resolved. The same disease has also been reported in North Korea and northern China.

7.2.3. Drunken Bread

Another human mycotoxicosis reported in the former Soviet Union is known as "drunken bread." This syndrome is apparently caused by consumption of bread made from rye grain infected with *F. graminearum*. The illness is milder than ATA and is a nonfatal self-limiting disorder. Symptoms associated with this disease are headache, dizziness, tinnitus, trembling, and shaking of the extremities with an unsteady or stumbling gait, hence the name. There is also flushing of the face and gastrointestinal symptoms, including abdominal pain, nausea, and diarrhea. Victims may appear to be euphoric and confused. The duration of the illness is 1–2 days after the consumption of the toxic food is ceased. The infected grain may seem normal or appear shriveled and light in weight with a white-to-pinkish coloration suggestive of *Fusarium* head blight, or scab.

7.2.4. Akakabi-byo

Akakabi-byo is also called scabby grain intoxication or red mold disease. It has been observed in Russia, Japan, and China. In China the disease is known as *mi-chum*. In all cases the disease is associated with eating bread made from scab-infested wheat, barley, or other grains infected with *F. graminearum*. Symptoms of this illness are anorexia, nausea, vomiting, headache, abdominal pain, diarrhea, chills, giddiness, and convulsions. Clinical signs are similar to those of the so-called drunken bread syndrome. No specific mycotoxins have been shown to cause this illness, but deoxynivalenol and nivalenol naturally occur in scabby grain from the endemic regions.

7.2.5. Foodborne Illness Outbreaks

Outbreaks of foodborne illness associated with *Fusarium* species have involved foods made from wheat or barley infected mainly with *F. graminearum*. These outbreaks resemble scabby grain intoxication or red mold disease and may be essentially the same thing. Foodborne illnesses have been reported in Japan, Korea, and China that have involved foods, particularly noodles, made from scabby wheat. The onset of illness is usually rapid, from 5 to 30 min, suggesting the presence of preformed toxin, most likely deoxynivalenol. The most common symptoms included nausea, vomiting, abdominal pain, diarrhea with headache, fever, chills, and throat irritation in some victims. An outbreak occurred in India that was characterized by an onset time of 15–60 min, abdominal pain, irritation of the throat, vomiting, and diarrhea, some with blood in the stools. In addition, some victims had a facial rash, nausea, and flatulence. The food involved was bread made from moldy wheat. Apparently, flour millers had mixed infected wheat with sound wheat and milled it into flour. *Fusarium* species were isolated from the wheat and flour, and trichothecene mycotoxins involving deoxynivalenol, acetyl-deoxynivalenol, nivalenol, and T-2 toxin were detected in many samples.

Outbreaks of foodborne illness have occurred in China that have involved corn and wheat contaminated with *Fusarium* species, deoxynivalenol, and zearalenone. An outbreak of precocious pubertal changes in thousands of young children occurred in Puerto Rico in which zearalenone or its derivatives or other exogenous estrogenic substances were the suspected cause. Affected persons experienced premature pubarche, prepubertal gynocomastia, and precocious pseudopuberty. Zearalenone or a derivative were found in the blood of some of the patients, and food was believed to be the source of the estrogenic substances.

7.2.6. Esophageal Cancer

Fusarium moniliforme has been associated with high rates of esophageal cancer in certain parts of the world, particularly in the Transkei region of South Africa, northeastern Italy, and northern China. In these regions, corn is a dietary staple and is the main or only food consumed. Corn and corn-based foods from these regions may contain significant amounts of fumonisins and possibly other metabolites of this fungus.

7.2.7. Immunotoxic Effects of *Fusarium* Toxins

The foregoing discussions of the human diseases that have been associated with various *Fusarium* species in grains and grain-based foods illustrate the range of illnesses caused by *Fusarium* toxins. Members of the genus *Fusarium* produce a diverse array of biologically active compounds. These differ in chemical structure and activity. However, it is believed that many of these compounds, especially the trichothecenes, can affect the immune system by suppressing immune functions. For example, T-2 toxin is known to be highly immunosuppressive. There are reports of so-called sick houses, that is, houses where individuals have contracted diseases such as leukemia, where *Fusarium* species have been detected in dust.

Immunotoxicity can cause two general types of adverse effects on the immune system. In the first type, a toxin or chemical suppresses one or more functions of the immune system. This can result in increased susceptibility to infection or neoplastic disease. Trichothecenes, for example, are known to inhibit protein and deoxyribonucleic acid (DNA) synthesis and to interact with cell membranes, causing weakening and damage. Exposure to trichothecenes can cause damage to bone marrow, spleen, thymus, lymph nodes, and intestinal mucosa. T-2 toxin and deoxynivalenol affect B-cell and T-cell mitogen responses in lymphocytes. Dietary exposures to deoxynivalenol at concentrations as low as 2 μg/g for 5 weeks or 5 μg/g for 1 week can cause decreased mitogen responses. T-2 toxin and diacetoxyscirpenol cause increased susceptibility to *Candida* infections, as well as to *Listeria*, *Salmonella*, *Mycobacterium*, and *Cryptococcus*, in experimental animals.

In the second type of immunotoxicity, the toxin or chemical may stimulate an immune function, resulting in autoimmune types of disorders. Dietary deoxynivalenol has been shown to stimulate immunoglobulin production, causing elevated immunoglobulin A (IgA) levels in mice. Among the harmful effects of this stimulation are kidney damage that is very similar to a common human kidney condition known as glomerulonephritis or IgA nephropathy. While the

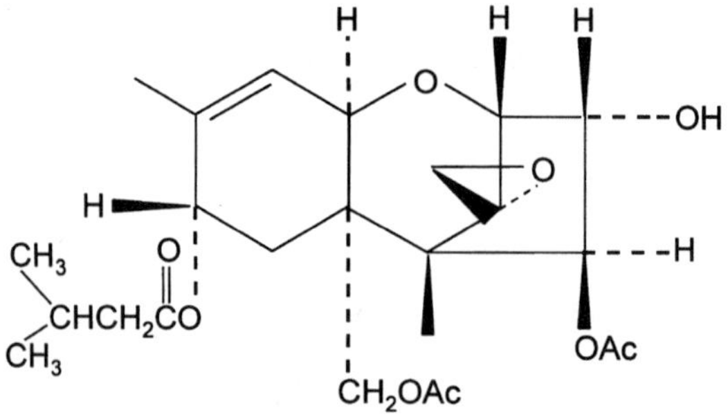

Figure 7.1. Chemical structure of T-2 toxin, a type A trichothecene.

cause of this condition in humans is unknown, there is an association with grain-based diets.

7.3. CHARACTERISTICS OF *FUSARIUM* TOXINS

Fusarium species produce several toxic or biologically active metabolites. The trichothecenes are a group of closely related compounds that are esters of sesquiterpene alcohols that possess a basic trichothecene skeleton and an epoxide group. The trichothecenes are divided into three groups: the type A trichothecenes, which include diacetoxyscirpenol, T-2 toxin (Fig. 7.1), HT-2 toxin, and neosolaniol; the type B trichothecenes, which include deoxynivalenol (Fig. 7.2), 3-acetyldeoxynivalenol, 15-acetyldeoxynivalenol, nivalenol, and fusarenon-X; and the type C or so-called macrocyclic trichothecenes known as satratoxins. Of these, the toxin most commonly found in cereal grains or most often associated with human illness is deoxynivalenol. T-2 toxin occurs rarely in grain in the United States but has been associated with ATA in Russia in 1940s and earlier and has been found in cereal grains in a number of other countries. Other *Fusarium* toxins associated with diseases are zearalenone and the fumonisins. Zearalenone (Fig. 7.3) is an estrogenic compound and is classified as an endocrine disrupter, causing estrogenic-like responses in animals. Fumonisins

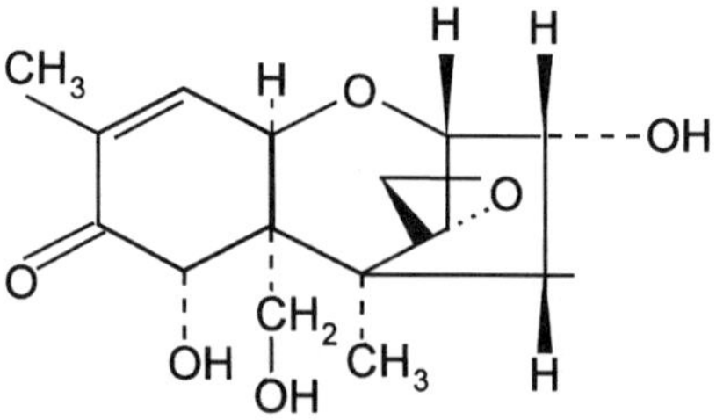

Figure 7.2. Chemical structure of deoxynivalenol (DON), a type B trichothecene.

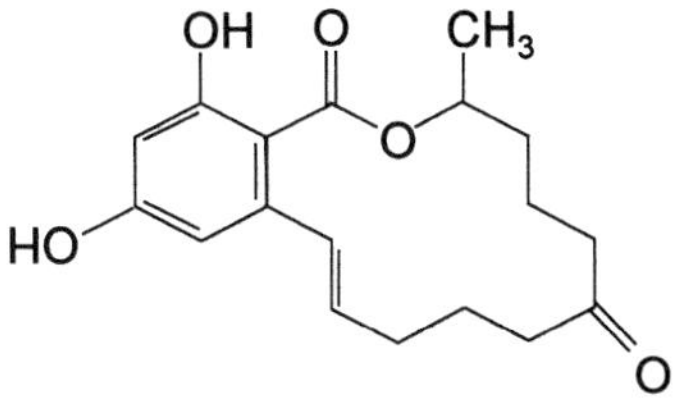

Figure 7.3. Chemical structure of zearalenone.

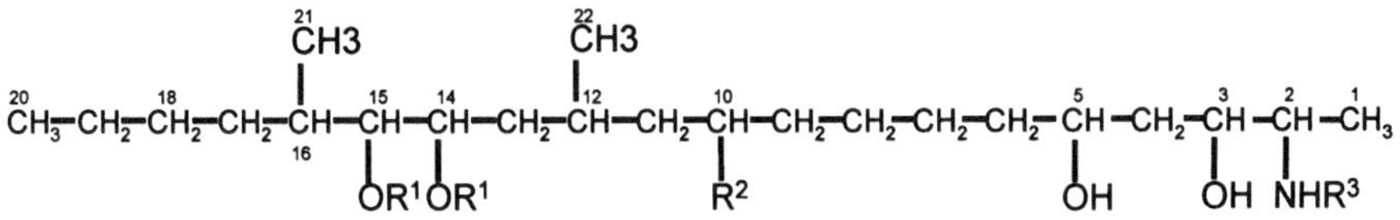

CH_3–CH_2–CH_2–CH_2–CH(CH_3)–CH(OR^1)–CH(OR^1)–CH_2–CH(CH_3)–CH_2–CH(R^2)–CH_2–CH_2–CH_2–CH_2–CH(OH)–CH_2–CH(OH)–CH(NHR^3)–CH_3

A_1: $R^1 = COCH_2CH(CO_2H)CH_2CO_2H$; $R^2 = OH$; $R^3 = C^{23}OC^{24}H_3$
A_2: $R^1 = COCH_2CH(CO_2H)CH_2CO_2H$; $R^2 = H$; $R^3 = COCH_3$
B_1: $R^1 = COCH_2CH(CO_2H)CH_2CO_2H$; $R^2 = OH$; $R^3 = H$
B_2: $R^1 = COCH_2CH(CO_2H)CH_2CO_2H$; $R^2 = R^3 = H$

Figure 7.4. Chemical structure of fumonisins.

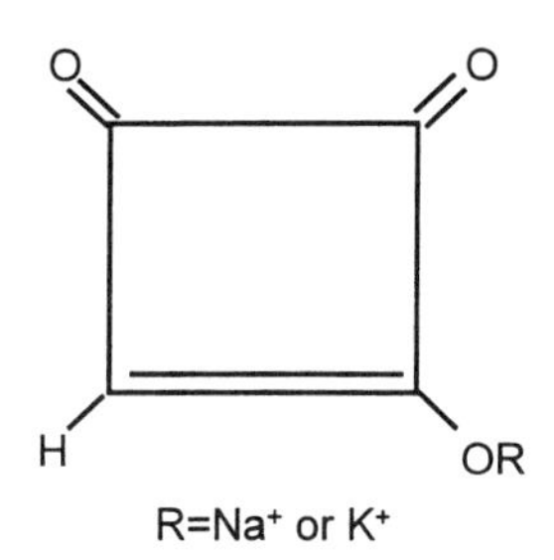

Figure 7.5. Chemical structure of moniliformin.

are a group of related toxins that consist of a long-carbon-chain backbone with two tricarboxylic acid side chains and an amino group (Fig. 7.4). These compounds resemble sphingosine and are known to disrupt sphingosine metabolism. Fumonisin B_1 (FB_1) is the most commonly occurring and most toxic and has been shown to cause cancer in male rats and female mice. It has also been shown to cause atherosclerotic lesions in a nonhuman primate and leukoencephalomalacia in horses and pulmonary edema in swine. Moniliformin (Fig. 7.5), fusarin C, and fusaric acid are also toxins of interest and concern but have not been shown to commonly occur or be specifically associated with diseases.

7.4. DETECTION, ISOLATION, AND IDENTIFICATION

Fusarium species are most often associated with cereal grains, seeds, milled cereal products such as flour and corn meal, barley malt, animal feeds, and necrotic plant tissue. These substrates may also contain or be colonized by many other microorganisms, and *Fusarium* species may be present in low numbers. To isolate *Fusarium* species from these products, it is necessary to use selective media. The basic techniques for detection and isolation of *Fusarium* employ plating techniques, either as plate counts of serial dilutions of products or by placement of seeds or kernels of grain directly on the surface of agar media in petri dishes, that is, direct plating.

Several culture media have been used to detect and isolate *Fusarium* species. These include Nash Snyder (NS) medium, modified Czapek Dox (MCZ) agar, Czapek iprodione dichloran (CZID) agar, potato dextrose iprodione dichloran (PDID) agar, and dichloran chloramphenicol peptone agar (DCPA). The NS medium and MCZ agar contain pentachloronitro-benzene, a known carcinogen, and are not favored for routine use in food microbiology laboratories. However, these media can be useful for evaluating samples that are heavily contaminated with bacteria and other fungi. The CZID agar is becoming a regularly used medium for isolating *Fusarium* from foods, but rapid identification of *Fusarium* isolates to species level is difficult, if not impossible on this medium. Isolates must be subcultured on other media such as carnation leaf agar (CLA) for identification. However, CZID agar is a good selective medium for *Fusarium*. While some other molds may not be completely inhibited on CZID, most are, and *Fusarium* species can be readily distinguished. It has been reported that PDID agar is as selective as CZID agar for *Fusarium* species, with the advantage that it supports *Fusarium* growth with morphological and cultural characteristics that are the same as on potato dextrose agar (PDA), which facilitates more rapid identification, since various monographs and manuals for *Fusarium* identification describe characteristics of colonies grown on PDA. Colony development by *Fusarium* on PDID and CZID agars is better than DCPA. Growth rates are much higher on DCPA, making colony counts more difficult. When modified with 0.5 μg/mL of crystal violet, DCPA increases selectivity by inhibiting *Aspergillus* and *Penicillium* species, but not *Fusarium* species.

Identification of *Fusarium* species is based largely on the production and morphology of macroconidia and microconidia. Identification keys rely heavily on the morphology of conidia and conidiophores, the structures on which conidia are produced. *Fusarium* species do not readily form conidia on all culture media, and conidia formed on high-carbohydrate media such as PDA are often more variable and less typical. A medium that supports abundant and consistent spore production is CLA. Carnation leaves from actively growing, disbudded, young carnation plants free of pesticide residues are cut into small pieces (5 mm^2), dried in an oven at 45–55°C for 2 h, and sterilized by irradiation. The CLA is prepared by placing a few pieces of carnation leaf on the surface of 2.0% water agar. *Fusarium* isolates are then inoculated on the agar and leaf interface, where they form abundant and typical conidia and conidiophores in sporodochia, rather than mycelia. Carnation leaf agar is low in carbohydrates and rich in other complex naturally occurring substances that apparently stimulate spore production.

Since many *Fusarium* species are plant pathogens and all are found in fields where crops are grown, these molds respond to light. Growth, pigmentation, and spore production are most typical when cultures are grown in alternating light and dark cycles of 12 h each from fluorescent light or diffuse sunlight from a north window. Fluctuating temperatures such as 25°C (day) and 20°C (night) also enhance growth and sporulation. For identification keys, refer to Nelson et al. (1983), Samson et al. (1999), Marasas et al. (1984), and Marasas (1991).

If present, *Fusarium* toxins are usually found at low levels in cereal grains and processed grain-based foods. Their concentrations may range from less than nanogram to microgram quantities per gram (ppb to ppm, respectively). *Fusarium* toxins vary in their chemical structures and properties, making it difficult to develop a single method for quantitating all toxins. The basic steps involved in detection of *Fusarium* mycotoxins are similar to those for other mycotoxins. These include sampling, size reduction and mixing, subsampling, extraction, filtration, clean-up, concentration, separation of components, detection, quantification, and confirmation.

The first problem encountered in the analysis of grains for *Fusarium* toxins is the same as for other mycotoxins, that is, sampling. Obtaining a representative sample from a large lot of cereal grain can be very difficult if the toxin is present in a relatively small percentage of the kernels, which may be the case with toxins such as deoxynivalenol and zearalenone. On the other hand, fumonisins appear to be more evenly distributed in grain such as corn. Processed grain-based foods may contain a more even distribution of toxins due to grinding and mixing. Samples are usually ground and mixed further, and a subsample of 50–100 g is taken for extraction. *Fusarium* toxins, like all mycotoxins, must be extracted from the matrix in which they are found.

Most mycotoxins are more soluble in slightly polar organic solvents than water. The most commonly used extraction solvents consist of combinations of water with organics such as methanol, acetone, and acetonitrile. Following extraction, the extract is filtered to remove solids and subjected to a clean-up step to remove interfering substances. Clean-up can be done in several ways, but the most common method used for *Fusarium* toxins is to pass the extract through a column packed with sorbant packing materials. In recent years, the use of small, prepacked, commercially available disposable columns or cartridges such as Sep-Pak, Bond Elut, and MycoSep has become more common. After the extract has been cleaned, the sample may need to be concentrated before analysis in order to detect the toxin. This may be accomplished by mild heating such as in a water bath, heating block, or rotary evaporator under reduced pressure or a stream of nitrogen. Detection and quantification of the toxins are done after they are separated from other components by chromatographic means. The most common chromatographic separation techniques used are thin-layer chromatography (TLC) and high-performance liquid chromatography (HPLC). Gas chromatography (GC) also has some applications, particularly when coupled with mass spectrometry (GCMS).

One method for quantitating deoxynivalenol is TLC. Gas chromatography is more sensitive than TLC but is also more laborious. While HPLC methods employing UV absorbance at 219 nm for detection are fairly sensitive, they require purification of deoxynivalenol using high-capacity activated charcoal columns. The method of choice for quantitation of zearalenone is HPLC with fluorescence detection. A TLC method for zearalenone has been tested collaboratively and is useful as a screening method. The methods most commonly used for T-2 toxin are GC methods. Because type A trichothecenes lack a UV chromophore and are not fluorescent, TLC and HPLC methods are unsuitable, resulting in the reliance on GC methods. The most widely used analytical methods for fumonisins are HPLC methods involving the formation of fluorescent derivatives. Methods have been developed using derivatizing agents, such as *o*-phthaldehyde (OPA), fluorescamine, and naphthalene dicarboxaldehyde. A TLC method for fumonisins has also been developed but is used mainly for screening. Methods for moniliformin include TLC and HPLC.

Immunoassays have been developed for *Fusarium* toxins. Enzyme-linked immunosorbant assay (ELISA) kits for *Fusarium* toxins are commercially available (Neogen, Lansing, MI). Qualitative kits for screening as well as kits for quantitative analyses are available for deoxynivalenol, zearalenone, T-2 toxin, and fumonisins. A rapid-screening TLC kit for deoxynivalenol is available from Romer Labs (Union, MO). This method uses a special clean-up column that requires only 10 s per sample. An antibody-based affinity column for fumonisins is also available (Vicam, Watertown, MA).

7.5. OCCURRENCE AND STABILITY OF TOXINS IN FOODS

Fusarium toxins, particularly deoxynivalenol and fumonisins, have been found in finished human food products. Various *Fusarium* toxins have been found naturally occurring in numerous cereal grains, but most of these grains have been destined for animal feed. Deoxynivalenol is the most common trichothecene found in commodity grains; therefore, the greatest potential exists for it to occur in finished foods. Food products such as bread, pasta, and beer may contain at least trace amounts of the toxin. Trace amounts (~5 ng/mL) of deoxynivalenol have been found in samples of commercially available domestic and imported beer in Canada. In the United States, deoxynivalenol has been found in breakfast cereals in average amounts of 100 ng/g, while corn syrup and beer samples were negative. The toxin was found in processed foods, including corn breakfast cereals, wheat flour muffin mixes, wheat- and oat-based cookies, crackers, corn chips, popcorn, and mixed-grain cereals in concentrations ranging from 4 to 19 μg/g and from 0.08 to 0.3 μg/g in wheat, flour, corn, corn meal, and snack foods. Thus there is evidence that deoxynivalenol is a contaminant of processed human food products and that levels sometimes exceed the U.S. government guideline of 1.0 μg/g in finished food products. Deoxynivalenol is quite heat stable, tolerating most thermal processes to some degree.

Fumonisins have also been found in processed or finished food products. Food products that have been examined include corn meal, corn grits, corn breakfast cereals, tortillas, tortilla chips, corn chips, popcorn, and hominy corn. The most consistently contaminated products with the greatest amounts of fumonisins are those foods that receive only physical processing such as milled products (e.g., corn meal and corn muffin mixes). Fumonisin levels in corn meal obtained from Canada, Egypt, Peru, South Africa, and the United States were highest in corn meal from Egypt and the United States. Corn meal from the United States contained fumonisins in concentrations less than 1.0 μg/g up to 2.8 μg/g of corn meal. In another survey, the highest levels of fumonisins in corn-based foods were found in corn meal and corn grits. Corn flakes and corn pops cereals, corn chips, and corn tortilla chips were negative for fumonisins, and very low levels were found in tortillas, popcorn, and hominy. Fumonisins have been detected in processed corn products in Germany, Italy, Japan, Spain, and Switzerland. Fumonisins have also been found to be very heat stable in thermally processed food products.

7.6. PREVENTION AND CONTROL

Preventing contamination of cereal grains with *Fusarium* species and their toxins is very difficult because contamination occurs in the field and is subject to weather, drought, and insect activity. The main preventative measure is to divert contaminated grain from being processed into human foods. The advisory level for deoxynivalenol is 1.0 μg/g in processed foods. An advisory or action level for fumonisin B_1 has not yet been set, but it is anticipated that it will be in the range of 1–3 μg/g. High-temperature processing systems such as extrusion cooking may offer some potential for reducing levels of *Fusarium* toxins in finished foods.

REFERENCES

Marasas, W. F. O. 1991. Toxigenic fusaria. In: J. E. Smith and R. S. Henderson (Eds.), *Mycotoxins and animal foods.* Boca Raton, FL: CRS Press, pp. 119–139.

Marasas, W. F. O., P. E. Nelson, and T. A. Tousson. 1984. *Toxigenic Fusarium Species: Identity and Mycotoxicology*. University Park, PA: Pennsylvania State University Press.

Nelson, P. E., T. A. Tousoun, and W. F. O. Marasas. 1983. *Fusarium Species. An Illustrated Manual for Identification*. University Park, PA: Pennsylvania State University Press.

Samson, R. A., E. S. Hoekstra, J. C. Frisvad, and O. Filtenborg (Eds.). 1995. *Introduction to Food-Borne Fungi*. Baarn, the Netherlands: Centraalbureau voor Schimmelcultures.

BIBLIOGRAPHY

Beardall, J. M., and J. D. Miller. 1994. Diseases in humans with mycotoxins as possible causes. In: J. D. Miller and H. L. Trenholm (Eds.), *Mycotoxins in Grain. Compounds Other than Aflatoxin*. St. Paul, MN: Eagan Press, pp. 487–539.

Bhat, R. V., S. R. Beedu, Y. Ramakrisna, and K. L. Munshi. 1989. Outbreak of trichothecene mycotoxicosis associated with consumption of mould-damaged wheat products in Kashmir Valley, India. *Lancet* 7(Jan.), 35–37.

Hayes, A. W. 1981. Involvement of mycotoxins in animal and human health. In: *Mycotoxin Teratogenicity and Mutagenicity*. Boca Raton, FL: CRC Press, pp. 11–40.

Joffe, A. Z. 1960. The mycoflora of overwintered cereals and its toxicity. *Bull. Res. Counc. Israel* **90**, 101–126.

Miller, J. D. 1995. Fungi and mycotoxins in grain: Implications for stored product research. *J. Stored Prod. Res.* **31**, 1–16.

Mills, J. T. 1989. Ecology of mycotoxigenic *Fusarium* species on cereal seeds. *J. Food Prot.* **52**, 737–742.

Nelson, P. E., A. E. Desjardins, and R. D. Plattner. 1993. Fumonisins, mycotoxins produced by *Fusarium* species: Biology, chemistry and significance. *Annu. Rev. Phytopathol.* **31**, 233–252.

Pestka, J. J., and G. S. Bondy. 1994. Immunotoxic effects of mycotoxins. In: J. D. Miller and H. L. Trenholm (Eds.), *Mycotoxins in Grain. Compounds Other than Aflatoxin*. St. Paul, MN: Eagan Press, pp. 339–359.

Pitt, J. I., and A. D. Hocking. 1985. *Fungi and Food Spoilage*. Sydney: Academic.

Saenz de Rodriguez, C. A. 1984. Environmental hormone contamination in Puerto Rico. *N. Engl. J. Med.* **310**, 1741–1742.

Saenz de Rodriguez, C. A., A. M. Bongiovanni, and L. Conde de Borrego. 1985. An epidemic of precocious development in Puerto Rican children. *J. Pediat.* **107**, 393–396.

Scott, P. M. 1984. Effects of food processing on mycotoxins. *J. Food Prot.* **47**, 489–499.

Sharma, R. P., and Y. W. Kim. 1991. Trichothecenes. In: R. P. Sharma, and D. K. Salunkhe (Eds.), *Mycotoxins and Phytoalexins*. Boca Raton, FL: CRC Press, pp. 339–359.

CHAPTER 8

Listeria monocytogenes

CATHERINE W. DONNELLY

8.1. INTRODUCTION

As a leading cause of death due to foodborne illness, *Listeria monocytogenes* continues to cause sporadic cases and outbreaks of illness linked to consumption of food products. *Listeria monocytogenes* was first discovered almost 100 years ago and prior to 1981 was recognized primarily as an animal pathogen. To prevent foodborne listeriosis, it is necessary to understand the disease, susceptible persons, distribution of the organism within the environment, and behavior of the organism in foods. This chapter is designed to summarize current knowledge with respect to the foodborne role of *L. monocytogenes*.

8.1.1. Characteristics of *Listeria*

Listeria monocytogenes is a small (1.0–2.0 × 0.5 μm), gram-positive, facultatively anaerobic, rod-shaped bacterium. The organism can exist in an intracellular state within monocytes and neutrophils and is accordingly named because large numbers of monocytes are often found in the peripheral blood of monogastric animals this organism infects (Gray and Killinger, 1966). *Listeria* is recognized as a causative agent of the disease listeriosis, a zoonotic illness that affects both animals and humans.

Reports of organisms resembling *Listeria* first appeared in the scientific literature in 1891 when Hayem of France observed small, gram-positive rod-shaped organisms in human tissue (Gray and Killinger, 1966). This observation was repeated two years later (1893) by Henle, who was working in Germany. Hulphers, working in Sweden in 1911, reported the presence of gram-positive rod-shaped organisms in the livers of rabbits and designated the organism *Bacillus hepatis*. Complete characterization of the organism now recognized to be *L. monocytogenes* was achieved by Murray et al. (1926). These investigators,

Guide to Foodborne Pathogens, Edited by Ronald G. Labbé and Santos García
ISBN 0-471-35034-6

working at Cambridge University, observed infection of an animal colony caused by an etiologic agent that induced pronounced monocytosis and caused hepatic lesions. The name given to this agent was *Bacterium monocytogenes.* Pirie (1940), working in South Africa one year later, isolated a small gram-positive bacillus from the livers of gerbils that he designated *Listerella hepatolitica,* in honor of Lord Lister (Seeliger, 1961). In 1940, the genus–species designation *Listeria monocytogenes* was proposed for this organism. A detailed historical chronology can be found in the classic review article on *Listeria* and listeric infection, published by Gray and Killinger (1966).

Included within the genus *Listeria* are the species *L. monocytogenes, L. innocua, L. seeligeri, L. welshimeri, L. ivanovii,* and *L. grayi* (Table 8.1). Deoxyribonucleic acid (DNA) base composition and DNA–DNA hybridization studies led Stuart and Welshimer (1974) to propose a new genus *Murraya* to include *M. grayi* and *M. grayi* subsp. *murrayi.* However, more recent examination of the genomic relatedness of *L. grayi* and *L. murrayi* based upon DNA–DNA hybridization and multilocus enzyme electrophoresis reveals that these organisms should be considered as members of the single species *L. grayi* (Rocourt et al., 1992). Members of the genus *Listeria* can be differentiated by the following biochemical reactions: reduction of nitrates to nitrites; beta-hemolysis; acid production from mannitol, L-rhamnose, and D-xylose; and the CAMP test (Table 8.1). *Listeria monocytogenes* and *L. innocua* are very closely related, and recent studies have confirmed that within the 16S ribosomal ribonucleic acid (rRNA), only 2 of 1281 base pairs differ between the two species (Czajka et al., 1993; Wang et al., 1991).

Of the six species within the genus *Listeria,* only *L. monocytogenes* is generally regarded as capable of causing illness in humans. Three reports of human infection caused by *L. ivanovii* exist in the scientific literature (Busch, 1971; Rocourt and Seeliger, 1985), along with one report of a case of human illness caused by *L. seeligeri* (Hof and Rocourt, 1992; Rocourt et al., 1986). *Listeria innocua* and *L. welshimeri* are not capable of causing illness. These nonpathogenic species are of interest from a food microbiology standpoint since their presence indicates the potential for the presence of the pathogenic *L. monocytogenes. Listeria monocytogenes* strains can be differentiated on the basis of serology, and to date, over 14 serotypes have been designated: 1/2a, 1/2b, 1/2c, 3a, 3b, 3c, 4a, 4b, 4bX, 4c, 4d, 5, 6a, and 6b (Graves et al., 1999). Despite the widespread occurrence of *Listeria* in nature, only three serotypes (4b, 1/2a, and 1/2b) account for 96% of human infections reported in the United States (Tappero et al., 1995). A similar survey of strains from cases of human listeriosis in Britain between 1967 and 1984 revealed 1/2, 3, and 4 as the predominant serogroups causing human infection (McLauchlin et al., 1986). Serotyping has poor discriminating power and is therefore of limited value as a subtyping method when compared to more advanced methods of genetic analysis. Of all available phenotypic and genotypic typing methods, pulsed-field gel electrophoresis (PFGE) is the preferred method for subtyping of *L. monocytogenes* strains (Brosch et al., 1996), offering a high degree of both discrimination of *Listeria* strains as well as reproducibility.

TABLE 8.1. Characterization of *Listeria* Species

Characteristic	*L. monocytogenes*	*L. ivanovii*	*L. seeligeri*	*L. innocua*	*L. welshimeri*	References
β-hemolysis	+	+	+	−	−	Farber and Peterkin, 1991
CAMP						Lovett, 1990
S. aureus	+	−	+	−	−	
Rhodococcus equi	−	+	−	−	−	
Fermentation of:						Lovett, 1990
Mannitol	−	−	−	−	−	
Xylose	−	+	+	−	+	
Rhamnose	+	−	−	±	±	
Pathogenic in humans	Yes	Extremely rare	Extremely rare	No	No	Busch, 1971; Rocourt et al., 1985; Rocourt et al., 1986

8.1.2. Distribution of *Listeria*

Listeria monocytogenes is very widely distributed in nature and can be readily isolated from soil, water, sewage, green plant material, decaying vegetation, and numerous species of birds and mammals, including humans (Gray and Killinger, 1966). Cattle, sheep, and goats are the domestic mammals most frequently afflicted by listeriosis (Dutta and Malik, 1981; Gray, 1960). A close relationship between onset of listeriosis in ruminants and feeding of contaminated silage has long been recognized (Gray, 1960; Gronstol, 1979; Kalac, 1982; Olafson, 1940). Palsson (1962) reported that in Iceland the relationship between silage feeding and onset of listeriosis is so strong that the disease has been referred to as "votheysveiki," or silage sickness. Presence of *Listeria* in silage is strongly influenced by pH, and samples having a pH in the range of 5.0–6.0 or above are far more likely to be sources of *L. monocytogenes* than silage where the pH is below 5.0 (Gronstol, 1979; Irvin, 1968; Perry and Donnelly, 1990; Ryser et al. 1997). The most common disease syndrome of listeriosis in ruminants is encephalitis, leading to observations of nervous system involvement in cattle and sheep. Infected animals become disoriented and circle endlessly in one direction or another depending upon the direction in which their head is drooped. For this reason, listeriosis in ruminants is often referred to as "circling disease" (Gill, 1933; Seeliger, 1961). Previous investigations have identified sheep as a major reservoir of *Listeria* in nature. In one study alone, 88% of tested sheep were identified as carriers of some member of the genus *Listeria* (Rodriguez et al., 1984). Gray (1960) used serotyping techniques to demonstrate the relatedness of isolates obtained from listeric sheep and the oat silage they consumed. More recent studies have used strain-specific ribotyping to support the link between on-farm sources of *Listeria* (silage) and subsequent contamination of dairy processing environments (Arimi et al., 1997). Infected animals displaying symptoms of listeric infection may excrete *L. monocytogenes* in milk, blood, and feces. High excretion rates of *L. monocytogenes* in milk from asymptomatic cows and goats have frequently been reported (Loken et al., 1982).

8.2. LISTERIOSIS IN HUMANS

8.2.1. Disease Characterization

Because of its ubiquity in the environment, humans frequently come into contact with *L. monocytogenes*. Exposure alone does not necessarily dictate that infection will result. *Listeria monocytogenes* is frequently shed in the stools of healthy humans who otherwise show no signs of the illness (Gellin and Broome, 1989). It is estimated that 5% of healthy humans harbor *L. monocytogenes* in their gastrointestinal tract. The first case of human listeriosis was described in 1929 (Nyfeldt, 1929), and since that time, listeriosis has been recognized as a rare but often fatal illness. In adults, the disease listeriosis is characterized by

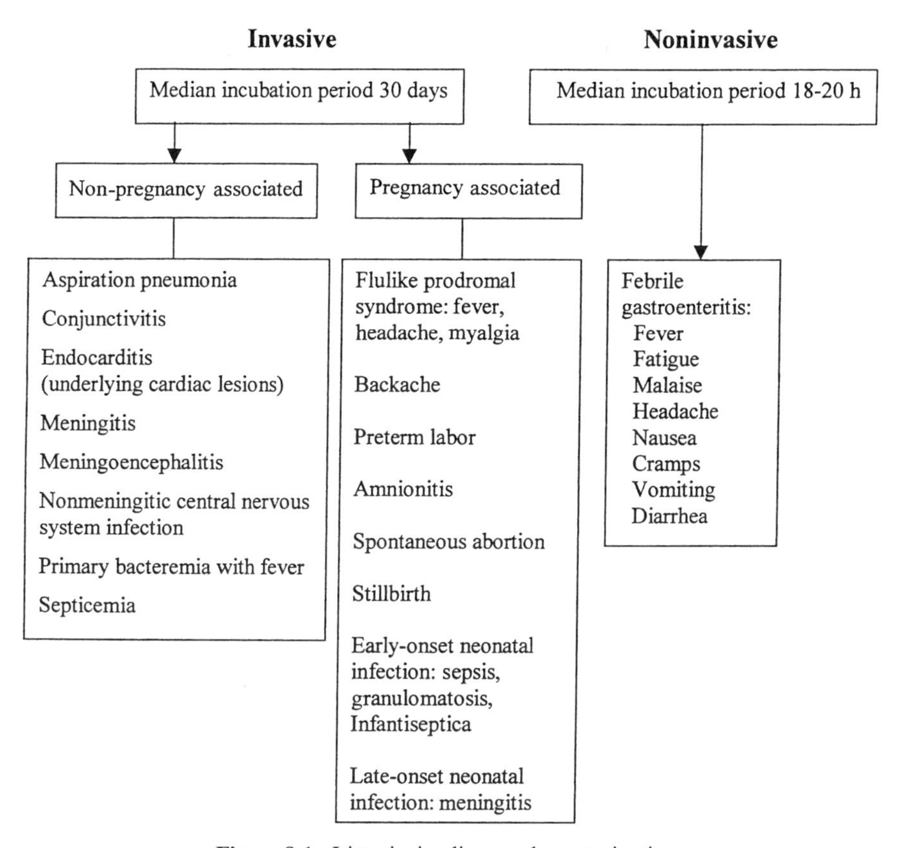

Figure 8.1. Listeriosis: disease characterization.

two primary syndromes, an invasive form of the illness versus a noninvasive form (Fig. 8.1). Invasive illness is characterized by the onset of severe symptoms, including meningitis, septicemia, primary bacteremia, endocarditis, nonmeningitic central nervous system infection, conjunctivitis, and flulike illness (fever, fatigue, malaise, nausea, cramps, vomiting, and diarrhea). The median incubation period for invasive illness prior to onset of symptoms is approximately 30 days [Centers for Disease Control and Prevention (CDC), 1998; Linnan et al., 1988]. FoodNet monitoring data from 1998 indicated that *Listeria* infections resulted in a higher rate of hospitalization (95% of infected patients hospitalized compared to 21% for *Salmonella* infections) than any other pathogen and caused approximately half of all reported deaths (CDC, 1998). Gastrointestinal symptoms are observed in approximately one-third of documented cases of listeriosis (Armstrong, 1995). A noninvasive form of listeriosis resulting in febrile gastroenteritis has been documented in several outbreaks (Dalton et al., 1997; Salamina et al., 1996) (Fig. 8.1). Unlike the invasive form, the median incubation period prior to onset of symptoms is

TABLE 8.2. Underlying Patient Conditions Causing Predispositions to Listeriosis

Condition	References
Cancer (leukemia, lymphoma, hematological, pulmonary)	Fleming et al., 1985; Gellin and Broome, 1989; Linnan et al., 1988; Slutsker and Schuchat, 1999.
Administration of steroids (corticosteroids)	Gellin and Broome, 1989; Slutsker and Schuchat, 1999.
Treatment with cytotoxic drugs	
Renal transplant recipients, renal dialysis, chronic renal disease	
Human immunodeficiency virus (HIV), acquired immunodeficiency syndrome (AIDS)	Gellin and Broome, 1989; Slutsker and Schuchat, 1999.
Pregnant women and neonates	Fleming et al., 1985; Gellin and Broome, 1989; Linnan et al., 1988; McLaughlin, 1990; Slutsker and Schuchat 1999.
Age > 60 years	Gellin and Broome, 1989.
Alcoholism, cirrhosis, liver disease, hepatitis	Fleming et al., 1985
Antacid, cimetidine use	Ho et al., 1986
Diabetes	Schuchat, 1992
Collagen vascular disease	Slutsker and Schuchat, 1999
Sarcoidosis	
Ulcerative colitis	
Aplastic anemia	
Intravenous drug abuse	
Conditions associated with iron overload	Mossey and Sondheimer, 1985

short, typically 18–20 h. The frequency of febrile gastroenteritis as a result of *L. monocytogenes* infection is undetermined, as are host characteristics associated with this syndrome. The infectious dose needed to cause symptoms of febrile gastroenteritis is not known but has been shown to be higher than that associated with the invasive form of disease (Dalton et al., 1997). When gastroenteritis due to listeriosis is suspected, clinicians and public health officials are advised to examine stool cultures for presence of *L. monocytogenes* (Slutsker and Schuchat, 1999).

Susceptible individuals typically have one or more underlying conditions that predispose these patients to acquiring listeriosis (Table 8.2). Humans shown to be at high risk for acquiring listeriosis include pregnant women, neonates, the elderly, organ transplant recipients, or those receiving immunosuppressive therapy. In the latter case, treatment of patients with corticosteroids or antimetabolites renders a suppressed immune system. Persons suffering from chronic disorders such as alcoholism, malignancy, diabetes, heart disease, or acquired immunodeficiency syndrome (AIDS) have also been shown to be at risk. Additional underlying factors that have been reported in association with listeriosis include sarcoidosis, chronic otitis, collagen-vascular disease, idiopathic thrombocytopenic purpura, asthma, ulcerative colitis, and aplastic anemia (Nieman and Lorber, 1980). Age has been shown to be a predisposing factor in listeriosis. An 11% case–fatality rate is documented in persons age 40 or under; a 63% case–fatality rate is recorded for persons over age 60 (Gellin and Broome, 1989). Age-related reasons for increased incidence of listeriosis may include a decline of the immune system as a function of age, increased prevalence of immunosuppressive disorders, and increased dependence on immunosuppressive medications. Although the above-listed conditions may predispose patients to acquiring listeriosis, it should be noted that persons showing no apparent immunocompromising conditions have been shown to acquire listeriosis. A recent study by Mead et al. (1999) has confirmed that while listeriosis is a rare human illness, it remains a leading cause of death from a foodborne pathogen with approximately 2518 cases and 499 deaths occurring annually in the United States. This illness has a case–fatality rate that ranges from 23 to 35% (Gellin and Broome, 1989; Schuchat et al., 1992).

8.2.2. Listeriosis in Immunocompromised Hosts

Cell-mediated immunity (CMI) plays an important role in dictating the resistance or susceptibility of a human host to infection by *L. monocytogenes* (Hahn and Kaufman, 1981; Lane and Unanue, 1972; Mackaness, 1962). Listeriosis occurs most often in those persons with impaired CMI. Cell-mediated immunity is dependent upon the activity of mononuclear phagocytes as early response, nonspecific effectors, and specific T-cells as a secondary response to infection. Alteration of T-cell or macrophage function due to immunosuppression would result in an impairment of CMI allowing a listeric infection to occur after the primary infection or after further contact with *Listeria*. Patients with cancer or undergoing treatment with steroids or cytotoxic drugs, pregnant women or neonates, renal transplant recipients, patients with AIDS, elderly or alcoholic patients, and those with diabetes are well known to be the primary targets for listeriosis (Gellin and Broome, 1989; Jensen et al., 1994; Slutsker and Schuchat, 1999). The risk of complications due to *L. monocytogenes* infection has long been recognized in CMI-compromised conditions such as Hodgkin's lymphoma and other hematological malignancies (Louria et al., 1967). North (1970) suppressed CMI of mice to *L. monocytogenes* by administration of an antimitotic drug and induced a listeric infection. Golnazarian et al. (1989) showed that *L. monocytogenes* infectious dose for mice immunocompromised by administration of hydrocortisone acetate was much lower than for normal resistant mice.

During pregnancy, selective factors of CMI become depressed to prevent rejection of the fetus by the mother. However, depression of these selective factors may result in decreased maternal resistance to *L. monocytogenes* infections and thereby increase the maternal or fetal risk to onset of listeriosis (Schlech 1988). Such selective factors include shifts in levels of hormones or serum factors that affect lymphocyte or macrophage synthesis, activation, or function during pregnancy (Weinberg, 1984). Plasma levels of hydrocortisone increase during pregnancy to levels three to seven times higher than those found in nonpregnant humans (Weinberg, 1984). Corticosteroids are known to suppress both lymphokine activation and phagocytic activity of macrophages. Low levels of immunoglobulin M (IgM) and decreased activity of the classic complement pathway during the neonatal period also occur and demonstrate the importance of opsonization in the immune response to *Listeria* (Gellin and Broome, 1989).

In humans, listeriosis occurs most often during the third trimester of pregnancy. Three outcomes are normally followed: an asymptomatic maternal infection and a resulting infected infant; a severely ill mother who enters premature labor and delivers a stillborn or severely ill infant; or an unaffected fetus with death of the mother (Weinberg, 1984). In most perinatal cases of listeriosis, the mother is usually mildly affected, exhibiting flulike symptoms, but neonatal morbidity and mortality are common. In early-onset neonatal listeriosis, transplacental infection results in a syndrome known as *granulomatosis infantisepticum*, a necrotic disease of the internal organs (Bojsen-Moller, 1972). Spontaneous abortion of the fetus and stillbirth of the neonate are most common, but if the fetus infected in utero is born alive, recovery is not likely (Bojsen-Moller, 1972; Gray and Killinger, 1966). Late-onset listeriosis occurs several days after birth, and infants are generally full term and healthy at birth. Late-onset listeriosis is more likely than early-onset listeriosis to present as meningitis, and case–fatality rates are lower than for early-onset infection. Of late-onset cases reported in Britain during 1967–1985, 93% of cases presented evidence of central nervous system infection (McLauchlin, 1990).

Buchdahl et al. (1990) described several cases in which complications due to listeriosis arose during the course of pregnancy. In one case, a mother at 32 weeks gestation developed a flulike illness and developed irregular uterine contractions. Spontaneous membrane rupture occurred with release of meconium-stained amniotic fluid. The infant, delivered by caesarean section, was found to have blood and cerebrospinal fluid contaminated with *L. monocytogenes*. Although the infant survived, neurological handicap was evident. The mother recalled consumption of a soft-ripened French cheese 9 days prior to delivery. Fortunately, perinatal listeriosis declined 63% between 1989 and 1997 in the United States due to peripartum use of the antibiotic ampicillin for group B *Streptococcus* prevention.

Listeriosis is well recognized as a complication of renal transplantation. Most patients become ill while they are receiving immunosuppressive therapy, which increases their susceptibility to listeriosis (Stamm et al., 1982). Meningitis is recorded as the most common presentation of listeriosis in renal transplant patients, and the fatality rate for listerial meningitis in this patient population is 38%. However, pneumonia due to *L. monocytogenes* has also been observed in renal patients, suggesting a possible respiratory route of transmission (Stamm et al., 1982). In a study of healthy renal transplant recipients, fecal carriage of *L. monocytogenes* in 8 of 37 patients was documented (Stamm et al., 1982).

In a review of 83 cases of listeriosis in renal patients, Stamm et al. (1982) found that one-third of patients had been treated for acute rejection. Nieman and Lorber (1980) have reported that hemodialysis is not a predisposing factor for most patients in acquiring listeriosis. However, Mossey and Sondheimer (1985) reported four cases of *L. monocytogenes* bacteremia associated with long-term hemodialysis and transfusional iron overload. None of these patients were receiving immunosuppressive therapy. Many surveys document a higher incidence of listeriosis in the months from July to October, and the same seasonal variation has been reported for renal transplant recipients (Stamm et al., 1982).

Patients with AIDS exhibit an impairment of T-cell-mediated immune response and therefore are at high risk for listeriosis. In early studies of the incidence of listeriosis in AIDS patients, *L. monocytogenes* was rarely implicated as an agent affecting persons with AIDS (Jacobs and Murray, 1986). Reasons given for this surprising finding included the fact that AIDS patients who displayed frequent gastrointestinal tract infections were given multiple courses of antibiotic therapy, thus decreasing exposure to *Listeria* (Jacobs and Murray, 1986; Mascola et al., 1988). Five cases of listeriosis in Los Angeles County between January 1985 and March 1986 occurred in patients with AIDS. Prior or concurrent gastrointestinal illness was recorded in three of the patients, and four patients had no history of prior antibiotic administration. This and subsequent investigations have shown that while listeriosis is a rare infection in patients exhibiting human immunodeficiency uirus (HIV) infection, persons with AIDS have a 300- to 1000-fold increased risk of acquiring listeriosis compared with the general population (Gellin and Broome, 1989; Jensen et al., 1994). Persons with AIDS are therefore advised to refrain from ingestion of food items associated with listeriosis (Mascola et al., 1988).

8.3. PATHOGENESIS

The production of sulfhydryl-activated hemolysin, listeriolysin O (α-listeriolysin) is associated with the pathogenic potential of *L. monocytogenes* (Gaillard et al., 1986, 1987; Geoffroy et al., 1989; Kathariou et al., 1987). Listeriolysin O is similar to streptolysin O and pneumolysin, and antigenic cross reactivity with these hemolysins as well as with hemolysins produced by *L. ivanovii* and *L. seeligeri* has been demonstrated (Geoffroy et al., 1989; Kraft et al., 1989). The virulence of *Listeria* species has also been associated with the ability to survive and grow intracellularly (Bortolussi et al., 1987; Kuhn et al., 1988).

Gaillard et al. (1986, 1987) studied the role of hemolysin (listeriolysin O) in pathogenicity of *L. monocytogenes*. Working with transposon Tn1545, these investigators inactivated a genetic determinant for hemolysin production and obtained nonhemolytic mutants from hemolysin-producing strains. The loss of hemolysin production was shown to be associated with loss of virulence in a mouse model. The spontaneous loss of the transposon restored hemolysin production and virulence to the revertants (Gaillard et al., 1986). Further studies by Gaillard et al. (1987) found that *L. monocytogenes* and *L. ivanovii* invaded a continuous gut epithelial cell line, whereas *L. seeligeri*, *L. innocua*, and *L. welshimeri* did not. A nonhemolytic mutant of *L. monocytogenes* invaded these gut enterocytes at the same rate as the hemolytic wild type. This finding, which has been corroborated by others in a fibroblast 3T6 continuous cell line (Kuhn et al., 1988), demonstrates that listeriolysin O is not involved in invasion. An extracellular protein (p60) may be involved in the process of attachment and invasion of *L. monocytogenes* (Kuhn and Goebel, 1989). Both listeriolysin and p60 are produced by all virulent *L. monocytogenes* strains and are involved in the virulence of *L. monocytogenes* (Kuhn et al., 1988; Kuhn and Goebel 1989). Under heat shock conditions, listeriolysin is synthesized, whereas production of p60 no longer occurs (Sokolovic and Goebel, 1989). Protein p60 is found both as a major secreted protein and on the cell surface of all *L. monocytogenes* isolates. In addition, this protein possesses murein hydrolase and is involved in cell division. Rough mutants of *L. monocytogenes* that lack p60 form long chains of cells that fail to separate. These mutants also show reduced uptake by 3T6 fibroblast cells (Kuhn and Goebel, 1989).

Factors other than hemolysin have recently been defined as essential virulence factors for *L. monocytogenes*. Hof and Rocourt (1992) found that a construct of a virulent *L. monocytogenes* EGD with selective blockade of phospholipase C production became avirulent. Tilney and Portnoy (1989) demonstrated that *L. monocytogenes* is capable of bypassing the humoral immune system by remaining in an intracellular state and spreading cell to cell. Following phagocytosis by host macrophages and escape from the phagocytic vacuole, *Listeria* species are coated with actin filaments, form a pseudopod, dissolve the phagocytic vacuolar membrane presumably by use of hemolysin, and repeat the cycle. It was postulated from this and other studies (Bortolussi et al., 1987; Kuhn et al., 1988; MacGowan et al., 1983,) that once *Listeria* enters macrophages, listeriolysin O is needed to lyse phagosomes, thereby releasing *Listeria* into the cytoplasm so that it can multiply. *Listeria* species that lack hemolysin fail to grow in vivo because of inability to dissolve the endosomal membrane and failure to escape from the endosome into the cytoplasm (MacGowan et al., 1983; Tilney and Portnoy, 1989).

Listeria monocytogenes can enter host cells in two distinct ways: through active ingestion by phagocytic cells such as macrophages or through the production of specific gene products that control ingestion by normally nonphagocytic cells. (Gaillard et al., 1987; Tilney and Portnoy, 1989). Studies conducted by Gaillard et al. (1991) identified a surface protein of *L. monocytogenes*, internalin, that mediates bacterial invasion of epithelial cells. Once internalized, the life cycle of *L. monocytogenes* within both phagocytic and nonphagocytic cells is similar. The majority of virulence genes that produce products associated with the intracellular life cycle of *L. monocytogenes* reside on a region of the chromosome known as the *PrfA*-dependent gene cluster (Kuhn and Goebel, 1999). This region is comprised of the genes *prfA*, *plcA*, *hly*, *mpl*, *actA* and *plcB*. The *prfA* gene a product is a positive regulatory factor, a 27-kDa regulatory factor that controls all virulence genes of the virulence gene cluster. The *plcA* product is phosphatidylinositol-specific phospholipase C (PI-PLC), which contributes to vacuole escape in cells such as bone marrow–derived macrophages. The *plcB* gene encodes for a phosphatidylcholine-specific phospholipase C (PC-PLC), which, together with metalloprotease, the *mpl* gene product, enables listeriolysin O–independent escape of *L. monocytogenes* from primary vacuoles in human epithelial cells. Metalloprotease (*mpl*) permits bacterial movement from the cytosol to the host surfaces and the ensuing cell-to-cell spread (Tilney and Portnoy, 1989). The *actA* gene locus is responsible for the accumulation of actin around *Listeria* in the cytosol. The lack of one of these determinants has been shown to interfere with the pathogenicity of *L. monocytogenes* (Rocourt et al., 1986).

Cowart (1987) showed that hemolysin activity is stimulated in iron-deprived medium. The cytolytic activity of hemolysin is maximally expressed at pH 5.5. Therefore, when *L. monocytogenes* are engulfed in phagosomes that do not contain iron and have a pH value around 5.5, hemolysin production is maximized, allowing destruction of internal membranes surrounding them. An additional mechanism of intracellular survival is dependent upon the ability of *L. monocytogenes* to resist killing by oxidizing agents produced by phagocytes. The production of superoxide dismutase (SOD) and catalase by *L. monocytogenes* has been factors associated with intracellular survival (Bortolussi et al., 1987; Welch et al., 1979). Bortolussi et al. (1987) demonstrated that resistance of *L. monocytogenes* to hydroxyl radical ($^{\bullet}OH$) during the log phase of growth was due to the production of sufficient amounts of catalase to inactivate this product. Welch et al. (1979) found that catalase-negative strains of *L. monocytogenes* possessed at least twofold greater SOD activities than catalase-positive strains.

Not all strains of *L. monocytogenes* are pathogenic. Further, within *L. monocytogenes* strains, there may be particular serotypes that possess enhanced virulence potential. A survey conducted by Pinner et al. (1992) showed that foods containing *L. monocytogenes* serotype 4b were four times as likely to contain strains identical to patient strains than were foods containing serotypes 1/2a or 1/2b. These and other observations suggest that serotype 4b strains may have an enhanced capacity to cause human disease (Pinner et al., 1992; Schlech, 1992). Wiedmann et al., 1997) characterized 133 isolates of *L. monocytogenes* according to ribotype and virulence gene analysis. These authors found that *L. monocytogenes* strains could be clustered into three distinct lineages. Within lineage 1 resided all strains isolated during outbreaks of listeriosis. In contrast, lineage 3 contained no human isolates, indicating that strains in this grouping may have reduced virulence.

8.4. FOODBORNE TRANSMISSION

8.4.1. Foodborne Disease Epidemics: North America

Listeria monocytogenes has emerged as a foodborne pathogen of major significance within the last 15 years. Most of our knowledge of routes of foodborne transmission of *Listeria* has been gained through study of epidemiological data from outbreak investigations of human listeriosis. While outbreaks have occurred worldwide, this chapter will focus primarily on listeriosis in North America, where six major outbreaks have occurred since 1979. (Table 8.3). In 1979, listeriosis was diagnosed in at least 23 hospitalized patients in the Boston, Massachusetts, area (Ho et al., 1986). The vehicle of infection in this outbreak was linked to hospital food, and patients who had consumed lettuce, carrots, and radishes were more likely to contract the illness. Isolates from 20 of 23 cases were identified as serotype 4b. Symptoms reported by the afflicted patients included bacteremia or meningitis. Fifty percent of the patients involved in this outbreak were immunosuppressed because of cancer, chemotherapy, or steroid treatment. Curiously, 60% of the afflicted patients had reported the use of antacids or the antiulcer medication cimetidine. Cimetidine is a histamine-2 antagonist that blocks the H_2 effects of histamine, thereby decreasing gastric acid secretion (Levine et al., 1979), whereas antacids neutralize gastric acid. It was found that patients who had consumed antacids or cimetidine were more likely to develop hospital-acquired infection caused by *L. monocytogenes.* It was hypothesized that gastric acid neutralization following use of antacids or cimetidine predisposed humans to acquiring listeriosis as a result of ingestion of this organism via a foodborne vector. This finding was corroborated in later studies, where patients were more likely than controls to have used antacids, laxatives, or H_2-blocking agents prior to onset of listeriosis (Schuchat et al., 1992).

In 1981, an outbreak of listeriosis occurred in the Maritime Provinces of Canada (Schlech et al., 1983). The vehicle of transmission was identified as commercially prepared cole slaw. Cabbages used to prepare cole slaw were traced to a sheep farm where an outbreak of listeriosis had killed several sheep. Use of manure from infected sheep was suspected as a factor in this outbreak. Thirty-four cases of listeriosis in pregnant women resulted in spontaneous abortions, still births, or live birth of ill infants. Seven nonpregnant adults who showed no evidence of immunosuppression had symptoms of meningitis, aspiration pneumonia, and sepsis. The overall mortality rate for this outbreak was 41%. All patient isolates were identified as serotype 4b, and *L. monocytogenes* isolates from unopened packages of cole slaw were also identified as serotype 4b.

In 1983, 49 patients in Massachusetts were diagnosed with listeriosis (Fleming et al., 1985). Epidemiological evidence pointed to a strong association between consumption of pasteurized whole and 2% milk and onset of the illness. Forty-two patients were characterized with underlying illnesses such as cancer or alcoholism, and several patients were undergoing corticosteroid therapy. Seven of the cases involved fetuses or infants. The overall case–fatality rate in this outbreak was 29%. Of 49 isolates available for serotyping, 32 were identified as serotype 4b. Despite numerous attempts, the epidemic serotype of *L. monocytogenes* responsible for this outbreak was never recovered from the incriminated milk.

In June of 1985, Jalisco-brand Mexican-style cheese was implicated as the

TABLE 8.3. Foodborne Listeriosis: North American Outbreaks, 1979–Present

Date and Location	Illness Presentation	Cases		Food Source	Epidemic Serotype	Risk Factors	Reference
		Number	(% Mortality)				
1979: Boston, MA	Bacteremia meningitis	23	15	Lettuce, carrots, radishes	4b	Cimetidine, antacid use	Ho et al., 1986
1981: Maritime Provinces, Canada	Meningitis, aspiration pneumonia sepsis abortion, stillbirth	41	41	Cole slaw	4b	Pregnancy	Schlech et al., 1983
1983: Boston, MA	Mengitis, septicemia, death in utero	49	29	Pasteurized whole and 2% milk	4b	Cancer, alcoholism corticosteroid therapy	Fleming et al., 1985
1985: Orange County, CA	Fever, vomiting, stillbirth gastroenteritis	142	33	Mexican-style cheese	4b	Pregnancy, cancer, steroid therapy, AIDS	Linnan et al., 1988
1994: Elizabeth, IL	Gastroenteritis with fever	45	0	Chocolate milk	1/2b	No chronic illness, pregnant female	Dalton et al., 1997
1998–1999: multistate	Severe febrile gastroenteritis	101 15 adult deaths, 6 miscarriages	21	Hot dogs	4b	Pregnancy, diabetes, kidney disease, lupus malignancy	CDC, 1999

vehicle of infection in an outbreak of listeriosis in southern California (Linnan et al., 1988) A total of 142 cases involving 93 pregnant women or their offspring and 49 nonpregnant immunocompromised adults were documented in Los Angeles County. Forty-eight deaths were recorded, accounting for a mortality rate of 33.8%. The majority of afflicted individuals (62%) were pregnant Hispanic women. Although an additional 160 cases occurred in other parts of California, for logistical reasons, the study reported by Linnan et al. (1988) was limited to Los Angeles County. In this outbreak, the cheese was most likely manufactured from a combination of raw as well as pasteurized milk, and the cheese plant that manufactured the incriminated cheese was found to harbor *Listeria* as an environmental contaminant. The epidemic strain in this outbreak was serotype 4b, and this serotype was recovered from unopened packages of Queso Fresco and Cotija Mexican-style cheese.

Dalton et al. (1997) reported an outbreak of listeriosis linked to consumption of chocolate milk served at a picnic during a Holstein cow show in Illinois in 1994. Forty-five individuals developed illness due to *L. monocytogenes*; however, unlike symptoms reported in previous outbreaks, illness in this outbreak was characterized by fever and gastroenteritis without progression to invasive disease. Additionally, none of the patients reported immune deficiency or chronic illness. One pregnant female patient delivered a healthy baby. This outbreak, in particular, illustrates how failure to adhere to good manufacturing practices can have severe human health consequences. In reviewing steps in the manufacture of this product, milk was pumped into a nonrefrigerated holding tank where a breach in the lining may have allowed milk to leak into the insulating jacket and back into the product. Plant inspections revealed that sanitizing solution sprayers were severely clogged. The product was subsequently pumped into a filling machine over a 7-h period. Milk left the filler at 45°F and was refrigerated but was transported for more than 2 h in an unrefrigerated truck the day before the picnic. The milk was refrigerated overnight. The next morning milk was placed in an unrefrigerated cooler and transported to the picnic. Most of the milk was consumed within the first hour of the picnic, but it remained available throughout the afternoon. Unopened cartons of milk contained *L. monocytogenes* at levels of 1.2×10^9 CFU/mL. The median dose of *Listeria* consumed by the afflicted individuals may have been as high as 2.9×10^{11} CFU/person. None of the samples taken from the holding tank yielded *Listeria*. However, isolates obtained from the floor drain and the valve connected to the chocolate milk pasteurizer yielded *L. monocytogenes* serotype 1/2b. Postpasteurization contamination was implicated as the cause of this outbreak. Proctor et al. (1995) used PFGE to link four additional sporadic cases of invasive listeriosis to recalled 1% low-fat chocolate milk responsible for this outbreak.

During the period of August 1998 to March 1999, 101 cases of listeriosis were reported in 22 states (CDC, 1999). Twenty-one fatalities (15 adult deaths, 6 miscarriages) were recorded in this outbreak. The outbreak strain was identified as *L. monocytogenes* serotype 4b (E_0, E_1, E_2 PFGE pattern), and this rare strain was isolated from packages of hot dogs as well as environmental samples taken from the hot dog manufacturing plant. The start of the outbreak coincided with the removal of a large refrigeration unit near a hot dog packaging line. Plant records revealed an increase in the incidence of psychrophilic organisms from product contact surfaces coincident with removal of the refrigeration unit from the plant. Samples of hot dogs cultured quantitatively for *L. monocytogenes* serotype 4b revealed contamination at extremely low levels of <0.3 CFU/g, suggesting the possibility of enhanced virulence of this particular strain of *L. monocytogenes* (Mead, 1999).

8.4.2. Sporadic Cases of Listeriosis

The vast majority of human cases of listeriosis are not outbreak related but rather occur as sporadic illnesses confined to a single individual (Gellin and Broome, 1989). Ongoing and active disease surveillance has confirmed that most sporadic listeriosis cases are the result of foodborne transmission. From September to June 1987, the CDC conducted a population-based active surveillance for *L. monocytogenes* infections (Schwartz et al. 1988). This surveillance involved 154 patients from six regions of the United States (New Jersey, Missouri, Oklahoma, Tennessee, Washington, and Los Angeles). From these data, it was estimated that approximately 1700 cases of listeriosis occurred in the United States in 1986, for an annual incidence rate of 7.1 cases per million persons. Epidemiological evidence suggested that consumption of contaminated foods accounted for 30 of the 154 cases (20%) of listeriosis reported. Two foodborne sources were epidemiologically linked with onset of illness, these being uncooked hot dogs and undercooked chicken. As a result of this active surveillance, a recall of turkey franks commenced after being linked with the death of a patient in Oklahoma (CDC, 1989). *Listeria monocytogenes* serotype 1/2a strains of identical isoenzyme types were isolated from the patient as well as unopened packages of turkey franks (CDC, 1989).

Between November 1, 1988, and December of 31, 1990, the CDC conducted a second major case–control study in order to identify dietary risk factors for sporadic listeriosis (Schuchat et al., 1992). The population base in this active surveillance was in excess of 18 million persons distributed within five geographic regions of the United States. Cases were enrolled from patients identified through active surveillance. Underlying patient conditions included pregnancy, steroid therapy, cancer, renal dialysis, diabetes, HIV infection, liver disease, and organ transplant recipients. Three hundred and one cases of listeriosis were confirmed in this study. Foods in refrigerators of patients were examined for presence of *Listeria*. Sixty-four percent of refrigerators (79 out of 123 examined) yielded an *L. monocytogenes* isolate. Of 2229 foods examined, 11% were positive for *L. monocytogenes* (Pinner et al., 1992). Serotypes 4b, 1/2a, and 1/2b accounted for 95% of *L. monocytogenes* isolates recovered from foods. Of the *L. monocytogenes*–positive foods, 33% matched the patient isolates. Sixty-seven percent of dairy isolates matched patient strains, implicating dairy products as sources of *L. monocytogenes*. Specific dairy product sources included Mexican-style cheese, Feta cheese, and commodity cheeses. Thirty-two percent of sporadic cases of listeriosis could be attributed to eating foods purchased from delicatessen counters, Mexican-style and feta cheeses, and undercooked chicken (Schuchat et al., 1992). Preliminary CDC data for 1991 suggested a decrease of 30–40% in the number of cases compared with 1989–1990. The annualized sporadic incidence for listeriosis was found to be 7.4 cases per million population, with an overall case–fatality rate of 23%. Serotypes 1/2a (23%) and 1/2b (36%) together accounted for 59% of the cases; serotype 4b was isolated from 37% of the patients (Schuchat et al., 1992).

Tappero et al. (1995) reported a 44% decrease in rates of invasive listeriosis and a 48% decrease in the numbers of deaths due to listeriosis in the United States from 1989 to 1993. There were 1092 cases of listeriosis reported in 1993, resulting in 248 deaths, for an overall annual incidence of 4.2 cases per million population. Case–fatality rates remained similar (25% in 1989 compared to 23% for 1993). The decreased incidence rate was attributed to food industry efforts, sustained prevention efforts, and continued active surveillance.

8.5. SOURCES OF *Listeria* IN FOODS AND FOOD-PROCESSING ENVIRONMENTS

A host of unique properties possessed by *Listeria* render this a difficult organism to control in foods. *Listeria* can grow over a wide range of temperatures (−1.5°C to 45–50°C; Hudson et al. 1994; Juntilla et al. 1988) and pH (4.3–9.6; Farber et al., 1989; Petran and Zottola, 1989), survives freezing (Golden et al. 1988; Palumbo and Williams, 1991), and is relatively resistant to heat (Bradshaw et al. 1985; Bunning et al. 1986, 1988; Doyle et al. 1987; Farber et al. 1987, 1992; Fernandez-Garayzabel et al., 1986; Knabel et al., 1990). Minimal water activity levels for growth of *L. monocytogenes* and *L. innocua* range from 0.90 to 0.97 (Miller, 1992; Nolan et al., 1992). Shahamat et al. (1980) reported survival of *L. monocytogenes* for 132 days at 4°C in trypticase soy broth containing 25.5% NaCl. *Listeria* is a psychrotrophic pathogen, and growth at temperatures as low as −0.1 to −0.4°C in chicken broth and pasteurized milk (Walker et al., 1990) and −1.5°C vacuum-packaged meat (Hudson et al., 1994) has been recorded.

The ability of *L. monocytogenes* to resist the heating temperatures used during milk pasteurization continues to be debated in the scientific literature (Bearns and Girard, 1958; Bradshaw et al., 1985; Bunning et al., 1986, 1988; Doyle et al., 1987; Farber et al., 1987, 1992; Fernandez-Garayzabel et al., 1986; Fleming et al. 1985; Knabel et al., 1990). Fleming et al. (1985), in their studies of an outbreak of listeriosis in Boston, Massachusetts, concluded that "intrinsic contamination of the milk and survival of some organisms despite adequate pasteurization is both consistent with the results of this investigation and biologically plausible" (pg. 407). Fleming et al. reached this conclusion based upon the fact that milk involved in the outbreak came from farms where outbreaks of listeriosis had occurred; there was no evidence of improper pasteurization or postpasteurization contamination of the processed milk; and evidence in the scientific literature demonstrated the ability of *L. monocytogenes* to survive pasteurization (Bearns and Girard, 1958; Doyle et al., 1987; Garayzabel et al., 1987).

Comprehensive studies conducted by the U.S. Food and Drug Administration (FDA) and the U.S. Department of Agriculture (Bunning et al., 1986, 1988) and by Health and Welfare Canada (Farber et al., 1992) have shown that *Listeria* is unable to survive normal conditions of milk pasteurization. Knabel et al. (1990) found that growing *L. monocytogenes* at 43°C prior to heat inactivation caused an increase in thermotolerance, but a study conducted by Farber et al. (1992) demonstrated that even under worst-case scenario conditions, which included cultivation of *L. monocytogenes* populations at 43°C prior to inactivation, pasteurization would render a 4.5- to 6.2-D (decimal reduction value) process. Lovett et al. (1987) and Beckers et al. (1987) estimate that extremely low levels of *L. monocytogenes* (0.5–1.0 *Listeria*/mL) exist in commercial bulk tank raw milk. Therefore, while populations of *L. monocytogenes* have been shown to survive minimum pasteurization treatments of 71.1°C/16 s in various laboratory studies, survival under actual conditions of commercial milk pasteurization and processing is unlikely. *Listeria* contamination of procesed dairy products is most likely a function of postpasteurization from the dairy plant environment, and numerous surveys (Charlton et al. 1990; FDA, 1987; Klausner and Donnelly, 1991) document presence of *Listeria* within the dairy plant environment. Sources of *Listeria* within the dairy plant environment include floors in coolers, freezers, and processing rooms, particularly entrances; cases and case washers; floor mats and foot baths; and the beds of paper fillers (Klausner and Donnelly, 1991). Pritchard et al. (1994), in a study of dairy processing facilities, found that those processing plants having a farm contiguous to the processing facilities had a significantly higher incidence of *Listeria* contamination than those farms without an on-site dairy farm. Arimi et al. (1997) used ribotype analysis to demonstrate the link between on-farm sources of *Listeria* contamination (dairy cattle, raw milk, and silage) and subsequent contamination of dairy processing environments. Raw milk is a well-recognized source of *Listeria*, and for this and numerous other microbiological reasons, consumption of raw milk should be avoided (Beckers et al., 1987; Lovett et al., 1987; Vogt et al., 1990).

Studies by Ryser and Marth (1987a,b, 1989) examined the fate of *L. monocytogenes* during the manufacture of Cheddar, Camembert, and brick cheese. Rapid growth of *Listeria* to populations of 5×10^7 CFU/mL is observed in Camembert cheese, which has a pH that increases during ripening, thereby creating a favorable growth environment for *Listeria* (Ryser and Marth, 1987b). In contrast, *Listeria* populations show a marked decline in viable population levels during ripening of Cheddar cheese. However, population levels do not decline to undetectable levels. Current U.S. regulations call for cheese made from raw or subpasteurized milk to be ripened at 1.7°C (35°F) for at least 60 days prior to sale. Ryser and Marth (1987a) have shown that aging alone will not ensure the production of *Listeria*-free Cheddar cheese.

Genigeorgis et al. (1991) evaluated the ability of 24 types of market cheeses to support growth of *L. monocytogenes*. Cheeses able to support growth of *L. monocytogenes* included soft Hispanic-type cheeses, ricotta, Teleme, Brie, Camembert, and cottage cheeses (pH range 4.9–7.7). Cheeses not supporting growth and that resulted in gradual death of *L. monocytogenes* included Cotija, cream, blue, Monterey Jack, Swiss, Cheddar, Colby, string, provolone, muenster, feta, and Kasseri (pH range 4.3–5.6). A correlation was observed between growth of *Listeria* in cheeses having a pH greater than 5.5 and in cheeses manufactured without a starter culture.

Ryser et al. (1985) examined the fate of *L. monocytogenes* during manufacture and storage of cottage cheese. *Listeria monocytogenes* survived in both creamed and uncreamed cottage cheese during 28 days of storage at refrigeration temperatures and was recovered in higher numbers from creamed (pH 5.32–5.45) versus uncreamed (pH 5.12–5.22) cottage cheese. Hicks and Lund (1991) examined the ability of *L. monocytogenes* to survive in creamed cottage cheese when stored at 4, 8, or 12°C for 14 days. The organism survived but failed to increase in numbers during storage in product with a pH range of 5.06–4.53. Chen and Hotchkiss (1992), however, found that *L. monocytogenes* grew in cottage cheese stored at 7°C for 16 days or 4°C for 63 days but would not grow under these conditions when modified CO_2 packaging conditions were used.

Conner et al. (1986) investigated the effects of temperature, NaCl, and pH on the growth of *L. monocytogenes* in cabbage. Results indicated that cabbage juice provided a good substrate for growth of *Listeria*. The organism was found to survive well at 5°C even in the presence of 5% NaCl, and the organism could grow and tolerate a pH of less than 5.6. This study, together with the findings of Schlech et al. (1983), confirms that cabbage can serve as a vector of transmission of *L. monocytogenes* to humans and demonstrates the potential for *L. monocytogenes* to persist and proliferate on vegetables and in brines used to ferment vegetables. Potatoes and radishes have been identified as sources of *L. monocytogenes* during retail food surveys (Heisick et al., 1989). Additional studies have confirmed growth and survival of *L. monocytogenes* on asparagus, broccoli, cauliflower, corn, green beans, lettuce, and radishes (Berrange et al., 1989; Hughey et al., 1989).

Meat and poultry products including ground beef, pork sausage, cooked beef, roast beef, ham, bologna, bratwurst, sliced chicken, and turkey have been identified as products that promote growth and can serve as sources of *L. monocytogenes* (Anonymous, 1997; Glass and Doyle, 1989). Farber and Daley (1994), however, found that when present initially on sliced ham, turkey breasts, wieners, and paté at very low levels and stored at 4°C, numbers of *L. monocytogenes* did not increase during storage. The U.S. Department of Agriculture Food Safety and Inspection Service (USDA-FSIS) microbiological monitoring data from 1993 to 1999 have pointed to hot dogs and luncheon meats as two products of concern that may serve as primary vehicles for foodborne transmission of *L. monocytogenes* (FSIS, 2000).

Raw poultry is a well-recognized source of *L. monocytogenes*, and numerous surveys have confirmed the presence of *L. monocytogenes* in retail poultry samples (Genigeorgis et al., 1989, 1990). Bailey et al. (1989) recovered *L. monocytogenes* from 23% of sampled broiler carcasses, the most prevalent serotype isolated being 1/2b. Ready-to-eat poultry products have been implicated as vectors of transmission of listeriosis to humans. Cooked-chilled chicken and turkey frankfurters were confirmed as vehicles of *Listeria* infection in England and the United States, respectively, during 1988 and 1989 (CDC, 1989; Kerr et al., 1988). Gilbert et al. (1989) confirmed the presence of *L. monocytogenes* in 12% of precooked, ready-to-eat poultry products collected from London-area retail establishments between mid-November 1988 and mid-January 1989. Survival of *L. monocytogenes* on chicken breasts processed by moist and dry heating methods has been demonstrated (Harrison and Carpenter, 1989).

Wenger et al. (1990) examined a turkey frank production facility in order to determine sources and incidence of contamination. *Listeria monocytogenes* was isolated from only 2 of 41 environmental samples from the plant, which included a cooler room floor and a conveyer belt attached to a peeler. Yet *L. monocytogenes* was isolated from six of seven retail lots of product produced over a 37-day production period. Product samples taken at the production line post peeler were more likely (12 of 14 samples positive) to be contaminated than samples from other production locations (2 of 40 samples positive). Therefore, product contamination was found to occur at a single point during the peeling process prior to packaging of product.

Seafood is recognized as a source of *L. monocytogenes*. Weagant et al. (1988) documented presence of *L. monocytogenes* in frozen seafood samples that included shrimp, crabmeat, lobster tail, fin fish, and surimi-based products. Farber (1991) isolated *L. monocytogenes* from ready-to-eat shrimp, crab, and smoked salmon, and further laboratory studies demonstrated growth at 4°C of *L. monocytogenes* on cooked crabmeat, lobster, shrimp, and smoked salmon. Jemmi (1990), upon examination of 377 samples of smoked and marinated fish, found *L. monocytogenes* in 47 samples. A survey in Newfoundland conducted by Dillon et al. (1992) also revealed the presence of *Listeria* in smoked seafood products. Crab and smoked fin fish accounted for the majority of seafood products involved in class 1 recalls during the 11-year period from 1987 to 1998 because of *L. monocytogenes* contamination (Jinneman et al., 1999).

8.6. DETECTION OF *Listeria* IN FOODS

8.6.1. Selective Enrichment and Enumeration

Detection of *L. monocytogenes* in food products or food-processing environments is accomplished by use of a variety of standard or rapid microbiological procedures. Among the most widely used are protocols devised by the USDA-FSIS for the detection of *Listeria* in meat and poultry products (Cook, 1999) and the FDA (Hitchins, 1995) for the detection of *Listeria* in dairy products, fruits, vegetables, and seafood products. A detection method widely used in Europe is the Netherlands Government Food Inspection Service (NGFIS) method developed by Netten et al. (1989). Foods are usually mixed with an enrichment broth and allowed to incubate for 24–48 h. Following incubation, a portion of the enrichment mixture is either again mixed with an enrichment broth or plated onto the final isolation agar. Enrichment broths are usually nutritious liquid media that employ various antimicrobial agents to which *L. monocytogenes* is resistant. The most common antimicrobial agents include nalidixic acid, acriflavin, and cycloheximide. Isolation agars include those used for direct plating (Cook, 1999; Curtis et al., 1989; Netten et al., 1989), although less selective agars have also been used successfully. Numerous studies have been conducted to compare efficacy of these and other widely used detection protocols. Hayes et al. (1992) compared three enrichment methods for examination of foods obtained from the refrigerators of patients with active clinical cases of listeriosis. This study examined 2229 foods, of which 11% were positive for *L. monocytogenes*. A comparative evaluation of three microbiological procedures was conducted on 899 of the examined foods. The FDA procedure detected *L. monocytogenes* in 65% of the foods shown to be positive, while the USDA-FSIS and NGFIS procedures detected *L. monocytogenes* in 74% of foods shown to be positive. Thus, none of the widely used conventional methods proved to be highly sensitive when used independently for analysis of *Listeria* contamination in foods. It was noted, however, that use of a combination of any two methods improved detectability from 65–74% (for individual protocols) to 87–91% for combined protocols.

The USDA-FSIS verifies the adequacy of hazards analysis and critical control point (HACCP) systems used by meat and poultry processing facilities in part through the collection and testing of ready-to-eat meat products for pathogenic bacteria such as *L. monocytogenes*. The use of HACCP strategies to effectively eliminate organisms such as *Listeria* from food-processing environments and therefore processed food products is predicated upon the use of sensitive, reliable, inclusive enrichment methods that facilitate the recovery of *Listeria*. However, since current procedures used by the USDA-FSIS to test for the presence of *Listeria* rely on the use of highly selective primary enrichment media, these media lack the sensitivity and reliability necessary to detect low-level contamination of *Listeria* in food products. Low levels of *Listeria* existing initially in food products that undergo long periods of refrigerated storage can multiply to dangerous levels that can ultimately cause human illness. As research cited above shows, existing regulatory procedures have only a 65–75% sensitivity and reliability rate. Simple modifications to existing regulatory protocols, such as those that utilize more than one primary enrichment broth, raise the sensitivity of detection to 90% (Pritchard and Donnelly, 1999).

Enumeration procedures developed by the USDA-FSIS (Cook, 1999) include a choice of two methods: a most probable number (MPN) technique and a direct plating method. The MPN procedure is described as the most sensitive, detecting 100 CFU/g or less of *L. monocytogenes*. This procedure utilizes University of Vermont medium (UVM) in a nine-tube series. The direct-plating method utilizes UVM as the diluent for homogenization of the sample, followed by direct plating onto modified Oxford medium (MOX). This method is described as being considerably less sensitive than the MPN method, possibly underestimating the actual number of cells present, particularly with respect to injured cells. Therefore, the direct-plating method would be utilized for samples anticipated to contain high levels of *L. monocytogenes* (Cook, 1999). Several rapid commercial methods are available for the detection of *Listeria* and are based on enzyme-linked immunosorbent assay (ELISA) and DNA probe and amplification technology. These procedures go beyond traditional identification methods through incorporation of genetic and immunological techniques to reduce identification time, and commercial kits are available for both (AOAC, 1995; Feldsine et al., 1997a,b; Feng, 1995; Hitchins, 1995; Knight et al., 1996). Rapid methods are faster than conventional methods and performance is similar. The main drawbacks to the rapid methods developed to date are cost, requirement for sufficient cell density to record positive results, and in certain instances inability to distinguish pathogenic from nonpathogenic species along with viable versus nonviable cells. Despite advances in *Listeria* detection methodology, all procedures developed to date lack sensitivity and reliability, and often positive samples escape detection.

8.6.2. Sublethal Injury

In addition to missing low-level contamination due to *Listeria*, highly selective enrichment procedures do not account for recovery of sublethally injured *Listeria* that can exist within heated, frozen, or acidified foods or heated, frozen, and sanitized areas within food-processing environments. It is well recognized that *Listeria* can be injured as a result of exposure to a variety of processing treatments, including sublethal heating and freezing, drying, irradiation, or exposure to chemicals (sanitizers, preservatives, acids). Under ideal conditions in food systems, injury is reversible and injured *Listeria* can repair sublethal damage. Refinement of existing *Listeria* recovery methods should consider the nutritional needs associated with specific genetic types, along with the physiological condition of *Listeria* isolates in foods. Archer (1996) has stated that the stresses to which bacteria in foods are subjected may result in increased expression of virulence in stressed pathogens and may promote adaptive mutations resulting in more virulent pathogens. As we employ processing procedures that tend to promote stress adaptation of organisms (extended refrigeration, cook/chill procedures, sous-vide), once repaired, we may generate organisms that become not only more adaptive to hostile environments but also more virulent. A host of genes control stress adaptation in *Listeria* through production of heat shock proteins, cold acclimating proteins, and acid tolerance response proteins. Hill and Gahan (2000) have shown that exposure of *L. monocytogenes* to pH 5.5 for a short period (30 min) can result in the subsequent survival of these cells to a normally lethal pH of 3.5. Bacteria, including foodborne pathogens, have developed a number of elaborate genetic systems that encode for the production of proteins that allow bacterial survival during lethal environmental challenges. For instance, the *rpos*-regulated proteins enhance acid tolerance and cross protect *Escherichia coli* O157:H7 against heat and salt challenges (Foster, 1995). The acid tolerance response ATR response encodes for the ability to withstand lethal pH conditions following adaptation to sublethal pH in *Listeria monocytogenes, Salmonella typhimurium, Escherichia coli,* and *Aeromonas hydrophila* (O'Driscoll et al., 1996). These mechanisms play a role in predicting the fate of pathogens in acidic foods. Roth and Donnelly (1996) assessed survival of acid-injured *Listeria* in acidic food products such as apple cider, yogurt, fresh cole slaw, and salsa. Temperature was found to greatly dictate survival of *Listeria* in salsa. When stored at 30°C, populations declined rapidly within 3 days. At 4°C, populations persisted for up to 17 days of storage. In further studies, the efficacy of two different enrichment media for recovery of acid-injured *Listeria* from acidic foods was assessed using *Listeria* repair broth (LRB) and UVM. At time points where differences were detected, LRB detected the organism in 22 of 54 samples, compared with UVM, which detected only 3 of 54 contaminated samples.

Ryser et al. (1996) evaluated the ability of UVM and LRB to recover different strain-specific ribotypes of *L. monocytogenes* from meat and poultry products. Forty-five paired 25 g retail samples of ground beef, pork sausage, ground turkey, and chicken were enriched in UVM and LRB (30°C/24 h) followed by secondary enrichment in Fraser broth (35°C/24 h) and plated on modified Oxford agar. A 3-h period of nonselective enrichment at 30°C was used with LRB (with tested food) to enable repair of injured *Listeria* prior to addition of selective agents. Of 180 meat and poultry products tested, LRB identified 73.8% (133/180) and UVM 69.4% (124/180). Although there was not a statistically significant difference in these results, combining results from UVM and LRB enrichment improved recovery rates to 83.3%. These results demonstrate the enhanced recovery of *Listeria* through the use of LRB for repair/enrichment of samples in conjunction with the USDA-FSIS method. After 24 h of incubation at 35°C, *Listeria* colonies were biochemically confirmed and selected isolates were ribotyped using the automated Riboprinter Microbial Characterization System (E. I. DuPont), an automated system that distinguishes genetically unique strains. A total of 36 different *Listeria* strains comprising 16 *L. monocytogenes* (including four known clinical ribotypes), 12 *L. innocua* and 8 *L. welshimeri* ribotypes were identified from selected positive samples (15 samples of each product type, 2 UVM and 2 LRB isolates per sample). Twenty-six of 36 (3 *L. monocytogenes*) *Listeria* ribotypes were observed using both UVM and LRB; whereas 3 of 36 (1 *L. monocytogenes*) and 7 of 36 (3 *L. monocytogenes*) *Listeria* ribotypes were observed using only UVM or LRB, respectively. Ground beef, pork sausage, ground turkey, and chicken

TABLE 8.4. Ribotypes of *Listeria* spp. Recovered from Ten Samples of Raw Chicken Following Primary Enrichment in UVM or LRB and Secondary Enrichment in Fraser Broth

Ribotype	*Listeria* spp.	Number of Isolates	
		UVM	LRB
1-909-3	*L. innocua*	0	1
5-418-3	*L. moncytogenes*	2	0
5-415-4	*L. innocua*	4	0
5-413-2	*L. monocytogenes*	2	0
2-864-3	*L. welshimeri*	2	0
1-916-1[a]	*L. monocytogenes*	3	3
5-408-1	*L. monocytogenes*	2	0
1-909-4	*L. innocua*	5	5
1-910-7	*L. innocua*	0	1
5-426-1	*L. innocua*	0	1
1-923-1[a]	*L. monocytogenes*	0	3
5-408-4	*L. monocytogenes*	0	2
1-907-1[a]	*L. monocytogenes*	0	1
1-919-2	*L. monocytogenes*	0	1
1-864-7	*L. monocytogenes*	0	1
1-915-7	*L. monocytogenes*	0	1

yielded 22 (8 *L. monocytogenes*), 21 (12 *L. monocytogenes*), 20 (9 *L. monocytogenes*), and 19 (11 *L. monocytogenes)* different *Listeria* ribotypes, respectively, with some *Listeria* ribotypes confined to a particular product. Many meat samples (pork sausage and ground beef) were found to harbor three or four different *Listeria* ribotypes in a single sample (Table 8.4). More importantly, striking differences in both the number and distribution of *Listeria* ribotypes were observed when 10 UVM and 10 LRB isolates from five samples of each product were ribotyped. When a third set of isolates were obtained using only one of the two primary enrichment media, UVM and LRB failed to detect *L. monocytogenes* (both clinical and nonclinical ribotypes) in two and four samples, respectively (Table 8.4). These findings stress the complex microbial ecology of *Listeria* in foods and the limitations of existing detection procedures to fully represent the total distribution of *Listeria* isolates in foods. Furthermore, two of the *L. monocytogenes* riboprints undetected using UVM were known clinical isolates of serotypes 1/2a, both of which were responsible for sporadic and epidemic cases of human listeriosis in England and Scotland (McLauchlin et al., 1986). These findings combined with reports of *L. innocua* being able to outgrow *L. monocytogenes* in UVM media suggest that genetically distinct strains of *L. monocytogenes* may vary somewhat in nutritional requirements or their ability to compete with other genetically distinct strains of *L. monocytogenes* or of other *Listeria* spp. (Curiale and Lewus, 1994; Petran and Swanson, 1993).

Characterization of *L. monocytogenes* isolates beyond the species level is primarily confined to epidemiological investigations where investigators attempt to confirm the vehicle of infection and conduct trace-back studies to a particular food-processing facility and/or environment. Methods for subtyping *Listeria* can be separated into two broad categories: (a) conventional methods that include serotyping (Schonberg et al., 1996), phage typing (Donnelly, 1998), and bacteriocin typing (Bannerman et al., 1996) and (b) molecular methods that encompass multilocus enzyme electrophoresis (MEE) (Caugant et al., 1996), chromosomal DNA restriction endonuclease analysis (REA) (Gerner-Smidt et al., 1996), ribotyping (Swaminathan et al., 1996), DNA macrorestriction analysis by PFGE (Brosch et al., 1996), random amplification of polymorphic DNA (RAPD) by PCR (Wernars et al., 1996), and DNA sequence-based subtyping (Graves et al., 1999; Rasmussen et al., 1995). In general, serotyping and phage typing are best suited as preliminary subtyping strategies, with phage typing now particularly popular in Europe for routine screening of isolates (Graves et al., 1999). Ribotyping and MEE lack sufficient discrimination to be used alone in epidemiological investigations and, with the exception of the RiboPrinter (a fully automated ribotyping system developed by Qualicon, Wilmington, DE), are also fairly labor intensive. At present, REA, PFGE, and RAPD are the preferred methods for subtyping *Listeria* because of their high discrimination and ease of use. Pulsed-field gel electrophoresis has been used in conjunction with phage typing since the late 1980s for routine screening of *Listeria* isolates in France (Seeliger et al., 1984). The CDC has utilized the power of PFGE establish PulseNet, a national network of public health laboratories that fingerprint foodborne bacteria using PFGE. Through use of standardized methods, genetic fingerprints can be compared with fingerprints maintained on a common database at the CDC to determine if bacteria isolated from ill persons originate from a common source. Use of PulseNet helped to identify a common source outbreak linked to hot dogs. Alternative tools such as riboprinting allow discrimination of distinct *Listeria* species and therefore trace-back to points of contamination within the processing plant environment. We face an interesting juxtaposition against the call for a relaxation of zero-tolerance standards by the food industry on the one hand, and availability of highly effective and discriminatory technology on the other hand, which, when coupled with epidemiological data, can effectively be used to trace back to contaminated food products causing listeriosis. The FDA is currently conducting a risk assessment for the purpose of determining the prevalence and extent of consumer exposure to *L. monocytogenes* in foods and to assess the resultant public health impact of this exposure (FDA/USDA-FSIS, 1999).

8.7. CONCLUSION

Despite reductions in disease incidence due to *L. monocytogenes*, this organism remains a leading cause of death due to a foodborne pathogen (Mead et al.,

1999). Recent multistate outbreaks of illness and death highlight the need for renewed collaboration among industry, university, and governmental agencies to control this dangerous but interesting foodborne pathogen. Improvements in testing methods are also needed to ensure adequate sensitivity of detection of regulatory procedures used to identify and ultimately control *Listeria*.

REFERENCES

Anonymous. 1997. *Microbiological Monitoring of Ready-to-Eat Products, 1993–1996.* Washington, DC: U.S. Department of Agriculture, Food Safety and Inspection Service.

AOAC Official Methods of Analysis, 16th ed. 1995. Subchapter 10: *Listeria*. Arlington, VA: AOAC International, p. 94A.

Archer, D. L. 1996. How you gonna keep em down on the farm? *IFT Ann. Mtg. Book Abstr.* **34-1**, p. 96.

Arimi, S. M., E. T. Ryser, T. J. Pritchard, and C. W. Donnelly. 1997. Diversity of *Listeria* ribotypes recovered from dairy cattle, silage and dairy processing environments. *J. Food Prot.* **60**, 811–816.

Armstrong, D. 1995. *Listeria monocytogenes*. In: G. L. Mandell, J. E. Bennett, and R. Dolin (Eds.), *Mandell, Douglas and Bennett's Principles and Practice of Infectious Diseases.* New York: Churchill Livingstone, pp. 1880–1885.

Bailey, J. S., D. L. Fletcher, and N. A. Cox. 1989. Recovery and serotype distribution of *Listeria monocytogenes* from broiler chickens in the southeastern United States. *J. Food Prot.* **52**, 148–150.

Bannerman, E., P. Boerlin, and J. Bille. 1996. Typing of *Listeria monocytogenes* by monocin and phage receptors. *Int. J. Food Microbiol.* **31**, 245.

Bearns, R. E., and K. F. Girard. 1958. The effect of pasteurization on *Listeria monocytogenes. Can. J. Microbiol.* **4**, 55–61.

Beckers, H. J., P. S. S. Soentoro, and E. H. M. Delfgou van Asch. 1987. The occurrence of *Listeria monocytogenes* in soft cheeses and raw milk and its resistance to heat. *Int. J. Food Microbiol.* **4**, 249–256.

Berrange, M. E., R. E. Brackett, and L. R. Beuchat. 1989. Growth of *Listeria monocytogenes* on fresh vegetables stored under controlled atmosphere. *J. Food Prot.* **52**, 702–705.

Bojsen-Moller, J. 1972. Human listeriosis. *Acta Pathol. Microbiol. Scand. Suppl.* **229**, 1–157.

Bortolussi, R., C. M. J. E. Vandenbroucke-Grauls, B. S. van Asbeck, and J. Verhoef. 1987. Relationship of bacterial growth phase to killing of *Listeria monocytogenes* by oxidative agents generated by neutrophils and enzyme systems. *Infect. Immun.* **55**, 3197–3203.

Bradshaw, J. G., J. T. Peeler, J. J. Corwin, J. M. Hunt, J. T. Tierney, E. P. Larkin, and R. M. Twedt. 1985. Thermal resistance of *Listeria monocytogenes* in milk. *J. Food Prot.* **48**, 743–745.

Brosch, R., M. Brett, B. Catimel, J. B. Luchansky, B. Ojeniyi, and J. Rocourt. 1996. Genomic fingerprinting of 80 strains from the WHO multicentre international typing study of *Listeria monocytogenes* via pulsed-field gel electrophoresis (PFGE). *Int. J. Food Microbiol.* **32**, 343–355.

Buchdahl, R., M. Hird, H. Gamsu, A. Tapp, D. Gibb, and C. Tzannatos. 1990. Listeriosis revisited: The role of the obstetrician. *Br. J. Obstet. Gynecol.* **97**, 186–189.

Bunning, V. K., R. G. Crawford, J. G. Bradshaw, J. T. Peeler, J. T. Tierney, and R. M. Twedt. 1986. Thermal resistance of intracellular *Listeria monocytogenes* cells suspended in raw bovine milk. *Appl. Environ. Microbiol.* **52**, 1398–1402.

Bunning, V. K., C. W. Donnelly, J. T. Peeler, E. H. Briggs, J. G. Bradshaw, R. G. Crawford, C. M. Beliveau, and J. T. Tierney. 1988. Thermal inactivation of *L. monocytogenes* within bovine milk phagocytes. *Appl. Environ. Microbiol.* **54**, 364–370.

Busch, L. A. 1971. New from the Center for Disease Control—Human listeriosis in the United States, 1967–1969. *J. Infect. Dis.* **123**, 328–332.

Caugant, D. A., F. E. Ashton, W. F. Bibb, P. Boerlin, W. Donachie, C. Low, A. Gilmour, J. Harvey, and B. Norrung. 1996. Multilocus enzyme electrophoresis for characterization of *Listeria monocytogenes* isolates: Results of an international comparative study. *Int. J. Food Microbiol.* **32**, 301.

Centers for Disease Control and Prevention (CDC). 1989. Listeriosis associated with consumption of turkey franks. *MMWR* **38**, 267–268.

Centers for Disease Control and Prevention (CDC). 1998. *FoodNet 1998 Annual Report:* http://www.cdc.gov/ncidod/dbmd/foodnet/ANNUAL/98_surv.htm.

Centers for Disease Control and Prevention (CDC). 1999. Update: Multistate outbreak of listeriosis–United States, 1998–1999. *MMWR* **47**(51), 1117–1118.

Charlton, B. R., H. Kinde, and L. H. Jensen. 1990. Environmental survey for *Listeria* species in California milk processing plants. *J. Food Prot.* **53**, 198–201.

Chen, J. H., and J. H. Hotchkiss. 1992. Growth of *Listeria monocytogenes* and *Clostridium sporogenes* in cottage cheese in modified atmosphere packaging. *J. Dairy Sci.* **76**, 972–977.

Conner, D. J., R. E. Brackett, and L. R. Beuchat. 1986. Effect of temperature, sodium chloride, and pH on growth of *Listeria monocytogenes* in cabbage juice. *Appl. Environ. Microbiol.* **52**, 59–63.

Cook, L. V. 1999. Isolation and identification of *Listeria monocytogenes* from red meat, poultry, egg and environmental samples (revision 2). In: *USDA/FSIS Microbiology Laboratory Guidebook*, 3rd ed. Chapter 8.

Cowart, R. E. 1987. Iron regulation of growth and haemolysin production by *Listeria monocytogenes*. *Ann. Inst. Pasteur/Microbiol.* **138**, 246–249.

Curiale, M. S., and C. Lewus. 1994. Detection of *Listeria monocytogenes* in samples containing *Listeria innocua*. *J. Food Prot.* **57**, 1048–1051.

Curtis, G. D. W., R. G. Mitchell, A. F. King, and E. J. Griffen. 1989. A selective differential medium for the isolation of *Listeria monocytogenes*. *Lett. Appl. Microbiol.* **8**, 95–98.

Czajka, J., N. Bsat, M. Piani, W. Russ, K. Sultana, M. Weidmann, R. Whitaker, and C. Batt. 1993. Differentiation of *Listeria monocytogenes* and *Listeria innocua* by 16S rRNA genes and intraspecies discrimination of *Listeria monocytogenes* strains by random amplified polymorphic DNA polymorphisms. *Appl. Environ. Microbiol.* **59**, 304–308.

Dalton, C. B., C. C. Austin, J. Sobel, P. S. Hayes, W. F. Bibb, L. M. Graves, B.

Swaminathan, M. E. Proctor, and P. M. Griffin. 1997. An outbreak of gastroenteritis and fever due to *Listeria monocytogenes* in milk. *N. Engl. J. Med.* **336**, 100–105.

Dillon, R., T. Patel, and S. Ratnum. 1992. Prevalence of *Listeria* in smoked fish. *J. Food Prot.* **55**, 866–870.

Donnelly, C. W. 1998. Conventional methods to detect and isolate *Listeria monocytogenes*. In: E. T. Ryser, and E. H. Marth (Eds.), *Listeria, Listeriosis and Food Safety*. New York: Marcel Dekker, p. 279.

Doyle, M. P., K. A. Glass, J. T. Berry, G. A. Garcia, D. J. Pollard, and R. D. Schultz. 1987. Survival of *Listeria monocytogenes* during high-temperature short-time pasteurization. *Appl. Environ. Microbiol.* **53**, 1433–1438.

Dutta, P. K., and B. S. Malik. 1981. Isolation and characterization of *Listeria monocytogenes* from animals and human beings. *Ind. J. Animal Sci.* **51**, 1045–1052.

Farber, J. M. 1991. *Listeria monocytogenes* in fish products. *J. Food Prot.* **54**, 922–924.

Farber, J. M., and E. Daley. 1994. Presence and growth of *Listeria monocytogenes* in naturally-contaminated meats. *Food Microbiol.* **22**, 33–42.

Farber, J. M., E. Daley, F. Coates, D. B. Emmons, and R. McKellar. 1992. Factors influencing survival of *Listeria monocytogenes* in milk in a high-temperature short time pasteurizer. *J. Food Prot.* **55**, 946–951.

Farber, J. M., G. W. Sanders, S. Dunfield, and R. Prescott. 1989. The effect of various acidulants on the growth of *Listeria monocytogenes*. *Lett. Appl. Microbiol.* **37**, 50–54.

Farber, J. M., G. W. Sanders, D. B. Emmons, and R. C. McKellar. 1987. Heat resistance of *Listeria monocytogenes* in artificially inoculated and naturally contaminated raw milk. *J. Food Prot.* **50**, 893.

Feldsine, P. T., A. H. Lienau, R. L. Forgey, and R. D. Calhoon. 1997a. Assurance polyclonal enzyme immunoassay for detection of *Listeria monocytogenes* and related *Listeria* species in selected foods: Collaborative study. *J. AOAC Int.* **80**, 775.

Feldsine, P. T., A. H. Lienau, R. L. Forgey, and R. D. Calhoon. 1997b. Visual immunoprecipitate assay (VIP) for *Listeria monocytogenes* and related *Listeria* species detection in foods: Collaborative study. *J. AOAC Int.* **80**, 791.

Feng, P. 1995. Rapid methods for detecting foodborne pathogens. In: *U.S. Food and Drug Administration. Bacteriological Analytical Manual*, 8th ed. Gaithersburg, MD: AOAC International, App. 1.01.

Fernandez-Garayzabel, J. F., L. Dominguez Rodriguez, J. A. Vazquez Boland, J. L. Balnco Cancelo, and G. Suarez Fernandez. 1986. *Listeria monocytogenes* dans le lait pasteurise. *Can. J. Microbiol.* **32**, 149–150.

Fleming, D. W., S. L. Cochi, K. L. MacDonald, J. Brondum, P. S. Hayes, B. D. Plikaytis, M. B. Holmes, A. Audurier, C. V. Broome, and A. L. Reingold. 1985. Pasteurized milk as a vehicle of infection in an outbreak of listeriosis. *N. Engl. J. Med.* **312**, 404–407.

Food and Drug Administration (FDA). 1987. *FDA Dairy Product Safety Initiatives. Second Year Status Report.* Washington, DC: Milk Safety Branch, Center for Food Safety and Applied Nutrition.

Food and Drug Administration/U.S. Department of Agriculture Food Safety and Inspection Service (FDA/USDA-FSIS). 1999. Structure and initial data survey for the risk assessment of the public health impact of foodborne *Listeria monocytogenes*: http://vm.cfsan.fda.gov/~dms/listrisk.html.

Food Service and Inspection Service (FSIS). 2000. Food safety and inspection service revised action plan for control of *Listeria monocytogenes* for the prevention of foodborne listeriosis. May 12 Anonymous, 1997. Microbiological monitoring of ready-to-eat products, 1993–1996. Washington, DC: U.S. Department of Agriculture, Food Safety and Inspection Service.

Foster, J. W. 1995. Low pH adaptation and the acid tolerance response of *Salmonella typhimurium*. *Crit. Rev. Microbiol.* **21**, 215–237.

Gaillard, J. L., P. Berche, and P. Sansonetti. 1986. Transposon mutagenesis as a tool to study the role of hemolysin in the virulence of *L. monocytogenes*. *Infect. Immun.* **52**, 50–55.

Gaillard, J. L., P. Berche, J. Mounier, S. Richard, and P. Sansonetti. 1987. In vitro model of penetration and intracellular growth of *Listeria monocytogenes* in the human enterocyte-like cell line Caco-2. *Infect. Immun.* **55**, 2822–2829.

Gaillard, J. L., P. Berche, C. Frehel, E. Gouin, and P. Cossart. 1991. Entry of *L. monocytogenes* into cells is mediated by internalin, a repeat protein reminiscent of surface antigen from Gram positive cocci. *Cell* **65**, 1127–1141.

Garayzabel, J. F. F., L. D. Rodriguez, J. A. V. Boland, E. F. R. Ferri, V. B. Dieste, J. L. B. Cancelo, and G. S. Fernandez. 1987. Survival of *Listeria monocytogenes* in raw milk treated in a pilot plant size pasteurizer. *J. Appl. Bacteriol.* **63**, 533–537.

Gellin, B. G., and C. V. Broome. 1989. Listeriosis. *JAMA* **261**, 1313–1320.

Genigeorgis, C., M. Carniciu, D. Dutulescu, and T. B. Farver. 1991. Growth and survival of *Listeria monocytogenes* in market cheeses stored at 4 to 30°C. *J. Food Prot.* **54**, 662–668.

Genigeorgis, C. A., D. Dutulescu, and J. F. Garayzabal. 1989. Prevalence of *Listeria* spp. in poultry meat at the supermarket and slaughterhouse level. *J. Food Prot.* **52**, 618–624.

Genigeorgis, C. A., P. Onaca, and D. Dutulescu. 1990. Prevalence of *Listeria* spp. in turkey meat at the supermarket and slaughterhouse level. *J. Food Prot.* **53**, 282–288.

Geoffroy, C., J. Gaillard, J. E. Alouf, and P. Berche. 1989. Production of thiol-dependent hemolysins by *Listeria monocytogenes* and related species. *J. Gen. Microbiol.* **135**, 481–487.

Gerner-Smidt, P., P. Boerlin, F. Ischer, and J. Schmidt. 1996. High-frequency endonuclease (REA) typing: Results from the WHO collaborative study group on subtyping of *Listeria monocytogenes*. *Int. J. Food Microbiol.* **32**, 313.

Gilbert, R. J., K. L. Miller, and D. Roberts. 1989. *Listeria monocytogenes* and chilled foods. *Lancet* **i**, 383–384.

Gill, D. A. 1933. Circling disease: A meningoencephalitis of sheep in New Zealand. Notes on a new species of pathogenic organism. *Vet. J.* **89**, 258–270.

Glass, K. A., and M. P. Doyle. 1989. Fate of *Listeria monocytogenes* in processed meat products during refrigerated storage. *Appl. Environ. Microbiol.* **55**, 1565–1569.

Golden, D. A., L. R. Beuchat, and R. E. Brackett. 1988. Inactivation and injury of *Listeria monocytogenes* as affected by heating and freezing. *Food Microbiol.* **5**, 17–23.

Golnazarian, C. A., C. W. Donnelly, S. J. Pintauro, and D. B. Howard. 1989. Comparison of infectious dose of *Listeria monocytogenes* F5817 as determined for normal versus compromised C57B1/6J mice. *J. Food Prot.* **52**, 696–701.

Graves, L. M., B. Swaminathan, and S. B. Hunter. 1999. Subtyping *Listeria monocytogenes*. In: E. T. Ryser, and E. H. Marth (Eds.), *Listeria, Listeriosis and Food Safety*. New York: Marcel Dekker, pp. 279–280.

Gray, M. L. 1960. Isolation of *Listeria monocytogenes* from oat silage. *Science* **132**, 1767–1768.

Gray, M. L., and A. H. Killinger. 1966. *Listeria monocytogenes* and listeric infections. *Bacteriol. Rev.* **30**, 309–382.

Gronstol, H. 1979. Listeriosis in sheep: *Listeria monocytogenes* excretion and immunological state in sheep in flocks with clinical listeriosis. *Acta Vet. Scand.* **20**, 417–428.

Hahn, H., and S. H. E. Kaufman. 1981. The role of cell mediated immunity in bacterial infections. *Rev. Infect. Dis.* **3**, 1221–1250.

Harrison, M. A., and S. L. Carpenter. 1989. Survival of large populations of *Listeria monocytogenes* on chicken breasts processed using moist heat. *J. Food Prot.* **52**, 376–378.

Hayes, P. S., L. M. Graves, B. Swaminathan, G. W. Ajello, G. B. Malcolm, R. E. Weaver, R. Ransom, K. Deaver, B. D. Plikaytis, A. Schuchat, J. D. Wenger, R. W. Pinner, C. V. Broome, and the *Listeria* study group. 1992. Comparison of three selective enrichment methods for the isolation of *Listeria monocytogenes* from naturally contaminated foods. *J. Food Prot.* **55**, 952–959.

Heisick, J. E., D. E. Wagner, M. L. Nierman, and J. T. Peeler. 1989. *Listeria* spp. Found on fresh market produce. *Appl. Environ. Microbiol.* **55**, 1925–1927.

Hicks, S. J., and B. M. Lund. 1991. The survival of *Listeria monocytogenes* in cottage cheese. *J. Appl. Bacteriol.* **70**, 308–314.

Hill, C., and C. Gahan. 2000. *Listeria monocytogenes*: Role of stress in virulence and survival in food. *Irish J. Agric. Food Res.* **39**, 195–201.

Hitchins, A. D. 1995. *Listeria monocytogenes*. In: *Food and Drug Administration Bacteriological Analytical Manual,* 8th ed. Gaithersburg, MD: AOAC International, Chapter 10, pp. 10.01–10.13.

Ho, J. L., K. N. Shands, G. Friedland, P. Eckind, and D. W. Fraser. 1986. An outbreak of type 4b *L. monocytogenes* infection involving patients from eight Boston hospitals. *Arch. Intern. Med.* **146**, 520–524.

Hof, H., and J. Rocourt. 1992. Is any strain of *Listeria monocytogenes* detected in food a health risk? *Int. J. Food Microbiol.* **16**, 173–182.

Hudson, J. A., S. J. Mott, and N. Penney. 1994. Growth of *Listeria monocytogenes, Aeromonas hydrophila,* and *Yersinia enterocolitica* on vacuum and saturated carbon dioxide controlled atmosphere-packaged sliced roast beef. *J. Food Prot.* **57**, 204–208.

Hughey, V. L., P. A. Wilger, and E. A. Johnson. 1989. Antimicrobial activity of hen white lysozyme against *Listeria monocytogenes* Scott A in foods. *Appl. Environ. Microbiol.* **55**, 631–638.

Irvin, A. D. 1968. The effect of pH on the multiplication of *L. monocytogenes* in grass silage media. *Vet. Rec.* **82**, 115–116.

Jacobs, J. L., and H. W. Murray. 1986. Why is *Listeria monocytogenes* not a pathogen in the acquired immunodeficiency syndrome? *Arch. Intern. Med.* **146**, 1299–1300.

Jemmi, T. 1990. Actual knowledge of *Listeria* in meat and fish products. *Mitt. Geb. Lebensmittel. Hyg.* **81**, 144–157.

Jensen, A., W. Frederiksen, and P. Gerner-Smidt. 1994. Risk factors for listeriosis in Denmark, 1989–1990. *Scand. J. Infect. Dis.* **26**, 171–178.

Jinneman, K. C., M. M. Wekell, and M. W. Eklund. 1999. Incidence and behavior of *Listeria monocytogenes* in fish and seafood. In: *Listeria, listeriosis and food safety,* 2nd ed. New York: Marcel Dekker, p. 608–609.

Juntilla, J. R., Niemelä, and J. Hirn. (1988). Minimum growth temperatures of *Listeria monocytogenes* and non-haemolytic *Listeria. J. Appl. Bacteriol.* **65**, 321–327.

Kalac, P. 1982. A review of some aspects of possible association between feeding of silage and animal health. *Br. Vet. J.* **138**, 314–315.

Kathariou, S., P. Metz, H. Hof, and W. Goebel. 1987. Tn916-induced mutations in the hemolysin determinant affecting virulence of *Listeria monocytogenes. J. Bacteriol.* **169**, 1291–1297.

Kerr, K. G., S. F. Dealler, and R. W. Lacey. 1988. Materno-fetal listeriosis from cook-chill and refrigerated foods. *Lancet* **ii**, 1133.

Klausner, R. B., and C. W. Donnelly. 1991. Environmental sources of *Listeria* and *Yersinia* in Vermont dairy plants. *J. Food Prot.* **54**, 607–611.

Knabel, S. J., H. W. Walker, P. A. Hartman, and A. F. Mendonca. 1990. Effects of growth temperatures and strictly anaerobic recovery on the survival of *Listeria monocytogenes* during pasteurization. *Appl. Environ. Microbiol.* **56**, 370–376.

Knight, M. T., M. C. Newman, M. J. Benzinger, Jr., and J. R. Agin. 1996. TECRA *Listeria* visual immunoassay (TLIVA) for detection of *Listeria* in foods: Collaborative study. *J. AOAC Int.* **79**, 1083.

Kraft, J., D. Funke, A. Haas, F. Lottspeich, and W. Goebel. 1989. Production, purification and characterization of hemolysins from *Listeria ivanovii* and *Listeria monocytogenes* Sv4b. *FEMS Microbiol. Lett.* **57**, 197–202.

Kuhn, M., and W. Goebel. 1989. Identification of an extracellular protein of *Listeria monocytogenes* possibly involved in intracellular uptake by mammalian cells. *Infect. Immun.* **57**, 55–61.

Kuhn, M., and W. Goebel. 1999. Pathogenesis of *Listeria monocytogenes*. In: E. T. Ryser and E. H. Marth (Eds.), *Listeria, Listeriosis and Food Safety,* 2nd ed. New York: Marcel Dekker, pp. 97–130.

Kuhn, M., S. Kathariou, and W. Goebel. 1988. Hemolysin supports survival but not entry of the intracellular bacterium *Listeria monocytogenes. Infect. Immun.* **57**, 55–61.

Lane, F. C., and E. R. Unanue. 1972. Requirement of thymus (T) lymphocytes for resistance of listeriosis. *J. Exp. Med.* **135**, 1104–1112.

Levine, B. A., K. R. Sirinek, C. G. McLeod, Jr., D. K. Teegarden, and B. A. Pruitt. 1979. The role of cimetidine in the prevention of stress induced gastric mucosal injury. *Surg. Gyn. Obsetr.* **148**, 399–402.

Linnan, M. J., L. Mascola, X. D. Lou, V. Goulet, S. May, C. Salminen, D. W. Hird, M. L. Yonekura, P. Hayes, R. Weaver, A. Audurier, B. D. Plikaytis, S. L. Fannin, A. Kleks, and C. V. Broome. 1988. Epidemic listeriosis associated with Mexican-style cheese. *N. Engl. J. Med.* **319**, 823–828.

Loken, T., E. Aspoey, and H. Gronstol. 1982. *Listeria monocytogenes* excretion and humoral immunity in goats in a herd with outbreaks of listeriosis and in a healthy herd. *Acta Vet. Scand.* **23**, 392–399.

Louria, D. B., T. Hensle, D. Armstrong, H. S. Collins, A. Blevins, D. Krugman, and

M. Buse. 1967. Listeriosis complicating malignant disease. A new association. *Ann. Intern. Med.* **67**, 261–281.

Lovett, J., and A. D. Hitchins. 1989. *Listeria* isolation. In: *Bacteriological Analytical Manual,* 6th ed., Supplement, Sept. 1987 (2nd printing 1989): 29.01. Arlington, VA: Association of Official Analytical Chemists.

Lovett, J., D. W. Francis, and J. M. Hunt. 1987. *Listeria monocytogenes* in raw milk: Detection, incidence and pathogenicity. *J. Food Prot.* **50**, 188–192.

MacGowan, A. P., P. K. Peterson, W. Keane, and P. G. Quie. 1983. Human peritoneal macrophage phagocytic, killing and chemiluminescent responses to opsonized *Listeria monocytogenes. Infect. Immun.* **40**, 440–443.

Mackaness, G. B. 1962. Cellular resistance to infection. *J. Exp. Med.* **116**, 381–406.

Mascola, L., L. Lieb, J. Chiu, S. L. Fannin, and M. J. Linnan. 1988. Listeriosis: An uncommon opportunistic infection in patients with acquired immunodeficiency syndrome. *Am. J. Med.* **84**, 162–164.

McLauchlin, J. 1990. Human listeriosis in Britain, 1967–1985, a summary of 722 cases. I. Listeriosis during pregnancy and in the newborn. *Epidemiol. Infect.* **104**, 181–189.

McLauchlin, J., A. Audurier, and A. G. Taylor. 1986. Aspects of the epidemiology of human *Listeria monocytogenes* infections in Britain 1967–1984; the use of serotyping and phage typing. *J. Med. Microbiol.* **22**, 367–377.

Mead, 1999. Multistate outbreak of listeriosis traced to processed meats, August 1998–March 1999. Written communication, May 27, pp. 1–11.

Mead, P. S., L. Slutsker, V. Dietz, L. F. McCaig, J. S. Bresee, C. Shapiro, P. M. Griffin, and R. V. Tauxe. 1999. Food-related illness and death in the United States. *Emerg. Infect. Dis.* **5**, 607–625.

Miller, A. J. 1992. Combined water activity and solute effects on growth and survival of *Listeria monocytogenes* Scott A. *J. Food Prot.* **55**, 414–418.

Mossey, R. T., and J. Sondheimer. 1985. Listeriosis in patients with long-term hemodialysis and transfusional iron overload. *Am. J. Med.* **79**, 397–400.

Murray, E. G. D., R. A. Webb, and M. B. R. Swann. 1926. A disease of rabbits characterized by a large mononuclear leucocytosis, caused by a hitherto undescribed bacillus *Bacterium monocytogenes* (n. sp.). *J. Pathol. Bacteriol.* **29**, 407–439.

Netten, P. Van, I. Perales, A. Van de Moosdijk, G. D. W. Curtis, and D. A. A. Mossel. 1989. Liquid and solid selective differential media for the detection and enumeration of *L. monocytogenes* and other *Listeria* spp. Int. *J. Food Microbiol.* **8**, 299–316.

Nieman, R. E., and B. Lorber. 1980. Listeriosis in adults: A changing pattern: Report of eight cases and review of the literature, 1968–1978. *Rev. Infect. Dis.* **2**, 207–227.

Nolan, D. A., D. C. Chamblin, and J. A. Troller. 1992. Minimal water activity levels for growth and survival of *Listeria monocytogenes* and *Listeria innocua. Int. J. Food Microbiol.* **16**, 323–335.

North, R. J. 1970. Suppression of infection by an antimitotic drug. Further evidence that migrant macrophages express immunity. *J. Exp. Med.* **132**, 535–545.

Nyfeldt, A. 1929. Etiologie de la mononucléose infectieuse. *C. R. Soc. Biol.* **101**, 590–592.

O'Driscoll, B., C. G. M. Gahan, and C. Hill. 1996. Adaptive acid tolerance response in *Listeria monocytogenes*: Isolation of an acid-tolerant mutant which demonstrates increased virulence. *Appl. Environ. Microbiol.* **62**, 1693–1698.

Olafson, P. 1940. *Listerella* encephalitis (circling disease) in sheep, cattle and goats. *Cornell Vet.* **30**, 141–150.

Palsson, P. A. 1962. Relation of silage feeding to listeric infection in sheep. In: M. L. Gray (Ed.), *Second Symposium on Listeric Infection.* Bozeman, MT: pp. 73–84.

Palumbo, S. A., and A. C. Williams. 1991. Resistance of *Listeria monocytogenes* to freezing in foods. *Food Microbiol.* **8**, 63–68.

Perry, C. M., and C. W. Donnelly. 1990. Incidence of *Listeria monocytogenes* in silage and its subsequent control by specific and nonspecific antagonism. *J. Food Prot.* **53**, 642–647.

Petran, R. L., and K. M. J. Swanson. 1993. Simultaneous growth of *Listeria monocytogenes* and *Listeria innocua*. *J. Food Prot.* **56**, 616–618.

Petran, R. L., and E. A. Zottola. 1989. A study of factors affecting growth and recovery of *Listeria monocytogenes* Scott A. *J. Food Sci.* **54**, 458–460.

Pinner, R. W., A. Schuchat, B. Swaminathan, P. S. Hayes, K. A. Deaver, R. E. Weaver, B. D. Plikaytis, M. Reeves, C. V. Broome, J. D. Wenger, and the *Listeria* study group. 1992. Role of foods in sporadic listeriosis II. Microbiologic and epidemiologic investigation. *JAMA* **267**, 2046–2050.

Pirie, J. H. H. 1940. *Listeria*: Change of name for a genus of bacteria. *Nature* **145**, 264.

Pritchard, T. J., and C. W. Donnelly. 1999. Combined secondary enrichment of primary enrichment broths increases *Listeria* detection. *J. Food Prot.* **62**, 532–535.

Pritchard, T. J., C. M. Beliveau, K. J. Flanders, and C. W. Donnelly. 1994. Increased incidence of *Listeria* species in dairy processing plants having adjacent farm facilities. *J. Food Prot.* **57**, 770–775.

Proctor, M. E., R. Brosch, J. W. Mellen, L. A. Garrett, C. W. Kaspar, and J. B. Luchansky. 1995. Use of pulsed-field gel electrophoresis to link sporadic cases of invasive listeriosis with recalled chocolate milk. *Appl. Environ. Microbiol.* **61**, 3177–3179.

Rasmussen, O. F., P. Skouboe, L. Dons, L. Rossen, and J. E. Olsen. 1995. Listeria monocytogenes exists in at least three evolutionary lines: Evidence from flagellin, invasive associated protein and listeriolysin O genes. *Microbiology.* **141**, 2053.

Rocourt, J., and H. P. R. Seeliger. 1985. Classification of a different *Listeria* species. *Zbl. Bakteriol. Hyg. A* **259**, 317–330.

Rocourt, J., P. Boerlin, F. Grimont, Ch. Jacquet, and J.-C. Piffaretti. 1992. Assignment of *Listeria grayi* and *Listeria murrayi* to a single species, *Listeria grayi*, with a revised description of of *Listeria grayi*. *Int. J. Syst. Bacteriol.* **42**, 69–73.

Rocourt, J., H. Hof, A. Schrettenbrunner, R. Malinverni, and J. Bille. 1986. Méningite purulente aiguë à *Listeria seeligeri* chez un adulte immunocompétent. *Schweiz. Med. Wschr.* **116**, 248–251.

Rodriguez, L. D., G. S. Fernandez, J. Fernandez, F. Garayzabel, and E. R. Ferri. 1984. New methodology for the isolation of *Listeria monocytogenes* from heavily contaminated environments. *Appl. Environ. Microbiol.* **47**, 1188–1190.

Roth, T. T., and C. W. Donnelly. 1996. Survival of acid-injured *Listeria monocytogenes* and comparison of procedures for recovery. *IFT Book of Abstracts* **35–3**, 75.

Ryser, E. T., and E. H. Marth. 1987a. Behavior of *Listeria monocytogenes* during the manufacture and ripening of Cheddar cheese. *J. Food Prot.* **50**, 7–13.

Ryser, E. T., and E. H. Marth. 1987b. Fate of *L. monocytogenes* during manufacture and ripening of Camembert cheese. *J. Food Prot.* **50**, 372–378.

Ryser, E. T., and E. H. Marth. 1989. Behavior of *Listeria monocytogenes* during manufacturing and ripening of brick cheese. *J. Dairy Sci.* **72**, 838–853.

Ryser, E. T., S. M. Arimi, M. M.-C. Bunduki, and C. W. Donnelly. 1996. Recovery of different *Listeria* ribotypes from naturally contaminated, raw refrigerated meat and poultry products with two primary enrichment media. *Appl. Environ. Microbiol.* **62**, 1781–1787.

Ryser, E. T., S. M. Arimi, and C. W. Donnelly. 1997. Effects of pH on distribution of *Listeria* ribotypes in corn, hay and grass silage. *Appl. Environ. Microbiol.* **63**, 3695–3697.

Ryser, E. T., E. H. Marth, and M. P. Doyle. 1985. Survival of *Listeria monocytogenes* during manufacture and storage of cottage cheese. *J. Food Prot.* **48**, 746–750, 753.

Salamina, G., E. D. Donne, A. Niccolini, G. Poda, D. Cesaroni, M. Bucci, R. Fini, M. Maldini, A. Schuchat, B. Swaminathan, W. Bibb, J. Rocourt, N. Binkin, and S. Salmosa. 1996. A foodborne outbreak of gastroenteritis involving *Listeria monocytogenes*. *Epidemiol. Infect.* **117**, 429–436.

Schlech, W. F. 1988. Virulence characteristics of *Listeria monocytogenes*. *Food Technol.* **42**, 176–178.

Schlech, W. F. III. 1992. Expanding the horizons of foodbonre listeriosis. *JAMA* **267**, 2081–2082.

Schlech, W. F. III, P. M. Lavigne, R. A. Bortolussi, A. C. Allen, E. V. Haldene, A. J. Wort, A. W. Hightower, S. E. Johnson, S. H. King, E. S. Nicholls, and C. V. Broome. 1983. Epidemic listeriosis: Evidence for transmission by food. *N. Engl. J. Med.* **308**, 203–206.

Schonberg, A., E. Bannerman, A. L. Cortieu, R. Kiss, J. McLauchlin, S. Shah, and D. Wilhelms. 1996. Serotyping of 80 strains from the WHO multicentre international typing study of *Listeria monocytogenes*. *Int. J. Food Microbiol.* **32**, 279.

Schuchat, A., K. A. Deaver, J. D. Wenger, B. D. Plikaytis, L. Mascola, R. W. Pinner, A. L. Reingold, C. V. Broome, and the *Listeria* study group. 1992. Role of foods in sporadic listeriosis 1. Case control study of dietary risk factors. *JAMA* **267**, 2041–2045.

Schwartz, B., C. A. Cielsielski, C. V. Broome, S. Gaventa, G. R. Brown, B. G. Gellin, A. W. Hightower, L. Mascola, and the Listeriosis Study Group. 1988. Association of sporadic listeriosis with consumption of uncooked hot dogs and undercooked chicken. *Lancet* **2**, 779–782.

Seeliger, H. P. R. 1961. *Listeriosis.* New York: Hafner.

Seeliger, H. P. R., J. Rocourt, A. Schrettenbrunner, P. A. D. Grimont, and D. Jones. 1984. *Listeria ivanovii* sp. *nov*. *Int J. Syst. Bacteriol.* **34**, 336.

Shahamat, M., A. Seaman, and M. Woodbine. 1980. Survival of *Listeria monocytogenes* in high salt concentrations. *Zbl. Bakteriol. Hyg. I Abt. Orig. A* **246**, 506–511.

Slutsker, L., and A. Schuchat. 1999. Listeriosis in humans. In: E. T. Ryser and E. H. Marth (Eds.), *Listeria, Listeriosis and Food Safety*, 2nd ed. New York: Marcel Dekker, p. 75.

Sokolovic, Z., and W. Goebel. 1989. Synthesis of listeriolysin in *Listeria monocytogenes* under heat shock conditions. *Infect. Immun.* **57**, 295–298.

Stamm, A. M., W. E. Dismukes, B. P. Simmons, C. G. Cobbs, A. Elliott, P. Budrich, and J. Harmon. 1982. *Rev. Infect. Dis.* **4**, 665–682.

Stuart, S. E., and H. J. Welshimer. 1974. Taxonomic reexamination of *Listeria pirie* and transfer of *Listeria grayi* and *Listeria murrayi* to a new genus, *Murraya. Int. J. Syst. Bacteriol.* **24**, 177–185.

Swaminathan, B., S. B. Hunter, P. M. Desmarchelier, P. Gerner-Smidt, L. M. Graves, S. Harlander, R. Hubner, C. Jacquet, B. Pedersen, K. Reineccius, A. Ridley, N. A. Saunders, and J. A. Webster. 1996. WHO-sponsored international collaborative study to evaluate methods for subtyping *Listeria monocytogenes*: Restriction fragment length polymorphism (RFLP) analysis using ribotyping and Southern hybridization with two probes derived from *L. monocytogenes* chromosome. *Int. J. Food Microbiol.* **32**, 263.

Tappero, J. W., A. Schuchat, K. A. Deaver, L. Mascola, and J. D. Wenger. 1995. Reduction in the incidence of human listeriosis in the United States—Effectiveness of prevention efforts? *JAMA* **273**, 1118–1122.

Tilney, L. G., and D. A. Portnoy. 1989. Actin filaments and the growth, movement, and spread of the intracellular bacterial parasite, *Listeria monocytogenes. J. Cell Biol.* **109**, 1597–1608.

Vogt, R. L., C. Donnelly, B. Gellin, W. Bibb, and B. Swaminathan. 1990. Linking environmental and human strains of *Listeria monocytogenes* with isoenzyme and ribosomal RNA typing. *Eur. J. Epidemiol.* **6**, 229–230.

Walker, S. J., P. Archer, and J. G. Banks. 1990. Growth of *Listeria monocytogenes* at refrigeration temperatures. *J. Appl. Bacteriol.* **68**, 157–162.

Wang, R.-F., W.-W. Cao, and M. G. Johnson. 1991. Development of a 16S rRNA-based oligomer probe specific for *Listeria monocytogenes. Appl. Environ. Microbiol.* **57**, 3666–3670.

Weagant, S. D., P. N. Sado, K. G. Colburn, J. D. Torkelson, F. A. Stanley, M. H. Krane, S. C. Shields, and C. F. Thayer. 1988. The incidence of *Listeria* species in frozen seafood products. *J. Food Prot.* **51**, 655–657.

Weinberg, E. D. 1984. Pregnancy-associated depression of cell-mediated immunity. *Rev. Infect. Dis.* **6**, 814–831.

Welch, D. F., C. P. Sword, S. Brehm, and D. Dusanic. 1979. Relationship between superoxide dismutase and pathogenic mechanisms of *Listeria monocytogenes. Infect. Immun.* **23**, 863–872.

Wenger, J. D., B. Swaminathan, P. S. Hayes, S. S. Green, M. Pratt, R. W. Pinner, A. Schuchat, and C. V. Broome. 1990. *Listeria monocytogenes* contamination of turkey franks: Evaluation of a production facility. *J. Food Prot.* **53**, 1015–1019.

Wernars, K., P. Boerlin, A. Audurier, E. G. Russell, G. D. W. Curtis, L. Herman, and N. van der Mee-marquet. 1996. The WHO multicentre study on *Listeria monocytogenes* subtyping: Random amplification of polymorphic DNA (RAPD). *Int. J. Food Microbiol.* **32**, 325.

Wiedmann, M., J. L. Bruce, C. Keating, A. E. Johnson, P. L. McDonough, and C. A. Batt. 1997. Ribotypes and virulence gene polymorphisms suggest three distinct *Listeria monocytogenes* lineages with differences in pathogenic potential. Infect. *Immunology* **65**, 2707–2716.

CHAPTER 9

Clostridium perfringens

NORMA L. HEREDIA and RONALD G. LABBÉ

9.1. INTRODUCTION

The genus *Clostridum* consists of a diverse group of bacteria that do not grow in the presence of oxygen and have the ability to form heat-resistant endospores. Many of these anaerobes are pathogenic for both human and animals. Historically it is best known as the predominant cause of gas gangrene in wound infections. The advent of antibiotics has greatly reduced the incidence of such infections.

The bacterial pathogen *Clostridium perfringens* is the most prolific toxin-producing species within the clostridial group. The toxins are responsible for a wide variety of human and veterinary diseases, many of which can be lethal. *Clostridium perfringens* causes two quite different human diseases that can be transmitted by food, one a common form of foodborne illness and the other necrotic enteritis (pig-bel), which is found only rarely. The foodborne illness caused by *C. perfringens* is among the most common caused by consumption of contaminated food. Although the association of *C. perfringens* with foodborne illness was first proposed about 100 years ago, it was not until the 1960s and 1970s that conclusive evidence had accumulated showing that an enterotoxin is associated with sporulation of the organism in the intestine of ill individuals.

The bacterium possesses several attributes that contribute significantly to its ability to cause foodborne illness: (1) a ubiquitous distribution throughout the natural environment, giving it ample opportunity to contaminate foods; (2) the ability to form heat-resistant spores, allowing it to survive incomplete cooking of foods or improper sterilization; (3) the ability to grow quickly in foods, allowing it to reach the high levels necessary for food poisoning; and (4) the ability to produce an intestinally active enterotoxin, responsible of the characteristic gastrointestinal symptoms of *C. perfringens* food poisoning.

Guide to Foodborne Pathogens, Edited by Ronald G. Labbé and Santos García
ISBN 0-471-35034-6

9.2. NATURE OF ILLNESS IN ANIMALS AND HUMANS

Clostridium perfringens enterotoxin (CPE) is traditionally recognized as a virulence factor responsible for the diarrheal and cramping symptoms associated with *C. perfringens* type A food poisoning. This toxin is a single 34-kDa polypeptide with a isoelectric point of 4.3, a 309 amino acid sequence, and a unique mechanism of action. It is produced intracellularly during the sporulation of the bacterium, and it is released together with the mature spore. Genetic studies indicate that the *cpe* gene can be either chromosomal or plasmid borne. Most, perhaps all, food-poisoning isolates carry the *cpe* gene on their chromosome while the gene is episomal in CPE-associated, non-foodborne human gastrointestinal disease isolates. Molecular epidemiological surveys suggest that only a low percentage (4–6%) of all *C. perfringens* isolates carry the gene. The role of CPE in the physiology of the bacterial cell is unknown.

Foodborne illness is produced 8–24 h after the ingestion of food contaminated with large numbers of vegetative bacteria ($>10^5$ CFU/g). Many of the ingested cells may die when exposed to stomach acidity, but if the food vehicle is sufficiently contaminated, some vegetative cells survive passage through the stomach and enter the small intestine, where they multiply, sporulate, and produce CPE. The possible role of ingested sporulating cells or preformed enterotoxin is unlikely, since studies with volunteers indicate that the amount of ingested CPE necessary to produce symptoms would require cell numbers that would impart adverse sensory qualities to such foods.

Once the CPE is released into the small intestine, the following sequence of events occur: (1) CPE binding to a 50-kDa protein receptor, forming a small complex of 90 kDa; (2) the development of a postbinding physical change to this small complex; that could represent either the insertion of CPE into the membrane or a conformational change to small complex; (3) an interaction between this physically changed small complex and a 70-kDa membrane protein, forming a large 160-kDa complex; and (4) the initiation of a series of biochemical events that alter the normal permeability of brush border membranes in the epithelial cells of the small intestine. This CPE-induced permeability change becomes cytotoxic and causes localized tissue damage (Fig. 9.1), which leads to a breakdown in normal fluid and electrolyte transport properties and hence to diarrhea. Treatment of CPE with trypsin increases its activity at least twofold, suggesting a possible role for the intestinal enzyme in cases of human illness.

The main symptoms of *C. perfringens* food poisoning include diarrhea and severe abdominal pain. Nausea is less common, and fever and vomiting are unusual. Death is uncommon but has occurred in debilitated or institutionalized individuals, especially the elderly. The cases are self-limited and antibiotic therapy is not recommended.

Animal carcasses and cuts of meat can become contaminated with *C. perfringens* from soil, animal feces, or handling during slaughtering and processing. Many organisms that compete with *C. perfringens* are killed when meat and poultry are cooked, but *C. perfringens* spores may be difficult to eliminate.

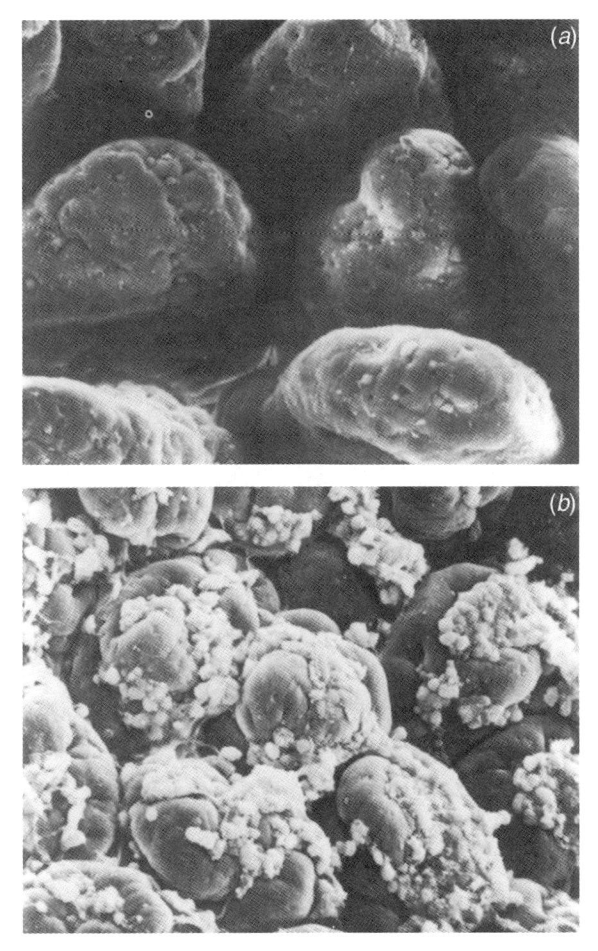

Figure 9.1. Effect of *C. perfringens* enterotoxin on rabbit villi: (*a*) control; (*b*) treated with enterotoxin for 90 min showing collapsed villus tips and protruding structures at the surface of epithelial cells. From McDonel, J., Chang, L., Pounds, J., and Duncan, C. The effects of *Clostridium perfringens* enterotoxin on rat and rabbit ileum. *Lab. Investig.* **39**, 210–218. Lippincot Williams & Wilkins, Baltimore MD, 1978. Reproduced with permission of publisher.

Heat activation of spores during the cooking process would facilitate germination when the temperature becomes favorable for growth. Also, during cooking E_h values drop to levels that favor subsequent multiplication of *C. perfringens*. Temperature abuse can occur during improper cooling, for example, at room temperature or by refrigeration of large portions that cool slowly or by improper holding temperatures. In such cases bacteria commence multiplication. If such foods are served without being reheated to a temperature sufficient to kill vegetative forms of *C. perfringens*, illness may result. Therefore, when foodborne outbreaks occur, one or more of the following events usually have

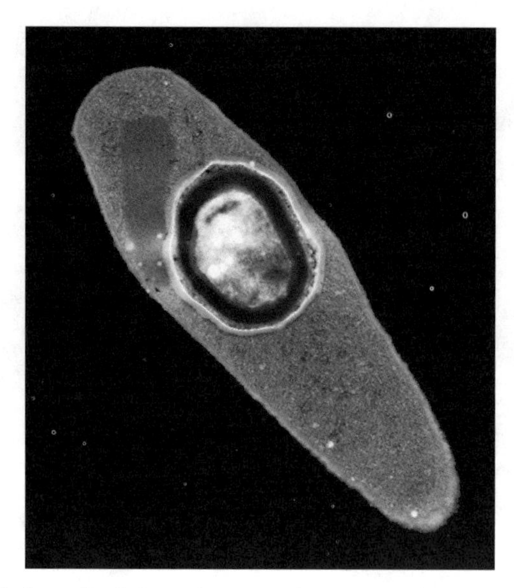

Figure 9.2. Sporulating cell of *C. perfringens*. The enterotoxin is released along with the spore upon lysis of the mother cell.

occurred: (a) improper cooling, (b) improper hot holding, (c) preparation of food a day or more in advance, or (d) inadequate reheating.

9.3. CHARACTERISTICS OF AGENT

Clostridium perfringens was first described as *Bacillus aerogenes* in 1892 and later called *Clostridium welchii*. It is an anaerobic, gram positive, rod-shaped, encapsulated, spore-forming bacterium (Fig. 9.2). Pulse-field gel electrophoresis revealed that strain CPN50 has a single circular 3.6-Mb chromosome. More than 100 restriction sites and almost 100 genetic loci have been located on the genome.

This organism is found in soil, water, air, and foods. The virulence of the microorganism results from it ability to produce numerous protein toxins (Table 9.1). It has been classified into five types (A–E) based on the production of extracellular toxins and hydrolytic enzymes, including lecithinase, hemolysins, hyaluronidase, collagenase, deoxyribonuclease (DNAse), and amylase. These allow it to scavenge required nutrients from its environment.

Foodborne illness is apparently caused only by type A strains. *Clostridium perfringens* types C and D also produced enterotoxin. However, there are no data implicating these strains in foodborne illness.

Types B, C, and D primarily occur in the intestines of animals and only occasionally in humans. The strains of these types have also been isolated from soils in areas where enteritis by the organisms was affecting a significant

TABLE 9.1. Toxins Produced by *C. perfringens*

C. perfringens Type	Toxins Produced										
	α	β	ε	ι	δ	θ	κ	λ	μ	ν	CPE
A	+++	−	−	−	−	+	+	−	+	+	+
B	+	++	+	−	+	+	+	+	+	+	+
C	+	++	−	−	+	+	+	−	+	+	+
D	+	−	++	−	−	+	+	+	+	+	+
E	+	−	−	+	−	+	+	+	−	+	+

number of animals and humans (Table 9.2). The pathogenicity of type E strains is not clear and has seldom been isolated.

Alpha toxin is produced by all five types and is a phospholipase C that can hydrolyze lecithin to phosphorylcholine and diglyceride. This toxin is believed to be a major factor responsible for the tissue pathology of the organism. The major lethal effects associated with this toxin are gas gangrene (clostridial myonecrosis) in humans and necrotic enteritis and enterotoxemia in animals.

Two toxins are known to be active in the human gastrointestinal tract, the β toxin and the CPE. The β toxin is produced by types B and C strains and is responsible for the lesions associated with necrotic enteritis also. The CPE toxin is more important in human illness and appears to be responsible for >10% of all foodborne illness in the United States.

Clostridium perfringens is different from many other clostridia in that it is nonmotile and in vitro forms spores only in specialized culture media. The organism is fermentative and grows rapidly in media containing carbohydrates. Under these conditions it produces copious amounts of H_2 and CO_2, which help to maintain an anaerobic environment. Owing to its rapid growth and relative aerotolerance, compared to other anaerobes, *C. perfringens* is relatively easy to work with in the laboratory.

TABLE 9.2. Disease Caused by *C. perfringens*

C. perfringens Type	Disease Produced
A	Gas gangrene (clostridial myonecrosis); food poisoning; necrotic enteritis of infants; necrotic enteritis of poultry; enterotoxemia in cattle and lamb; possibly colitis
B	Lamb dysentery; enterotoxemia of sheep, foals and goats; hemorrhagic enteritis in neonatal calves and foals
C	Enterotoxemia of sheep (struck); necrotic enteritis in animals; human enteritis necroticans (pig-bel); acute enterotoxemia in adult sheep
D	Enterotoxemia of sheep (pulpy kidney disease), lamb, and calves
E	Enterotoxemia of rabbits; canine and porcine enteritis

9.4. EPIDEMIOLOGY

Meat and poultry are the most common food vehicles for *C. perfringens* type A food poisoning with beef being the vehicle in nearly 30% of all outbreaks. Turkey and chicken together account for another 15%. Mexican foods containing meat are emerging as another important vehicle for *C. perfringens* type A food poisoning. Fish are not commonly involved in food-poisoning outbreaks due to this organism.

In the United States, for the periods 1983–1997, *C. perfringens* has ranked second or third as the causal agent of confirmed outbreaks of foodborne disease of bacterial origin. The number of individuals affected per outbreak is typically between 40 and 70. Nevertheless, like other agents of foodborne illness, it is widely underreported in part because of the anaerobic conditions required for its cultivation and the perceived mild nature of the illness. In fact, it is not one of the target foodborne pathogens in the recently instituted U.S. Foodborne Disease Active Surveillance Network (FoodNet). Its omission from the target group will reduce its relative prevalence as a foodborne pathogen. By contrast, in the United Kingdom, with its relatively smaller population but similar food habits, the number of outbreaks and cases caused by *C. perfringens* is dramatically greater than in the United States due to more diligent monitoring.

As mentioned above, the most common vehicles are meat and poultry-based products, spreading from intestinal contents during slaughter. The presence of this organism in the spore state on herbs and spices (as well as in meat and poultry) can create a hazard since they can survive the cooling process and resume growth in temperature-abused or poorly cooled foods, resulting in the large number of cells required for an outbreak. The organism's fastidious growth requirements are easily met in meat and poultry items.

Temperature abuse of cooked foods is invariable involved in outbreaks, reflecting the organism's relatively high optimum growth temperature [43–46°C (110–115°F)]. This and its spore-forming ability are the two outstanding characteristics contributing to its etiological role in foodborne illness.

Food service establishments such as restaurants, hospitals, prisons, schools, and caterers are the most likely sites for acquiring the illness.

9.5. DETECTION OF ORGANISM

Requirements for implicating *C. perfringens* in an outbreak include one of the following: (a) $>10^6$ spores/g feces from ill individuals, (b) $>10^5$ cells/g of incriminated food, or (c) direct detection of CPE in outbreak stools. Previous recommendations for identifying common serotypes in implicated foods and stools have not been widely adopted because many isolates cannot be serotyped. All responsible strains are assumed to be type A.

In North America and the United Kingdom the recommended selective plating medium for enumerating *C. perfringens* from food and feces is tryptose–

sulfite–cycloserine (TSC) agar, which is commercially available. The selective antibiotic cycloserine must be added separately. Normal stool samples typically contain 10^3–10^4 *C. perfringens* spores/g. Stool samples are heated before dilution and plating. Colonies appear black on this medium. Typically 10 are confirmed using lactose–gelatin and motility nitrate medium. Details of the procedure are described in the *Compendium of Methods for the Microbiological Examination of Food* (American Public Health Association).

Certain otherwise healthy elderly individuals may carry high spore levels of *C. perfringens*, limiting the elevated spore count as an implicating criterion. This has promoted the direct detection of CPE in outbreak stools as a useful technique for establishing this organism as the etiological agent. Several enzyme-linked immunosorbent assay (ELISA) procedures that have been developed for this purpose have sensitivities of several nanograms per gram of feces. Stools from healthy individuals contain undetectable levels of CPE compared with outbreak stools, which contain >1 μg/g. A reversed passive latex agglutination (RPLA) with similar sensitivity has been developed. Both RPLA and ELISA assay kits are commercially available.

Detection of CPE in laboratory cultures require the sporulation of such isolates. Many isolates sporulate in vitro only reluctantly, and numerous media have been proposed for this purpose. Because of such difficulties, molecular methods to detect the CPE have been developed. These have been useful since the *cpe* gene (chromosomal and episomal) is highly conserved. These techniques include simple polymerase chain reaction (PCR), nested PCR, randomly amplified polymorphic DNA (RAPD), ribotyping, and digoxigenin-labeled *cpe* gene probes. However, a disadvantage of these techniques is that most require pure cultures. An exception is the PCR. Multiplex PCR methods have also been developed that permit the detection of CPE as well as the four extracellular toxins used in toxigenic typing of the organism and should prove to be useful in research and epidemiological studies.

9.6. PHYSICAL METHODS FOR DESTRUCTION

As with most foodborne pathogens pasteurization temperatures [72°C (161°F)] for 5–10 min and routine cooking procedures readily inactivate vegetative cells of this organism. Of far greater concern is the heat resistance of their spores, which varies depending on the strain. This led to the designation of strains as being "heat resistant" or "heat sensitive." For example, in one study the decimal reduction values at 95°C (D_{95}) for the former were between 17.6 and 63 min compared with between 1.3 and 2.8 min for the latter. Not surprisingly the so-called heat-resistant strains were more often associated with cases of foodborne illness. However, determination of this phenotypic characteristic is no longer sought for purposes of laboratory confirmation.

Extrinsic factors that may affect spore heat resistance include sporulation temperatures and the menstrum in which spores are heated. The D values for a number of strains have been summarized elsewhere. These two groups also vary in their radiation resistance. Not surprisingly, prior irradiation has a sensitizing effect on spores that are subsequently heated.

Unlike vegetative cells of other species, *C. perfringens* is unusually sensitive to refrigerated and frozen storage. Because of this, food samples to be tested for *C. perfringens* vegetative cells should be analyzed immediately or otherwise refrigerated and tested as soon as possible but not frozen. For storage of more than 48 h, samples should be treated with a glycerol salt solution to give a 10% final concentration of glycerol and stored frozen at −60°C until analysis. By contrast, spores, for example in fecal samples, are unaffected by frozen storage at −20°C. Unlike *Staphylococcus aureus*, enterotoxin CPE is not detected preformed in foods implicated in outbreaks. Nevertheless, studies have demonstrated that it is relatively heat sensitive with substantial inactivation occurring within a couple of minutes at 60°C (140°F).

9.7. PREVENTION AND CONTROL

Clostridium perfringens is commonly found in retail meat and poultry at levels of perhaps 10–100 g^{-1}. It is not reasonable to expect such products to be free of this organism. Rather the goal is to prevent multiplication to levels sufficient to cause illness. If present as spores, the organism can survive cooking procedures and is able to multiply rapidly between 37 and 45°C (98 and 112°F). In virtually all outbreaks the most significant contributing factor is the failure to properly cool previously cooked foods, especially when prepared in large portions. Cooking procedures also drive off oxygen, promoting anaerobic conditions. Cooked, chilled foods should be reheated to a minimal internal temperature of 75°C before serving in order to inactivate vegetative cells. The standard dictum of holding cooked foods outside the range of 5°C (40°F) to 60°C (140°F) will effectively prevent the growth of this organism in hazardous foods.

All humans carry the organism in the large intestine, presumably enterotoxin-negative strains. Regardless, it is not possible to prevent carriers from handling food. Hence preventative measures depend largely of food service workers' knowledge of proper food preparation and storage techniques, in particular temperature control.

BIBLIOGRAPHY

Ando, Y., T. Tsuzuki, H. Sunagawa, and S. Oka. 1985. Heat resistance, spore germination, and enterotoxigenicity of *Clostridium perfringens*. *Microbiol. Immunol.* **29**, 317–326.

Anonymous. 1990. *Surveillance for Foodborne Disease Outbreaks, 1983–1987.* Atlanta, GA: U.S. Centers for Disease Control and Prevention.

Anonymous. 1996. *Surveillance for Foodborne Disease Outbreaks, 1988–1992.* Atlanta, GA: U.S. Centers for Disease Control and Prevention.

Anonymous. 2000. *Surveillance for Foodborne Disease Outbreaks, 1993–1997.* Atlanta, GA: U.S. Centers for Disease Control and Prevention.

Birkhead, G., T. R. Vog, E. Heun, J. Snyder, and B. McClane. 1988. Characterization of an outbreak of *Clostridium perfringens* food poisoning by quantitative fecal culture and fecal enterotoxin measurement. *J. Clin. Microbiol.* **26**, 471–474.

Bryan, F. L. 1980. Foodborne diseases in the United States associated with meat and poultry. *J. Food Protect.* **43**, 140–150.

Heredia, N. L., R. G. Labbé, and J. S. Garcia-Alvarado. 1998. Alteration in sporulation, enterotoxin production and protein synthesis by *Clostridium perfringens* type A following heat shock. *J. Food Protect.* **61**(9), 1143–1147.

Labbé, R. G. 1989. *Clostridium perfringens.* In: M. P. Doyle (Ed.), *Foodborne Bacterial Pathogens.* New York: Marcel Dekker, pp. 191–234.

Labbé, R. G. 1991. *Clostridium perfringens. J. Assoc. Off. Anal. Chem.* **74**, 711–714.

Labbé, R. 2001. *Clostridium perfringens.* In: F. Downes and K. Ito (Eds.), *Compendium of Methods for the Microbiological Examination of Foods*, 4th ed. Washington, DC: American Public Health Association, pp. 325–330.

McClane, B. A. 1996. An overview of *Clostridium perfringens* enterotoxin. *Toxicon* **34**, 1335–1343.

McClane, B. A. 1997. *Clostridium perfringens.* In: M. P. Doyle, L. R. Beuchat, and T. J. Montville (Eds.), *Food Microbiology, Fundamentals and Frontiers.* Washington, DC: ASM Press, pp. 305–326.

Meer, R., and J. G. Songer. 1997. Multiplex polymerase chain reaction assay for genotyping *Clostridium perfringens. Am. J. Vet. Res.* **58**, 702–705.

Rodriguez-Romo, L. A., N. L. Heredia, R. L. Labbé, and J. S. Garcia-Alvarado. 1988. Detection of enterotoxigenic *Clostridium perfringens* in spices used in Mexico by dot blotting using a DNA probe. *J. Food Protect.* **61**(2), 201–204.

Rood, J. 1998. Virulence genes of *Clostridium perfringens. Annu. Rev. Microbiol.* **52**, 333–360.

Rood, J., and S. Cole. 1991. Molecular genetics and pathogenesis of *Clostridium perfringens. Microbiol. Rev.* **55**, 621–648.

Stringer, M., P. Turnbull, and R. Gilbert. 1990. Application of serological typing to the investigation of outbreaks of *Clostridium perfringens* food poisoning. 1970–1978. *J. Hyg.* **84**, 443–456.

Tood, E. C. 1995. Worldwide surveillance of foodborne disease: The need to improve. *J. Food Protect.* **59**, 82–92.

Van Damme-Jongsten, M., J. Rodhouse, R. J. Gilbert, and S. Notermans. 1990. Synthetic DNA probes for detection of enterotoxigenic *Clostidium perfringens* strains isolated from outbreaks of food poisoning. *J. Clin. Microbiol.* **28**, 131–133.

CHAPTER 10

Escherichia coli

PETER FENG

10.1. INTRODUCTION

Escherichia coli is a normal inhabitant of the digestive tract of all animals, including humans. It serves a beneficial function in the body by synthesizing useful vitamins and by competing and suppressing the growth of pathogenic bacteria that may be present or ingested with food and water. With the exception of anaerobic bacteria, *E. coli* is one of the dominant species in the human feces; therefore, it has been used as an indicator of fecal contamination for most of this century. The concept of indicators is based on the premise that if *E. coli* is detected in food and water, it is indirect evidence that the product has been contaminated with sewage and that pathogenic bacteria may also be present. Over the years, the use of *E. coli* as an indicator of fecal contamination has been challenged and criticized as being unreliable; however, no adequate substitute has been proposed, and as a result, it continues to be used as indicator worldwide.

Escherichia coli is a member of the family Enterobacteriaceae and the genus is composed of gram-negative, aerobic, facultatively anaerobic, non-spore-forming rods. They have the ability to ferment a variety of sugars; but the fermentation of lactose with production of acid and gas is characteristic, as it is for the coliform group. The genus is classified based on biochemical reactions or biotypes, where typical biotype I *E. coli* consisted of strains that gave $++--$ on the IMViC tests (indole, methyl red, Voges–Proskauer, and citrate). Other distinguishing biochemical traits include presence of ß-glucuronidase activity but absence of urease, phenylalanine deaminase, and H_2S production. *Escherichia coli* is serologically complex, as it is serotyped based on the 173 somatic (O), 56 flagellar (H), and 80 capsular (K) antigens. However, serotyping remains to be a useful tool for the classification of the many isolates that exists in the genus.

Guide to Foodborne Pathogens, Edited by Ronald G. Labbé and Santos García
ISBN 0-471-35034-6

Escherichia coli is generally regarded as being harmless; but when intestinal barriers are compromised, especially in debilitated or immunosuppressed hosts, even typical *E. coli* may cause infections; hence, they are referred to as opportunistic pathogens. There is, however, another group of *E. coli*, which have acquired unique virulence factors and have the ability to cause diarrheal disease even in healthy human hosts. These are known as pathogenic *E. coli* (EC) and there are six major groups: enteropathogenic (EPEC), enterotoxigenic (ETEC), enterohemorrhagic (EHEC), enteroinvasive (EIEC), enteroaggregative (EAEC), and diffusely adherent (DAEC). Some type strains in these pathogenic *E. coli* groups, such as O157:H7, are widely recognized by their serotype; but, since these are grouped on the basis of pathogenicity, analysis for these characteristic virulence properties is required for definitive identification of each group.

This chapter examines the general characteristics, disease, and epidemiology of pathogenic *E. coli* groups. It also includes discussions on methods used for detection as well as preventive measures to inactivate and control these pathogens in foods. Of the pathogenic *E. coli* groups, EHEC, especially serotype O157:H7, has been implicated often in outbreaks of gastrointestinal illness worldwide; hence, more research has focused on this serotype, and this is also reflected in the contents of this chapter.

10.2. NATURE OF ILLNESS AND EPIDEMIOLOGY

All infections with pathogenic *E. coli* begin with the colonization of intestinal mucosa followed by elaboration of distinctive virulence factors, which causes the symptoms or gastrointestinal illness, that may be characteristic for that group. The illnesses resulting from pathogenic *E. coli* infections can range from asymptomatic or mild diarrhea in healthy adults to fairly severe complications in the young, the elderly, and immunocompromised patients that can result in fatalities. The importance of pathogenic *E. coli* groups also varies, as some are well-recognized pathogens and implicated in foodborne diseases worldwide while others only cause sporadic cases of gastrointestinal illness and often with unknown food source.

10.2.1. EPEC

Enteropathogenic EC was first characterized as a pathogen in 1955, and it is most commonly associated with diarrhea in infants, especially those under 2 years of age. Infantile diarrhea caused by EPEC is a prevalent problem in developing countries, and it sometimes surpasses in frequency to diarrheal infections caused by rotavirus. In developed countries or those with good hygienic standards, EPEC has become less significant and only causes sporadic cases of infection. A variety of foods (raw beef and chicken, cold pork, meat pie, coffee substitute drink, etc.) have been implicated in EPEC outbreaks worldwide, but water is also a common vehicle of infection. The use of contaminated water in the preparation of infant formulas is suspected to be a source of EPEC infection in children. The infectious dose of EPEC for infants and immunocompromised individuals is unknown; but in healthy adults, it is very high and estimated to be in the range of 10^8–10^{10}. After ingestion of the contaminated source, there is an average incubation period of 36 h, but it can range from 17 to 72 h, before the onset of illness. The predominant symptom of EPEC infection is severe diarrhea with watery stools that contains large amount of mucus, but it is seldom bloody. Other symptoms include fever, vomiting, nausea, and abdominal cramps. Infected children and adults exhibit similar symptoms, but these may be less severe and shorter lasting in adults. The illness is usually over in a few days; but in severe cases of infantile diarrhea, it can last up to 14 days. Enteropathogenic EC infections are best treated by preventing dehydration associated with the diarrheal symptoms. Some antibiotics have been shown to be effective against EPEC, but antibiotic-resistant strains are also known to exist. Infection or exposure to EPEC does build up immunity, which may account for the lower incidences of EPEC infections in older children.

10.2.2. ETEC

Enterotoxigenic EC is best known as the causative agent of traveler's diarrhea, but it is also an important diarrheal pathogen in infants. Enterotoxigenic EC infections are endemic in many developing countries in the tropics or in areas with poor hygenic standards, and infections are especially prevalent in the warm, wet months. The onset of illness is usually 26 h after ingestion of contaminant, but it can range from 8 to 44 h. Enterotoxigenic EC infection is characterized by sudden onset of watery diarrhea, without blood or mucus, and it is rarely accompanied by fever or vomiting. Illness is usually self-limiting, mild and brief, and lasting only a few days; but in some severe forms, ETEC infections can persist for up to 19 days and can resemble cholera, with up to five or more daily passages of rice water–like stools that results in severe dehydration. Antibiotic treatment is usually not required in ETEC infections but seems to be effective in reducing the duration and severity of illness. The infectious dose for ETEC is high, and data from volunteer feeding studies estimate it to be 10^8–10^{10} cells.

Often ETEC is found in feces of asymptomatic carriers; hence, humans are the most likely source of this pathogen. Enterotoxigenic EC infection does not appear to be transmitted by person-to-person contact, but some hospital infections have occurred and were probably caused by insanitary conditions. Consumption of contaminated food or water accounts for most ETEC outbreaks. In 1975, a large outbreak affecting 2000 people was traced to sewage-contaminated water at a national park. Contaminated well water in Japan and water supply aboard cruise ships have also been implicated in several ETEC outbreaks. Foodborne outbreaks of ETEC have occurred in restaurants and at various food service catered settings and implicated foods such as Brie cheese, curried turkey mayonnaise, crabmeat, deli food, salads, and others. In most of these cases, foods became contaminated with ETEC via infected food handlers or through the use of contaminated water during food preparation.

10.2.3. EHEC

The principal illness caused by EHEC is hemorrhagic colitis (HC), which is characterized by acute abdominal cramps and bloody diarrhea. In some cases, the diarrhea symptoms may be very severe, consisting of all blood and occurring every 15–30 min. Other symptoms of the disease include vomiting, but fever does not occur frequently. The incubation period before onset of illness is usually 3–4 days but can range from 1 to 9 days and the illness can last from 2 to 9 days. Approximately 3–7% of the HC infection may progress to more serious complications such as hemolytic uremic syndrome (HUS) and thrombotic thrombocytopenic purpura (TTP). Hemolytic uremic syndrome is a life-threatening sequela that affects all ages but seems to be more prevalent in children. It is characterized by acute renal failure and has a mortality rate of 3–5% or less; but many survivors suffer permanent disabilities, such as renal insufficiency and neurological deficits. Thrombotic thrombocytopenic purpura is a variant of HUS, with similar effects, but more commonly occurs in adults than in children. It also affects the central nervous system and therefore is often accompanied by fever and other neurological disorders. Treatment for EHEC infection consists mostly of administering fluids and salts to prevent dehydration, but in severe cases or HUS, dialysis and blood transfusions may be required. Antibiotic therapy has had mixed results in treating EHEC infections and, in some instances, seems to increase the patient's risk of developing HUS. One speculation for the latter may be the fact that Shiga toxin (Stx) produced by EHEC is mainly accountable for the illness, and antibiotics can lyse the bacteria to release more Stx into the host.

There are over 200 serotypes of *E. coli* that produces Stx, but not all are pathogenic for humans or regarded as EHEC (also see C.3). Various Stx-producing *E. coli* (STEC) serotypes, other than the well-known pathogenic serotypes, have been isolated from feces of healthy humans and are also found with frequency in fresh meat, poultry, seafood, and other foods that were not implicated in illness; hence the health significance of STEC is uncertain. Of those serotypes that are known to be pathogens and regarded as EHEC, serotype O157:H7 is the type strain and the one most often implicated in infections worldwide; other serotypes such as O26:H11, O111:NM, and several others have also been implicated in illness. The EHEC infections by non-O157:H7 serotypes vary from mild, nonbloody diarrhea to hemorrhagic colitis (HC), which can also progress to HUS, and in some countries, they surpass O157:H7 in frequency of infection and account for the majority of the HUS cases reported. Infections by these pathogens are mostly foodborne or waterborne or transmitted by person-to-person contact, but in many cases the source of infection could not be determined.

Serotype O157:H7 was first isolated in the United States in 1975 but was not recognized as a foodborne pathogen until 1982. Since O157:H7 accounts for almost 75% of the sporadic and outbreak cases of EHEC infections worldwide, many speculate that this serotype may be more virulent and more easily transmitted than other EHEC serotypes. Infections by O157:H7 are most commonly caused by the consumption of undercooked, contaminated ground beef or beef products, but illness caused by contaminated drinking or recreational water, raw milk, and person-to-person contact have also been well documented. Among foodborne vehicles, beef products still account for most of the O157:H7 cases; however, other foods like salad vegetables, fruits, alfalfa and radish sprouts, unpasteurized apple cider, mayonnaise, yogurt, and salami have also been implicated in recent major outbreaks, some of which affected close to 10,000 people. Many of the implicated products listed above are acidic foods that have a pH of less than 4.0, which is normally sufficient to control and inhibit bacterial growth. However, O157:H7 serotype has been shown to be fairly resistant to moderate acidity and can persist in acidity for extended periods. Cattle appear to be the primary reservoir for serotype O157:H7; however, it is also found in deer, sheep, and other animals. The infectious dose of EHEC O157:H7 in humans is estimated to be very low and in the range of 10–100 cells.

10.2.4. EIEC

Enteroinvasive EC causes bacillary dysentery, and its symptoms closely resemble that of shigellosis caused by *Shigella* species. Enteroinvasive EC infections exhibit symptoms such as chills, fever, headache, and abdominal cramps, but they are characterized by profuse, watery diarrhea that in rare cases has contained traces of blood. Symptoms usually begin 11–18 h after ingestion of contaminated vehicle, but the incubation period may be as short as a few hours up to 48 h.

Enteroinvase EC is most often implicated in outbreaks; but sporadic infections also occur, as it remains an important cause of endemic diarrhea in South America, Mexico, and some eastern European countries. First recognized in the 1940s and documented to be a diarrheal pathogen, EIEC caused an outbreak in 1947, when 47 children were infected after consuming canned salmon. It has since caused several outbreaks around the world, implicating water and various food vehicles such as imported cheese, vegetables, and potato salad. Enteroinvasive EC infections are infrequent in the United States, and it is not regarded as an important pathogen; however, one of the largest outbreaks occurred in Texas in 1992, when 370 people became infected after consuming guacamole that was prepared by a caterer. There is no known animal reservoir for EIEC; hence, humans are suspected to be the primary source. Also, the incidence of asymptomatic infections or carriers seems to be rare with EIEC; therefore, contamination of foods is most likely via infected food handlers. Several cases of EIEC infection have also been reported to occur via person-to-person transmission. Although EIEC very closely resembles *Shigella* in pathogenesis, the infective dose is higher than the 10–100 cells estimated for *Shigella*. Some human feeding studies showed that as much as 10^8 cells were required for infection, but most likely it is around 10^6. Such a high infectious dosage required tends to suggest that contact transmission may not be a common mode of spread for EIEC infections.

10.2.5. EAEC and DAEC

Enteroaggregative EC has been implicated in diarrheal illness in several countries around the world, especially in association with cases of persistent diarrhea syndrome with ≥14 days in duration. The common symptoms of EAEC infections are watery diarrhea without gross blood, afebrile or with a low-grade fever, and with little or no vomiting. There are exceptions, however, as bloody stools have been reported occasionally, and infected infants can also exhibit mucoid, liquid green stools. Enteroaggregative EC causes mostly sporadic infections, but several outbreaks have been reported in restaurants and in hospital environments. Four foodborne outbreaks of EAEC in England caused a total of 133 cases. The source of infection was traced to a meal in a restaurant, but the implicated food was not identified. Also, EAEC outbreaks of persistent diarrhea in a hospital malnutrition ward in Mexico resulted in a few fatalities, and in an outbreak in a hospital nursery in Serbia 19 children were affected. In most of these, the illness lasted only 3–9 days, but in some cases it persisted for 18–20 days. The sources of infection in these hospital outbreaks were never determined, but there seems to be a high prevalence of symptomatic carriers of EAEC. Also, like other infections, immunocompromised patients seem most susceptible.

Very little is known about the pathogenesis or illness caused by DAEC, and there is also discrepancy as to whether DAEC causes diarrheal illness. Studies performed in Chile, Brazil, and Thailand showed little correlation between the presence of DAEC in stools and illness, but other surveys done in Mexico and Bangladesh showed that DAEC was more prevalent in stool of diarrheal patients as compared to healthy controls. Another study of hospitalized patients in France showed that DAEC accounted for many of the diarrheal cases where no other gastrointestinal pathogens were found and hence suggested that DAEC was an important diarrheal pathogen in humans. The common clinical syndrome of DAEC infection is watery diarrhea without blood; however, others showed that vomiting may be more significantly linked to DAEC infections than diarrhea. Only a limited number of cases have been published on DAEC infections, and no food vehicles have been associated with illness.

10.3. DETECTION OF ORGANISM AND PATHOGENICITY TESTING

Because the various pathogenic *E. coli* groups are phenotypically diverse, it is difficult to develop microbiological methods that are specific for each group. As a result, few tests and no standard methods are available for detecting for pathogenic *E. coli* in foods. However, EHEC of serotype O157:H7 is an exception. Stimulated by the 1993 outbreak with hamburgers in the northwestern United States, great advances in assay development for O157:H7 have been made. This section examines microbiological methods, rapid methods, serological typing methods, and some pathogenicity tests for pathogenic *E. coli*. Regardless of the method format, however, it is important that the isolates are first identified biochemically as *E. coli* before they are serotyped and tested for the virulence factors that are crucial for the identification each pathogenic groups. Also included is a section that describes how the problems of testing for pathogenic *E. coli*, especially O157:H7, are being complicated by the emergence of variants that do not exhibit characteristic traits.

10.3.1. Microbiological Methods for Detecting Pathogenic *E. coli*

Like typical *E. coli*, strains in most pathogenic groups ferment lactose and are not affected by the elevated incubation temperatures; hence, they can still be distinguished from other enteric bacteria using the standard most probable number (MPN) method for *E. coli*. Exceptions, however, are EIEC, which do not ferment lactose, and EHEC O157:H7, which can be inhibited by the elevated incubation temperature used in the MPN confirmatory procedure. Elevated temperature and the detergent sodium lauryl sulfate, used in the lauryl tryptose broth, can also cause the loss of virulence-associated plasmids, which will complicate the subsequent pathogenicity testing required for identification. The MPN method will only identify the isolates as *E. coli*; hence, extensive serotyping and virulence analysis of colonies from Levine's eosin methylene blue (L-EMB) agar plate may be require to determine the pathogenic group.

Another method used for isolating pathogenic *E. coli* groups in general uses culture enrichment and selective plating. Food samples are preenriched in brain heart infusion broth for 3 h at 35°C to resuscitate stressed cells, then selectively enriched in tryptone phosphate broth at 44°C for 20 h. After enrichment, samples are plated on L-EMB or MacConkey (MAC) agar for differential isolation. For EIEC, Hektoen enteric (HE), *Salmonella–Shigella*, and MAC may be used; however, HE and MAC appear least inhibitory for the isolation of EIEC. Some pathogenic *E. coli* strains may not produce typical colonies on selective media; hence, both typical and atypical colonies need to be picked for biochemical identification that they are *E. coli*. Like the MPN method, extensive serotyping and virulence analysis of colonies from selective agar plates may be required for identification.

As with the other pathogenic *E. coli* groups, EHEC strains of serotypes other than O157:H7 are phenotypically diverse; hence the general procedures outlined above may have to be used to isolated these from foods. However, EHEC of serotype O157:H7 is phenotypically unique in that it does not ferment sorbitol (SOR) in 24 h and is negative for β-glucuronidase (GUD) activity; hence, these traits, especially the absence of SOR fermentation, are used extensively for the detection of O157:H7 in foods. There are a number of sorbitol-containing media for O157:H7; most are commercially available or they can be made from individual ingredients. Sorbitol MacConkey (SMAC) agar is the first differential plating medium developed for EHEC O157:H7 and is used most often in the analysis of clinical specimens. Prompt plating of patient's bloody stool onto SMAC has been shown to be very effective in isolating O157:H7. However, since SMAC is not very selective, it is of limited use in food analysis, where high levels of normal bacterial flora in foods often mask the presence of O157:H7 on this medium. Another sorbitol-containing, direct-plating medium developed for foods is the hemorrhagic colitis (HC) agar. Incubated at 43°C to provide some selectivity, HC agar also contains the fluorogenic substrate 4-methyl-umbelliferyl ß-D-glucuronide (MUG), which measures GUD activity and gives the medium additional differentiation. But, like SMAC, high levels of normal flora in foods can mask O157:H7 colonies on HC. Recently, the SMAC agar was made more selective with the inclusion of potassium tellurite and the antibiotic cefixime. This medium, known as TC-SMAC or CT-SMAC, is fairly selective for O157:H7 and effectively inhibits the growth of other enteric bacteria. Despite the added selectivity, however, food samples still need to be enriched prior to plating onto selective agars for isolation of O157:H7. Enrichment media often used are modified EC (mEC) or modified tryptic soy broth (mTSB) that are supplemented with novobiocin or acriflavin. But even with enrichment, foods with very high levels of normal flora can still cause overgrowth and masking problems on some selective plating agars. Another medium, the EHEC enrichment broth (EEB), developed for the analysis of carcasses uses the antibiotics vancomycin, cefixime, and cefsulodin to inhibit gram-positive bacteria, aeromonads, and *Proteus* species, respectively. The EEB is highly selective for O157:H7, especially in foods that contain high levels of normal flora. However, preliminary evidence suggests that in less contaminated foods the level of cefixime in EEB may need to be reduced or may inhibit the growth of O157:H7. Other enrichment media used include modified buffered peptone water (mBPW), which contains lactose and seems to show promise as an effective enrichment broth for isolating O157:H7 from various foods.

Sorbitol fermenting bacteria appear as large pink colonies on SMAC or TC-SMAC. The O157:H7 strains do not ferment sorbitol; so typical colonies appear pale or neutral/grey with a smoky center and are 1–2 mm in diameter. In the family Enterobacteriaceae, however, many genera include species that do not ferment SOR, including about 20% of wild-type *E. coli*, and these resemble O157:H7 colonies on sorbitol-containing media. Hence, it is important to pick at least 5–10 typical colonies and identify these biochemically as *E. coli* using miniaturized kits or conventional procedure before serotyping for the O157 and H7 antigens and for pathogenicity testing.

Recently, a number of specialty plating media have also become commercially available. These media, such as Rainbow, BCMO157, Chromagar, and others, use chromogenic substrates to detect enzymatic activities of GUD, ß-galactosidase, SOR fermentation, or a combinations of these, so that O157:H7 can be distinguished from other bacteria based on the color of the colonies. These specialty agars are useful but can be expensive, and few have been tested by collaborative studies or examined for analysis using food samples.

10.3.2. Rapid Methods for Detecting Pathogenic *E. coli*

Advancements in biotechnology have introduced many technologies that are currently being used in assays to detect for pathogenic *E. coli*. Unlike microbiological methods that use media to grow and isolate pathogens in foods, these "rapid methods" use antibodies or deoxyribonucleic acid (DNA) to detect for antigens or genes that are specific for a particular pathogenic *E. coli* group. These range from very simple latex agglutination assays that are useful for quick serological typing of pure culture isolates for the O157 and the H7 antigens to more sophisticated assays that use polymerase chain reaction (PCR) to amplify specific DNA targets.

Serotyping. Since pathogenic *E. coli* is classed based on trait virulence factors, serotyping in general does not provide a reliable or definitive identification of its pathogenic group. However, serotyping does provide useful epidemiological data, and in the case of EHEC O157:H7, typing for the somatic O157 and flagellar H7 antigens are crucial for the identification of this pathogen. A review by Nataro and Kaper (1998) shows a comprehensive listing of the major O serotypes with its associated H antigen types for each pathogenic *E. coli* group.

Sera for typing the various O and H serotypes are commercially available from several sources (Table 10.1); however, there are no specific serotyping kits for EIEC, ETEC, EPEC, EAEC, and non-O157:H7 EHEC strains. Sera are also available for EHEC O157:H7; in addition, a number of latex agglutination kits can be used to rapidly test pure culture isolates for the O157 antigen and, in some kits, the H7 antigen as well. Basically, a colony from an agar plate is used to make a suspension in a drop of anti-O157 bound, colored latex bead reagent. If the bacteria have the O157 antigen, the antibody–latex agglutinates with cells to form visible clumps within minutes. Although latex agglutination assays are rapid and simple to perform, it is important to follow manufacturer's instructions and use proper controls to eliminate incidences of autoagglutination. The use of heavy inoculum instead of a single colony can also cause false agglutinations in these assays.

In serotyping for O157:H7, a positive agglutination reaction with an anti-O157 reagent does not indicate that the isolate is O157:H7, as the reaction only indicates the presence of the O157 antigen. Many *E. coli* carry the somatic O157 antigen but are not of H7 flagellar type, and these do not produce Stx and are regarded as nonpathogenic. Furthermore, due to the similarity in the lipopolysaccharide (LPS) structure, many polyclonal anti-O157 sera can also cross react with isolates of *Escherichia hermanii*, *Salmonella* group N, *Citrobacter*, and others; therefore, it is important that O157 isolates are further tested for the H7 antigen. Sera for the H antigen are available (Table 10.1), and a few kits also provide anti-H7 latex reagent along with the anti-O157 latex. It should be cautioned, however, that H antigen expression in bacteria can be variable, and sometimes isolates from clinical or environmental samples may be nonmotile and will need to be induced for motility before H typing. Also, anti-H7 latex reagent should not be used independently of the anti-O157 reagent, as

TABLE 10.1. Selected Commercially Available Reagents and Assays for Serotyping and Detecting Pathogenic *E. coli* Groups

Target	Trade Name	Assay Format[a]	Application	Manufacturer
O antigen	—	Sera	Serotyping	Denka Seiken
		Sera	Serotyping	Difco
		Sera	Serotyping	Murex
O157	RIM	LA	Serotyping	REMEL
	E. coli O157	LA	Serotyping	Oxoid
	Prolex	LA	Serotyping	PRO-LAB
	Ecolex O157	LA	Serotyping	Orion Diagnostica
	Wellcolex	LA	Serotyping	Murex
	E. coli O157	LA	Serotyping	TechLab
	Bactrace	Serum[b]	Serotyping	Kirkegaard Perry Laboratory
		Serum	Serotyping	Difco
		Serum	Serotyping	Denka Seiken
H antigen	—	Sera	Serotyping	Denka Seiken
		Sera	Serotyping	Difco
H7	RIM	LA	Serotyping	REMEL
	Wellcolex	LA	Serotyping	Murex
		Serum	Serotyping	Difco
E. coli	Gene Trak	DNA probe	Detection	Gene Trak Systems
O157	PetrifilmHEC	Blot-ELISA	Detection	3M
	EZ COLI	Tube-ELISA	Detection	Difco
	Dynabeads	Ab-beads	Detection	Dynal
	Assurance[a]	ELISA	Detection	BioControl
	HECO157	ELISA	Detection	3M Canada
	TECRA	ELISA	Detection	TECRA
	E. coli O157	ELISA	Detection	LMD Lab
	Premier O157	ELISA	Detection	Meridian
	VIDAS	AutoELISA	Detection	bioMerieux
	EIA FOSS	AutoELISA	Detection	Foss Electric
	VIP[c]	Ab-ppt	Detection	BioControl
	Reveal	Ab-ppt	Detection	Neogen
	NOW	Ab-ppt	Detection	Binax
	QUIX	Ab-ppt	Detection	Universal Health Watch
O157:H7	BAX	PCR	Detection	Qualicon
	Probelia	PCR	Detection	Sinofi-Pasteur
	NOW	Ab-ppt	Detection	Binax
	Immunocard	Ab-ppt	Detection	Meridian
O157:H7, O26:H11	EHEC-TEK	ELISA	Detection	Organon-Teknika

[a] LA, latex agglutination; ELISA, enzyme-linked immunosorbent assay; Ab-ppt, immunoprecipitation; Ab-beads, immunomagnetic beads; auto-ELISA, automated ELISA; PCR, polymerase chain reaction.

[b] Available as unlabeled, fluorescein labeled or enzyme conjugate.

[c] Approved official first action, AOAC International.

Source: Modified from Feng (1998).

false-positive H7 agglutinations can also occur, and there are many *E. coli* that carry the H7 antigen but are not of the O157 serotype.

Detection. Pathogenic *E. coli* in the EIEC, ETEC, EPEC, and EAEC groups are not as frequently implicated in foodborne illness as EHEC; hence, few methods have been developed and there are no commercially available rapid methods for detecting these pathogenic groups in foods. Similarly, though EHEC of serotype O111, O48, O26, and others have caused foodborne illnesses worldwide, O157:H7 remains to be the predominant serotype implicated in outbreaks; hence, most detection assays target this serotype (Table 10.1). Feng (1997) reviewed the various formats used in rapid detection methods and include DNA-based tests such as PCR and DNA probes that target specific genetic sequences of O157:H7. There are also many antibody-based tests such as enzyme-linked immunosorbent assays (ELISA) and variations of ELISA that are automated or done on membranes, tubes, or dipsticks. There are immunomagnetic beads that can selectively capture O157 strains from enrichment and immunoprecipitation assays that are modeled after the home pregnancy test and can be done in minutes after enrichment and with minimal manipulation. With few exceptions, most of the antibody assays are specific only for the O157 antigen and, hence, will also detect O157, non-H7 serotypes that can be found in foods. Also, regardless of format, all detection assays for O157:H7 require cultural enrichment of food samples in broth media prior to analysis. However, even with the enrichment step, rapid methods can shorten analysis by a few days, are more sensitive and easier to perform than conventional microbiological methods, and hence are useful for screening of foods for toxins or pathogens. In screening, a negative result by rapid method is definitive, but a positive is regarded only as presumptive and must be confirmed by standard methods. Furthermore, since different rapid methods use different technologies and formats, their detection efficiency can vary from food to food. Hence, it is important that methods are validated prior to use in food testing. In the United States, only methods that have been collaboratively examined by the Association of Official Analytical Chemists (AOAC) International are regarded as official, standard methods. All rapid methods should be performed according to the manufacturers' instructions, including the use of the specified enrichment medium for the assay. Any deviations from the instructions or from the described methodology in the Official Method of Analysis of AOAC International invalidate the data or the claim made by the manufacturers. A partial listing of various rapid methods for pathogenic *E. coli*, mostly EHEC O157:H7, is given in Table 10.1.

10.3.3. Pathogenicity and Testing

With the exception of O157:H7, which is recognized worldwide based on serology alone as the prototype EHEC, definitive identification of pathogenic *E. coli* groups requires testing for the specific virulence factors characteristic of the groups. Virulence properties such as invasion, attachment, and effacing factors and production of toxins can be assayed phenotypically using tissue culture cells or antibodies or they can be tested genotypically using DNA probe and PCR assays to determine the presence of virulence genes. It should be pointed out that a positive genotypic test indicates that the bacteria carry genetic sequences for that particular virulence factor but does not indicate that the factor is actually expressed. Bacteria are known to carry genes that are not expressed due to genetic defects or physiological factors. Most tests for pathogenic *E. coli* are not commercially available except for a few antibody-based kits for testing for toxins produced by ETEC and EHEC strains.

EPEC. A model proposed for EPEC pathogenesis consists of three stages: localized adherence to epithelial cells, probably via a fimbriae structure known as the bundle-forming pilus (BFP). Adherence triggers signal transduction activity, which is encoded by genes, including *eae*, on the locus for enterocyte effacement (LEE) pathogenicity island. Then, the expression of intimin by *eae* causes intimate adherence of EPEC to cells, and the resulting attachment and effacement (A/E) lesion also causes accumulation of polymerized actin at the site of attachment. A typical EPEC, therefore, is defined as a strain that exhibits the A/E phenotype, does not produce Stx, and carries the EPEC adherence factor (EAF) plasmid, which carries the *bfpA* gene that encodes for the BFP.

The A/E phenotype can be assayed by electron microscopy or by the fluorescence actin staining (FAS) test, which stains the filamentous actin that concentrates in the HEp-2 or HeLa cells, directly beneath the sites of EPEC attachment. The *eae* gene, which encodes for intimin, is an essential component for the expression of the A/E phenotype by EPEC and also by some strains of EHEC. The intimin protein of EPEC and EHEC are 83% homologous overall, but there is more diversity in the receptor binding regions at the C terminus. The *eae* of EPEC is clustered with other virulence-associated genes in the LEE pathogenicity island on the chromosome. Various *eae*-specific probe and PCR assays have been developed, but these will often detect both EPEC and EHEC strains.

Since some EHEC strains also have *eae* and exhibit the A/E lesion, the absence of Stx production by EPEC is an important characteristic for distinguishing EPEC from EHEC. There are exceptions, however, as few rare EPEC isolates of O55:H7 serotype have been found to produce Stx2. The Stx genes in EHEC are phage encoded and potentially can be transferred laterally to other enteric bacteria (see Section 10.3.4). Procedures for testing for Stx or the *stx* genes are discussed below under EHEC.

The EAF plasmid, which encodes for BFP, is essential for the initial localized adherence of EPEC to cells. This phenotype can be assayed with the HEp-2 adherence assay, which is also the standard method for differentiating EPEC (localized adherence) from EAEC (aggregative adherence) and DAEC (diffusely adherent) strains. The presence of the EAF plasmid in bacteria can also be determined genotypically using DNA probes that are specific for the *bfpA* gene or to other markers on the EAF plasmid.

ETEC. The virulence factors that characterize the ETEC group are the production of enterotoxins and colonization factor antigens (CFAs), all of which are encoded by plasmids. The pathogenicity of ETEC resembles that of *Vibrio cholerae*, where the bacteria attaches to the intestinal mucosa via fimbrae or CFA, then produces heat-labile (LT) and/or heat-stable (ST) enterotoxin that causes the watery diarrhea symptoms. The LT is a large protein of 86 kDa and is easily inactivated at 65°C for 30 min. Within the LT class, there are two serologically distinct types. The first, LT-I, is important in causing illness in humans and animals and has two closely related subtypes or variants, designated LTh-I for human origin and LTp-I for pig origin. The other type, LT-II, is produced mostly by animal isolates of *E. coli* and has not been associated with illness. In contrast, the ST enterotoxin is a small peptide of about 2 kDa and is stable to heating at 100°C for 30 min. But like LT, there are also two distinct types of ST: STa is produced by ETEC and a few other pathogens and include the human (STh) and pig (STp) variants and STb is produced only by ETEC and limited mostly to those isolated from pigs.

The production of enterotoxins by ETEC strains used to be assayed mostly by in vivo tests. The LT toxin was detected with the Y1 adrenal cell test, which measured the rounding of the monolayer cells caused by the toxic effects of LT. Easier tests, such as the Biken test, uses anti-LT sera to cause the formation of a precipitin line when the antibody reacts with the LT secreted by the cell. Some antibody-based assays are now available commercially (Table 10.2); but since LT shares 80% homology in protein sequence to the cholera toxin (CT) of *V. cholerae*, the assays often will recognize and react with both toxins. The ST toxin used to be detected with the ligated rabbit ileal loop assay, but it was later replaced with the suckling mouse test, which measures intestinal fluid accumulations caused by the effects of the ST. Antibody-based assays, such as ELISA, are also available now to test for the production of ST (Table 10.2). The genes

TABLE 10.2. Selected Commercially Available Reagents and Assays for Detecting Toxins Produced by Pathogenic *E. coli*

Organism	Target	Trade Name	Assay Format[a]	Manufacturer
EHEC	Stx1, Stx2	VEROTEST	ELISA	MicroCarb
		Premier EHEC	ELISA	Meridian
		Verotox-F[b]	RPLA	Denka Seiken
			ELISA	LMD Laboratories
			Serum	Toxin Technology
ETEC	Labile-toxin	VET	RPLA	Denka Seiken
	Stable-toxin	ST	ELISA	Denka Seiken

[a] ELISA, enzyme-linked immunosorbent assay; RPLA, reverse passive latex agglutination.

[b] Differentiates between Stx1 and Stx2.

Source: Modified from Feng (1998).

encoding for LT and ST have been sequenced; hence DNA probes and PCR assays have also been developed to test isolates for the presence of LT and ST genes.

EHEC. The pathogencity of EHEC begins with the intimate attachment of bacteria to the intestinal epithelial cells, mediated by the intimin protein encoded by the chromosomal *eaeA* gene, but probably also involves factors encoded by a 60-mDa EHEC plasmid or pO157. Once attached, the bacteria produces potent cytotoxins, known as Shiga toxins (Stx), which inhibits protein synthesis by interfering with the functions of the 23S ribosomal ribonucleic acid (rRNA), causing cell death. The pO157 plasmid also encodes for enterohemolysin, but the role of this hemolysin in pathogenesis is uncertain.

Of the major EHEC virulence factors mentioned above, the production of Stx, especially Stx1 and Stx2, is probably the most distinguishing characteristic of EHEC that affects humans. Epidemiological evidence further shows that Stx2 is most often implicated in cases of HUS; hence, it appears to be more important in human infections. The Stx of EHEC are also known as verotoxins (VTs), because of the cytotoxic effects on Vero and HeLa cells. They were also known as Shiga-like toxins (SLTs) due to the homology of Stx1 to the Stx of *Shigella dysenteriae* type I. In EHEC, both Stx1 and Stx2 are phage encoded, but the two toxins only share 55% homology; hence antibodies to one toxin often will not recognize the other. There are also several variants of Stx2 (e.g., Stx2c, Stx2v, Stx2e), but most of these are produced by animal or environmental isolates of *E. coli*, and it is not certain that they causes illness in humans. It is estimated that there are over 200 *E. coli* serotypes that produce Stx, but many of these can be found in feces of healthy adults and have not been implicated in outbreaks. Common EHEC serotypes that cause infections include O157:H7, O111:H8, O26:H11, O103:H2, and nonmotile (NM) variants of these; but the type strain O157:H7 is still the predominant pathogen and accounts for most of the foodborne outbreaks worldwide.

Testing for Stx production or the presence of the *stx* genes is the most reliable means to identify EHEC. Stx used to be assayed by observing for its cytotoxic effects on HeLa or Vero cells; but anti-Stx antibodies and *stx*-specific probes or PCR assays are now available to test isolates for toxigenic potential. Similarly, the A/E lesion caused by intimin and encoded by the *eae* gene can now be assayed by PCR instead of having to use HEp-2 or HeLa cell monolayers. At least a dozen PCR primers and methods have been published for detecting *stx1* and *stx2* genes. Some of these are multiplex PCR assays, which combine primers specific for the *stx* genes with those of *eae* (intimin), *ehxA* (enterohemolysin), pO157, or mutation in *uidA* gene unique to O157:H7, to allow simultaneous detection of serotype and/or several virulence factors. These assays are very effective in testing pure culture isolates; however, they are less efficient in the analysis of foods, due to interference by food matrices and the presence of inhibitors and normal microbial flora.

There are also several commercially available kits that use anti-Stx antibodies to detect for the production of Stx1 and Stx2. Some are simple, reverse passive latex agglutination (RPLA) tests that can differentiate the two toxins and determine the titer of the toxin produced. But most use sandwich ELISA format, where anti-Stx and globotriaosyl ceramide (Gb3), the cellular receptor for Stx, are used to capture or immobilize Stx, followed by detection using an enzyme-conjugated anti-Stx antibody (Table 10.2). These assays that target Stx or *stx* genes will identify all EHEC strains that produce Stx or are potentially toxigenic; hence, they are not specific for O157:H7 or any other EHEC serotype. Also, assays that are specific for *stx1* or Stx1 will also react with the Stx or the gene of *S. dysenteriae* type I.

EIEC. Enteroinvasive EC isolates are characterized by their ability to invade epithelial cells. This pathogenic *E. coli* group closely resembles *Shigella*, and both also share similar virulence properties. The invasion phenotype is encoded by a plasmid, designated pInv; but its expression appears to also require regulatory genes that reside on the chromosome. The size of pInv can vary from 120 mDa in *Shigella sonnei* to 140 mDa in EIEC and other *Shigella* species, but the invasion gene on the plasmid is identical. The pathogenesis of EIEC and *Shigella* begins with cellular invasion via endocytic vacuoles; once internalized, the vacuoles lyse and the bacteria multiplies intracellularly to spread laterally to other cells. Both EIEC and *Shigella* also produce an enterotoxin, which may be involved in causing the watery diarrhea that precedes the dysentery symptoms associated with these pathogens. Since the invasion phenotype and genotype in EIEC and *Shigella* species are identical, the HeLa cell assays and the Sereny or keratoconjunctivitis test that uses guinea pig eye can be used to assay for the invasion phenotype in both pathogens. Similarly, *inv*-specific DNA probes or PCR assays will detect both but will not differentiate between the two pathogens. It is therefore important that isolates are identified biochemically as *E. coli*, before testing for invasiveness. The absence of Stx production by EIEC can also be a useful marker, but limited only for distinguishing EIEC from *S. dysenteriae* type I, which produces Stx. All other *Shigella* species are invasive but do not produce Stx.

EAEC and DAEC. The pathogenesis of diarrheal illness for these *E. coli* groups is not well understood. But since both are defined based on cellular adherence patterns, the HEp-2 adherence assay is the standard phenotypic test to distinguish these groups. The aggregative adherence of EAEC to HEp-2 cells is mediated by an adhesin, known as aggregative adherence fimbrae I (AAF/I). Adherence results in increased mucous secretion, which forms a biofilm to entrap the bacteria onto the small intestinal epithelium. The AAF/I factor appears to be encoded by large plasmid and probes, and PCR assays specific for the plasmid have been developed that can be used to identify EAEC isolates. The EAEC strains also produce a cytotoxin and an ST-like toxin designated EAST1. The role of these toxins in EAEC pathogenesis remains to be determined.

The diffusely adherent pattern of DAEC on HEp-2 cells is also mediated by a fimbria, designated F1845, but the genetic locus seems to be both chromosomal and plasmid encoded. A *daaC* gene, which encodes for a component essential in the expression of the F1845 fimbrae, has been identified and a *daaC* gene-specific DNA probe has been shown to react with 75% of the DAEC isolates tested.

10.3.4. Variants of Pathogenic *E. coli*

Previous sections described various methods used to test for pathogenic *E. coli*. Most of these assays target virulence factors or, in the case of EHEC O157:H7, specific phenotypic characteristics of this pathogen. However, reliance on few traits, such as the absence of SOR fermentation or GUD activity, for selection of O157:H7 can be risky, as many enteric bacteria, such as *E. hermanii* and *E. hafnia*, do not ferment SOR and are also GUD negative and hence resemble O157:H7 colonies on SOR media. Furthermore, SOR-positive and GUD-positive phenotypic variants of O157:H$^-$ serotype, which produce Stx2, have been isolated from HUS patients in Germany and are fairly prevalent in Central Europe. These phenotypic variants of O157:H7 are not easily detected using media that differentiate based on these traits.

Serotyping for the O157 and the H7 antigens are crucial for the identification of the O157:H7 serotype, but this is complicated by the existence of motility and serotypic variants in the environment. The mechanism of motility in *E. coli* is under complex regulatory control and is affected by environmental factors; hence, strains may become NM due to physiological or genetic factors. The isolation rate for NM strains of O157 serotype has increased from 6% in 1990 up to 47% in 1996. Since NM strains cannot be typed for the H7 antigen, their health significance is uncertain, as some may be NM variants of O157, non-H7 strains that are not pathogenic. However, genetic analysis of O157 NM strains from around the world showed that the majority of O157 NM variants examined that produced Stx are actually of the pathogenic H7 serotype and therefore important in disease. Similarly, an O rough strain of O157:H7 that does not make the O157 antigen has been isolated from beef in Asia. This serotypic variant was genetically identical to O157:H7, carried all the trait virulence factors, and even had genetic sequences essential for the synthesis of the O157 antigen; however, it did not express the O157 antigen. These motility and serotypic variants are not identified by routine serological assays.

Lastly, variants of O157:H7 that do not produce Stx have also been isolated from various sources worldwide. The genes for Stx-1 and Stx-2 are phage encoded, and in fact, O157:H7 is thought to have emerged evolutionarily from its closest genetic relative, EPEC O55:H7, via infection of *stx*-bearing phages and the acquisition of other virulence markers. Since phages may be acquired, lost, or exchanged within a population, it is not surprising that non-*stx*-producing variants of O157:H7 exist. This would also account for the emergence of *stx2*-producing strains of *Citrobacter freundii* and *Enterobacter aerogenes* that have been implicated in illness. Sewage has been shown to contain high titers of *stx2*-bearing phages and may be the reservoir for these toxin-encoding phages that infects EHEC and the enteric population.

10.4. PHYSICAL METHODS FOR DESTRUCTION

Generally, pathogenic *E. coli* has similar growth and survival characteristics as nonpathogenic *E. coli* and hence are inactivated by similar disinfectants and physical control measures. Because of the importance of EHEC O157:H7 in foodborne illness, most inactivation studies have been done with this pathogen. EHEC O157:H7 seeded into ground beef has the ability to survive up to 9 months in frozen storage at −20°C. In meats, including poultry, *Salmonella* and *Campylobacter* are usually inhibited by 2–4% NaCl; but EHEC O157:H7 will tolerate up to 8% salt, and its survival and growth at low temperatures may even be enhanced by some of the commonly used food additives. However, *E. coli* is not particularly heat resistant, and thermal inactivation studies showed that EHEC O157:H7 in ground beef had a D value of only 9.6 s at 64.2°C and was even more easily inactivated than *Salmonella*. Cooking and pasteurization of foods therefore remain the most effective measures to control pathogenic *E. coli* in foods.

Although the organism may be heat sensitive, the toxin produced by EHEC is more stable. Studies showed that purified Stx1 can withstand heating for 60 min at 75°C and up to 5 min at 80°C, suggesting that the toxin can survive the cooking conditions prescribed for ground beef. There is, however, little evidence that Stx are produced in foods or that ingestion of preformed Stx in foods can causes illness.

Other studies have examined the use of modified atmosphere packaging or the use of pH extremes to control growth of pathogenic *E. coli* in foods. Fresh produce seeded with EHEC O157:H7 and stored at various temperatures and under a combinations of mixtures of CO_2, O_2, and N_2 showed that a modified atmosphere had no inhibitory effect. The pathogen grew at 10 and 20°C and survived refrigerated storage at 5°C. At 8% concentration, trisodium phosphate (TSP) has a pH of >12 and the high alkalinity effectively lyses gram-negative bacteria. Trisodium phosphate is approved by the U.S. Department of Agriculture (USDA) for use as a postchill antimicrobial treatment for raw poultry, but it has also been shown to be effective in reducing EHEC O157:H7 populations in beef carcasses and potentially may be applicable to other foods. Traditionally, acidity has also been used as an effective means to control microbial growth in foods. However, some foodborne pathogens such as *Shigella* have been shown to be very resistant to acidity. Acid tolerance is thought to be an important trait of pathogenic bacteria as it enables them to pass through the gastric barrier in the digestive tract to cause disease. Little information is available on the acid resistance properties of most pathogenic *E. coli* groups; however, EIEC and EHEC are known to tolerate acidity. Several EHEC serotypes have been found to be acid resistant, but the ability of serotype O157:H7 to survive in moderate acidic conditions is especially well documented as it has been implicated in many outbreaks involving acidic foods, such as apple cider and juices, mayonnaise, yogurt, and fermented sausages. Enterohemorrhagic EC O157:H7 can survive up to 56 days at pH ≥ 4.0; hence, acidity is not effective in controlling the growth of this pathogen in foods. Likewise, the use of hot acid sprays has been ineffective in removing EHEC O157:H7 from beef carcasses. Bacterial acid resistance is genetically encoded, and cells can be induced to acid resistance by exposure to low-pH conditions. Properties of acid tolerance also seem to be linked to other factors, as induction to acid resistance also increases the organism's ability to resist other stress factors such as antimicrobial agents, heat, UV light, and irradiation.

Like other bacteria, pathogenic *E. coli* are relatively sensitive to irradiation and can be eliminated from foods by low-dose treatments. Studies showed that a dosage of 2–3 kGy of gamma irradiation is sufficient to decontaminate raw meats and poultry of all foodborne pathogens, including EHEC O157:H7. However, the effectiveness and the dosage of irradiation required may vary, depending on the temperature of the product during irradiation (i.e., fresh, refrigerated, or frozen) as well as the physiological state of the organism (i.e., induced to acid resistance state).

10.5. PREVENTION AND CONTROL

Pathogenic *E. coli* is transmitted via the fecal–oral route; hence, most outbreaks are caused by the consumption of contaminated food and water. A large variety of foods have been implicated in infections of EPEC, ETEC, EIEC, and EHEC worldwide. Little is known about the source of EAEC and DAEC, and no food vehicles have been reported to be associated with infections, but these two groups are thought to be disseminated by contact. Transmission of secondary infections via person-to-person contact has also been reported to occur for other pathogenic *E. coli* groups and is well documented for EHEC. A number of cases in outbreaks have been traced to contact transmission, and such infections appear to be especially common in nursing homes and day care centers, where good personal hygiene is not always observed. In these settings, incidences of pathogenic *E. coli* infection by contact transmission can be greatly reduced by emphasizing good personal hygiene and proper hand-washing techniques. Children or patients with symptoms of diarrhea should be closely monitored to prevent cross-contamination.

Enterohemorrhagic EC O157:H7 has been isolated from the intestinal tract of cattle, sheep, deer, and other animals, but cattle appear to be the primary reservoir for this pathogen. No animal reservoirs have been reported for other pathogenic *E. coli* groups; hence, infected patients and human carriers are believed to be the main source of these other pathogenic *E. coli* in infections. Foods can become contaminated with pathogens in the field or from improper sanitation, food handling, or processing. The soil flora, the use of manure or sewage-contaminated water for irrigation of vegetable fields, and droppings from wild animals and birds have all been suspected as sources of bacteria and pathogens on fresh produce. Several outbreaks of pathogenic *E. coli* have been linked to the consumption of produce such as lettuce, sprouts, and other vegetable. Proper washing and rinsing of produce prior to consumption may not eliminate the risk entirely but have been shown to greatly reduce bacterial populations on produce. Meat and dairy products can become contaminated with EHEC O157:H7 through contact with manure during slaughter or milking. If the food is not properly processed or handled, the pathogen may cross-contaminate other products. Likewise, in the preparation of foods, infected and/or asymptomatic food handlers with poor personal hygiene can contaminate foods. Hence, to prevent potential transmission of pathogenic *E. coli* from carrier to foods or from food to food, consumers and food handlers at institutions or restaurants should know how to properly handle and store foods and should thoroughly wash hands with hot soapy water prior to handling foods. Counter tops, cutting boards, utensils, and dishes should be cleaned and washed in between foods, especially in between preparation of raw meats and salad vegetables, which are often consumed without cooking. Also, food handlers that exhibit gastrointestinal symptoms should not be preparing meals, serving foods, or handling clean plates and utensils that are used in serving.

Lastly, infections of pathogenic *E. coli*, especially EHEC O157:H7, have most often been associated with the consumption of inadequately heat processed foods, such as undercooked meats and unpasteurized milk and fruit juices or by the cross-contamination of cooked foods with underprocessed foods. To reduce these incidences, cooked foods should never be placed on plates that previously held raw meat, poultry, or seafood or come in contact with raw meat juices. More importantly, since pathogenic *E. coli* is not particularly resistant to heat, it is essential that foods are adequately cooked at the proper temperature. After the 1993 hamburger outbreak in the northwestern United States, federal guidelines were modified to require that ground beef be cooked at 160°F and also recommended the use of instant-read thermometers to ensure that the ground meat has been adequately processed before consumption. Similarly, the majority of juices in the United States are now pasteurized, and those that are not are required to carry a label warning that the product is not heat processed and may have some risk from microbial contamination.

REFERENCES

Feng, P. 1997. Impact of molecular biology on the detection of foodborne pathogens. *Mol. Biotechnol.* **7**, 267–278.

Feng, P. 1998. Rapid methods for detecting foodborne pathogens. In: *FDA Bacteriological Analytical Manual*, 8th ed. (rev. A). Gaithersburg, MD: AOAC International. pp. 1.01–1.16.

BIBLIOGRAPHY

Aleksic, S., H. Karch, and J. Bochemuhl. 1992. A biotyping scheme for *Shiga*-like (Vero) toxin-producing *Escherichia coli* O157 and a list of serological cross-reactions between O157 and other gram-negative bacteria. *Zentr. Bakteriol.* **276**, 221–230.

Ewing, W. H. 1986. *Edwards and Ewing's Identification of Enterobacteriaceae*, 4th ed. New York: Elsevier.

Doyle, M. P., and V. V. Padhye. 1989. *Escherichia coli*. In: M. P. Doyle (Ed.), *Foodborne Bacterial Pathogens*. New York: Marcel Dekker.

Feng, P. 1995. *Escherichia coli* serotype O157:H7: Novel vehicles of infection and emergence of phenotypic variants. *Emerg. Infect. Dis.* **1**, 47–52.

Feng, P., K. A. Lampel, H. Karch, and T. S. Whittam. 1998. Phenotypic and genotypic changes in the emergence of *Escherichia coli* O157:H7. *J. Infect. Dis.* **177**, 1750–1753.

Griffin, P. M., and R. V. Tauxe. 1991. The epidemiology of infections caused by *Escherichia coli* O157:H7, other enterohemorrhagic *E. coli*, and the associated hemolytic uremic syndrome. *Epidemiol. Rev.* **13**, 60–98.

Hill, W. E., A. R. Datta, P. Feng, K. A. Lampel, and W. L. Payne. 1998. Identification of foodborne bacterial pathogens by gene probes. In: *FDA Bacteriological Analytical Manual*, 8th ed. (rev. A). Gaithersburg, MD: AOAC International, pp. 24.01–24.33.

Hitchins, A. D., P. Feng, W. D. Watkins, S. R. Rippey, and L. A. Chandler. 1998. *Escherichia coli* and the coliform bacteria. In: *FDA Bacteriological Analytical Manual*, 8th ed. (rev. A). Gaithersburg, MD: AOAC International, pp. 4.01–4.29.

Jay, J. M. 1996. Foodborne gastroenteritis caused by *Escherichia coli*. In: J. M. Jay (Ed.), *Modern Food Microbiology*. New York: Chapman & Hall, Chapter 24.

Johnson, R. P., R. C. Clarke, J. B. Wilson, S. C. Read, K. Rahn, S. A. Renwick, K. A. Sandhu, D. Alves, M. A. Karmali, H. Lior, S. A. Mcewen, J. Spika, and C. L. Gyles. 1996. Growing concern and recent outbreaks involving non-O157:H7 serotypes of verotoxigenic *Escherichia coli*. *J. Food Prot.* **59**, 1112–1122.

Karmali, M. A. 1989. Infection by verocytotoxin-producing *Escherichia coli*. *Clin Microbiol Rev* **2**, 15–38.

Meng, J.-H., P. Feng, and M. P. Doyle. 2001. Pathogenic *Escherichia coli*. In: *Compendium of Methods for the Microbiological Examination of Foods*. Washington, DC: American Public Health Association.

Nataro, J. P., and J. B. Kaper. 1998. Diarrheagenic *Escherichia coli*. *Clin. Microbiol. Rev.* **11**, 142–201.

Orskov, F., and I. Orskov. 1992. *Escherichia coli* serotyping and disease in man and animals. *Can. J. Microbiol.* **38**, 699–704.

Paton, J. C., and A. W. Paton. 1998. Pathogenesis and diagnosis of Shiga toxin-producing *Escherichia coli* infections. **11**, 450–479.

Tarr, P. I. 1995. *Escherichia coli* O157:H7: Clinical, diagnostic, and epidemiological aspects of human infections. *Clin. Infect. Dis.* **20**, 1–10.

CHAPTER 11

Salmonella

JEAN-YVES D'AOUST

11.1 INTRODUCTION

Salmonella spp. occur widely in the natural environment and in different sectors of the global food chain. The ability of these microorganisms to survive under adverse conditions and to grow in the presence of low levels of nutrients and at suboptimal temperatures and pH values presents a formidable challenge to the agricultural and food-processing industries in marketing safe products. The continued prominence of raw meats, eggs, and dairy products as the principal vehicles of human foodborne salmonellosis arises from major difficulties to coordinate sectorial control efforts within each industry. New foods such as vegetable sprouts, fresh fruits, and fruit juices have been incriminated in recent years as vehicles of human *Salmonella* infections. The problem of salmonellosis is further compounded by the massive and unrestricted movement of foods in international trade, the national disparities in the hygienic agricultural and aquacultural production of foods, and the nonuniform government and industry food safety controls during the processing, distribution, and marketing of fresh and processed food products. The emergence and persistence of highly virulent and antibiotic-resistant *Salmonella* strains in recent years are major public health concerns. The relentless human pandemic of *S. enteritidis* from the consumption of poultry meat and shell eggs and the increasing prevalence of the pentavalent antibiotic-resistant *S.* Typhimurium DT 104 in meat animals and in humans are disquieting. The use and misuse of antibiotics in clinical medicine, in prophylactic drug regimens in meat animals and aquacultural species, and in selected horticultural crops have led to a plethora of antibiotic-resistant *Salmonella* in human populations and in the global food chain.

Guide to Foodborne Pathogens, Edited by Ronald G. Labbé and Santos García
ISBN 0-471-35034-6

TABLE 11.1. Taxonomy of *Salmonella*

Species	Subspecies	Number of Serovars
S. enterica	*enterica*	1443
	salamae	488
	arizonae	94
	diarizonae	323
	houtenae	70
	indica	11
S. bongori		20
Total		2449

Source: Ref. 46.

11.2. NATURE OF ILLNESS

The genus *Salmonella* is divided into two species, *S. enterica* and *S. bongori*, and currently encompasses a total of 2449 serovars (Table 11.1). *Salmonella enterica* is further subdivided into six subspecies where serovars commonly isolated from humans, agricultural products, and foods generally belong to the subspecies *enterica*. The type strain for the genus is *S. enterica* subsp. *enterica* serovar Typhimurium LT2. All *Salmonella* strains are potentially pathogenic to humans.

Two clinical syndromes are associated with human salmonellosis, enteric fever and enterocolitis. The ingestion of viable salmonellae from food or water constitutes the principal route of human and animal salmonellosis. Enteric fever follows from the invasion of *S.* Typhi or *S.* Paratyphi A, B, or C in human host tissues and is characterized by watery diarrhea, prolonged and spiking fever, nausea, and abdominal cramps. The clinical management of these symptoms, which appear 7–28 days following exposure to the infectious agent, requires agressive medical therapy to circumvent serious to life-threatening sequelae. A chronic carrier state can follow the acute phase of the disease whereby the organism lodges for lengthy periods of time in the gall bladder, liver, or biliary tree and escapes inactivation by host humoral and cell-mediated defense mechanisms. The intermittent shedding of viable salmonellae in the stools of chronic carriers potentiates secondary human infections and cross-contamination of foods by infected food handlers. Diagnosis of enteric fever relies on symptomatology and isolation of the pathogen from blood, bone marrow, or stool specimens. The efficacy of traditional ampicillin (A), chloramphenicol (C), or trimethoprim sulfamethoxazole (SXT) treatments of enteric fever is rapidly falling into disfavor because of the increasing resistance of salmonellae to these antibiotics (Table 11.2). Although a gradual shift in therapy toward third-generation cephalosporins and fluoroquinolones has tempered the severity of this global medical problem, increasing reports on the resistance of *Salmonella* and other bacterial pathogens to fluoroquinolones (Table 11.2) are disquieting. This emerging problem is compounded by the observation that a

TABLE 11.2. Antibiotic Resistance in Human Isolates of *Salmonella* spp.

Country	Period	Number of Strains Tested	Serovar	Antibiotic Resistance (%)[a]							Reference
				A	C	TP[b]/SXT[c]	Su	G	Cip	Multiple[d]	
Crete	1992–1994	491	Multiple	36.4	4.7	3.5[c]	NL	0.4	NL	NL	51
Denmark	1993	228	*S.* Typhimurium	12.0	8.0	4.0[c]	12.0	<1.0	0.0	12.3	53
England/Wales	1995	3837	*S.* Typhimurium	87.5	87.2	33.1[b]	87.2	NL	6.2	87.2	55
Ethiopia	1993–1996	110	Multiple	36.4	35.5	51.8[c]	NL	20.0	NL	41.8	57
France	1997	2153	Multiple	42.3	27.1	7.0[c]	NL	1.5	0.4	NL	18
Iran	1994	28	*S.* Typhi	89.3	67.9	64.3[c]	NL	NL	NL	85.7	11
Ireland (Republic)	1996	181	Multiple	28.0	24.0	11.0[b]	41.0	NL	NL	NL	21
Italy	1989–1994	242	*S.* Typhimurium	28.9	19.3	9.4[b]	NL	5.7	0.0	NL	50
Thailand	1994	710	Multiple	19.3	20.4	31.1[c]	NL	13.2	NL	NL	16
Tunisia	1989–1995	46	*S.* Enteritidis	6.5	2.1	0.0[c]	NL	0.0	NL	NL	12
		18	*S.* Wien	100.0	94.4	38.8[c]	NL	94.4	NL	NL	
United States	1995	4042	Multiple	15.0	8.0	3.0[c]	24.0	3.0	0.0	23.0	2
Vietnam	1993–1994	294	*S.* Typhi	72.2	80.0	74.0[c]	NL	NL	NL	NL	37

[a] Ampicillin (A); chloramphenicol (C); sulfonamides (Su); gentamicin (G); ciprofloxacin (Cip). NL = not listed.

[b] Trimethoprim.

[c] Trimethoprim sulfamethoxazole.

[d] Resistant to ≥ 2 antibiotics that may not be listed above.

bacterial strain that acquires resistance to one quinolone may also exhibit cross-resistance to other quinolones. The present lack of effective replacement drugs for fluoroquinolones could rapidly lead to a serious international medical crisis. The emergence of ciprofloxacin-resistant *S.* Typhimurium DT 104 in England and Wales in recent years enhances public health concerns because this prominent epidemic strain commonly carries resistance determinants for ampicillin, chloramphenicol, streptomycin, sulfonamides, and tetracycline. The prominence of antibiotic resistance in human *Salmonella* isolates (Table 11.2) and rapid increases in the number of multiple antibiotic-resistant salmonellae likely follow from the overprescription of drugs by physicians, the over-the-counter public access to antibacterials in many countries, the incomplete patient use of prescribed drugs and storage of unused portions for future use, and the acquiescence of family practitioners to parent pressures to treat children with antibiotics. Human foodborne infections with antibiotic-resistant *Salmonella* also contribute to this alarming epidemiological trend. The widespread occurrence and interspecies exchange of bacterial resistance (R) plasmids that encode for single or multiple resistance factors in carrier strains heighten public health concerns.

Enterocolitis is usually a self-limiting clinical condition that appears 8–72 h following exposure to nontyphoid salmonellae with remission within 4–5 days following the onset of overt disease. Newborns, the elderly, and individuals with immune deficiencies are particularly susceptible to *Salmonella* infections, which can degenerate into serious systemic infections or fatal outcomes. The propensity for the rapid spread of salmonellae in hospital nurseries, intensive care units, geriatric wards, and nursing homes is notable. Enterocolitis is usually characterized by severe abdominal pain, diarrhea, vomiting, and fever. Treatment of uncomplicated cases of salmonellosis is limited to fluid and electrolyte replacement. Antibiotic therapy is contraindicated because it tends to prolong the convalescent carrier state that commonly follows the acute phase of the disease. Nonetheless, nontyphoid *Salmonella* strains can be highly invasive and engender serious systemic infections that require timely and effective antibiotic treatment. Human nontyphoid salmonellosis can also lead to chronic autoimmune diseases that appear weeks to months after the acute phase of infection and the destruction of viable salmonellae by host defense mechanisms. For example, reactive arthritis involves the chronic inflammation of articulated joints in the upper and lower limbs, a condition that is putatively triggered by fragments of *Salmonella* cells that lodge in the articulations and induce a local inflammatory response. There is evidence that humans carrying the human leucocyte antigen (HLA) B27 may be predisposed to reactive arthritis and other chronic arthropathies.

Recent years have witnessed the commercial development of several human typhoid vaccines that are gradually replacing the classical heat-inactivated, phenol-preserved, whole-cell vaccines. The Vivotif Berna oral typhoid vaccine is manufactured in Switzerland and is based on an avirulent strain of *S.* Typhi lacking the capsular Vi antigen and other important genetic determinants for bacterial virulence. The vaccine consists of three enteric-coated capsules each taken on alternate days and provides protection for up to 7 years with booster doses recommended at 3-year intervals. An injectable phenol-preserved Vi capsular vaccine manufactured in France under the trade name Typhim Vi protects against *S.* Typhi for 3 years with low to moderate levels of adverse reactions. The growing importance of *Salmonella* as a human disease agent and a major contaminant of the global food chain supports the development of new human and animal vaccines using attenuated *Salmonella* strains that are auxotrophic in biosynthetic pathways or that lack key virulence determinants. However, the great diversity and rapid succession of serovars in foods and the limited specificity of vaccines for a single or a few closely related serovars pose a major challenge to the development of effective vaccines for meat animals. Performance-tested vaccines have been proposed for the abatement of *S.* Typhimurium in bovine and *S.* Enteritidis, *S.* Gallinarum, and *S.* Heidelberg in poultry.

Virulence factors encoded in the *Salmonella* chromosome or in plasmids contribute to strain pathogenicity. Such factors protect against the acidicity of the human stomach and the cytoplasm of phagocytic leucocytes and favor the bacterial colonization of the intestinal tract and the spread of invasive salmonellae to extraintestinal tissues through vascular and lymphatic conduits. Although a detailed discussion of the molecular intricacies of these virulence determinants is beyond the scope of this chapter, a brief description of key factors is indicated. A region of the *Salmonella* chromosome (pathogenicity island 1) contains gene clusters encoding determinants for the invasion of intestinal epithelial cells (enterocytes). Of these, the invasion (*inv*) genome, consisting of no fewer than 15 genes (*invA* to *invO*), encodes structural and enzyme proteins that facilitate the *Salmonella* penetration of enterocytes. Upon the intimate contact of *Salmonella* with a targeted enterocyte, the *invC*, *invG*, *invI*, and *invJ* genes are activated toward the formation of essential but transient surface appendages (invasomes). Concurrently, the *invE* gene encodes an unknown factor (possibly a secretory protein) that directs enterocyte uptake of lumenal Ca^{2+}. These events lead to the evagination of the enterocyte outer membrane (membrane ruffles), which engulfs and internalizes *Salmonella* by macropinocytosis. The foregoing is an example of signal transduction where molecules secreted by the invasive microorganism direct profound structural changes in targeted host cells. Another important *Salmonella* virulence factor is the two-component chromosomal *phoP*/*phoQ* transcriptional regulator system that encodes factors for protection against host defensins (bactericidal cationic peptides) and for enhanced bacterial survival in unfavorable acidic environments. The *phoQ* gene product is a protein located in the outer bacterial membrane that detects hostile acidic conditions in the bacterial environment and triggers the transcription of genes encoding protective outer membrane proteins (OMPs). Virulence plasmids, autonomous deoxyribonucleic acid (DNA) organelles that occur in low copy numbers in the bacterial cytoplasm, encode factors that promote the systemic spread and survival of salmonellae in host tissues. The distribution of virulence plasmids within *Salmonella* spp. is limited and notably is absent in the highly infectious and host-adapted *S.* Typhi. Of the many genetic loci found on this organelle, the *Salmonella* plasmid virulence (*spv*) region is noteworthy because it affords protection against the hostile cytoplasm of phagocytic leucocytes and enables salmonellae to cope with conditions of iron and nutrient deprivation, elevated temperatures, and highly acidic environments. The armamentarium of *Salmonella* virulence factors also includes siderophores that compete with host cells for the limiting amounts of essential iron in host tisssues, OMPs that restrict the entry of antibiotics into the bacterial cytoplasm and help maintain homeostasis under potentially lytic osmolar conditions, and the capsular Vi antigen in typhoid and paratyphoid C strains that protect against lysis by the host complement system. Toxins also figure prominently as virulence determinants. The diarrheagenic *Salmonella* enterotoxin is a thermolabile protein encoded by the chromosomal *stn* gene that is carried by most strains of *S. enterica* but not *S. bongori*. Thermolabile cytotoxins are located in the bacterial outer membrane and putatively contribute to virulence through inhibition of protein synthesis and lysis of host cells, thereby facilitating the spread of the pathogen into deeper host tissues. *Salmonella* endotoxin is also located in the bacterial outer membrane and consists of polysaccharide and lipid A moieties. The lytic release of lipid A into host cells triggers a leucocyte-dependent inflammatory response.

11.3. CHARACTERISTICS OF *Salmonella* spp.

11.3.1. Growth

Salmonella grows in the temperature range of 2–47°C with rapid growth between 25 and 43°C. The minimum temperature for growth prevails at neutral pH and increases sharply with increasing acidity or alkalinity of the suspending medium. The ability of salmonellae to grow at common refrigerator temperatures (4–10°C) underlines the importance of rapid turnover of perishable foods and the need to separate raw from cooked foods during refrigerated storage. Some food products are packaged under vacuum or gaseous mixtures of CO_2, N_2, and O_2 to increase product quality and extend shelf life. The benefits of modified atmosphere packaging (MAP) stem from the inhibition of endogenous spoilage microflora and enhanced organoleptic stability of the product. However, MAP inhibition of putrefactive microorganisms also removes the natural indicators of product abuse and potential human health hazards. The high concentrations of CO_2 (>50% v/v) commonly used in MAP inhibit the growth of *Salmonella* spp. but exert little or no effect on survival. At concentrations of $\geq$3% (w/v), NaCl generally inhibits the growth of salmonellae. Increasing temperatures in the range of 10–30°C increase the salt and acid (pH $\geq$ 5.0) tolerance of the organism, whereas cold temperatures reinforce salt- and acid-dependent bacteriostasis. *Salmonella* also grows at low water activities ($a_w \geq 0.93$) where the response is strain and food dependent. *Salmonella* actively grows in the pH range of 3.6–9.5 and optimally at near-neutral pH values. Bacteriostasis is acidulant dependent, as evidenced by the greater inhibition with acetic than with equimolar citric or lactic acids. Although fermentation of dairy and meat products enhances food safety by inhibiting the growth or eliminating bacterial pathogens through the formation of organic acids by starter cultures or native microflora, slow product acidification could induce an adaptive pathogen response toward greater acid tolerance and possible growth under normally inhibitory acidic conditions.

11.3.2. Survival

The propensity of *Salmonella* spp. to survive for long periods of time in hostile environments impacts greatly on the public health safety of the global food supply. Human salmonellosis from the ingestion of very few viable cells contrasts sharply with the need for high numbers of other foodborne bacterial pathogens to engender human disease. The persistence of *Salmonella* at freezer temperatures is strain dependent and varies with food composition. Freezer storage at −20°C or lower favors greater bacterial survival than storage at near-freezing temperatures. The protective role of induced cold-shock proteins in temperature-stressed salmonellae has yet to be fully elucidated. The survival of *Salmonella* in low-moisture foods increases with decreasing water content (a_W) and decreases with increasing storage temperature.

The ability of salmonellae to trigger protective responses in acidic, acidified, and fermented foods is of major concern in food safety assessments. Although bacterial lethality generally increases with increasing acidity, slow acidification can condition salmonellae to a state of enhanced acid tolerance and survival. Cold temperatures tend to reduce the antibacterial action of acids as evidenced by the greater lethality of *Salmonella* in acidified mayonnaise stored at room than at refrigerator temperatures. Cells grown at 37–43°C are more acid resistant than cells grown at ambient temperature. In addition to a constitutive homeostasis system that maintains cytoplasmic pH between 6.0 and 7.0 at external pH values of $\geq$4.0, *Salmonella* also possesses other mechanisms collectively termed acid tolerance responses (ATRs) for protection against highly acidic environments through the induced synthesis of acid-shock proteins. Different ATR responses are activated in the logarithmic and stationary phases of growth. Short exposure of log-phase cells to pH 4.3 triggers either a transient or a sustained ATR response, enabling salmonellae to survive under normally bactericidal conditions (pH 3.0). The sustained response affords a more effective and rigorous protection of cells. Two other ATR responses that operate in stationary-phase cells are responsible for the synthesis of up to 15 acid-shock proteins. In addition to a general acid-independent stress response that is activated when cells enter the stationary phase of growth, another highly protective, acid-induced mechanism operates under extreme acidic conditions. Interestingly, these acid tolerance responses also confer cross-protection against thermal, osmotic, and oxidative stress conditions. Salmonellae are highly susceptible to extreme alkaline conditions (pH $>$ 10.0) where elevated temperatures amplify the bactericidal effect.

11.4. EPIDEMIOLOGY

The wide variations in the national prevalence of human salmonellosis likely arise from inequalities in national reporting systems, the endemicity of salmonellae in certain countries, poor hygienic practices in third world countries, and lack of adequate medical and epidemiological resources to limit the spread of human disease outbreaks. The last decade has witnessed marked increases in the incidence of human *Salmonella* infections. For example, rates per 100,000 population increased from 27.0 to 70.3 in Denmark (1985–1995), 103.0 to 175.5 in Germany (1989–1993), and 32.2 to 69.0 in Sweden (1980–1991). In the mid-1980s, *S.* Enteritidis, which had already pervaded the poultry and egg industries in many countries, challenged the prominence of *S.* Typhimurium as the most common agent of human salmonellosis. The continuing human pandemic of *S.* Enteritidis stems, in part, from its natural transmission from the reproductive tissues of layer hens into the magma of the egg prior to shell deposition. Although the incidence of internally contaminated shell eggs is generally low (<1.0%), the consumption of cooked eggs whose yolk is still liquid or the use of raw eggs in uncooked and in partially cooked foods remains an important cause of human salmonellosis. Recent years have witnessed the worlwide emergence of the highly virulent *S.* Typhimurium DT104 in European and North American countries where strains can carry resistance determinants to seven medically important antibiotics, including ampicillin, chloramphenicol, sulfonamides, streptomycin, trimethoprim, and ciprofloxacin. Several factors, including tourist travel, international movement of foods and food ingredients, animal feed trade, and the importation of infected animal replacement stocks, contribute to the national succession of *Salmonella* serovars in human populations and in the food chain. The prominence of agricultural products as major reservoirs of *Salmonella* spp. arises from the ubiquity of the microorganism in the natural environment, the intense on-farm husbandry practices that favor the spread of salmonellae among reared animals, the use of untreated sludge to fertilize agricultural land, and the rendering of animal offals into frequently contaminated feed proteins. Animal slaughtering practices favor the spread of salmonellae from the external surfaces of live animals to consumer meat products in the following decreasing order of importance: chicken $\gg$ pork > sheep/lamb > beef (Table 11.3). The rate of isolation of salmonellae from turkey, duck, and other poultry carcasses is similar to the $\geq$20% levels of contamination in eviscerated chicken (Table 11.3). Several slaughtering operations contribute to the spread of salmonellae in raw eviscerated poultry. These include the immersion of fecally contaminated birds in a common scald tank operated at 50–60°C to soften surface tissues and facilitate the removal of feathers by rotary drum pickers, the automated removal of bird viscera, and the chilling of eviscerated birds in large water baths.

Fresh fruits and vegetables are also potential sources of *Salmonella* and other pathogenic microorganisms. Spraying or irrigation of field crops with contaminated waters, fertilization of soil with untreated sludge, cross-contamination of crops by infected wildlife, manual handling of fresh fruits and vegetable by

TABLE 11.3. Prevalence of *Salmonella* spp. in Raw Meats

		Number of Samples	
Product	Country	Tested	Percent Positive
Beef	Denmark, 1995[a]	2,559	1.3
	Germany, 1991[b]	18,242	5.1
	United States, 1993[b]	2,112	2.7
Pork	Canada, 1985[b]	448	10.0
	Mexico, 1994[a]	50	76.0
	Portugal, 1987[b]	405	5.4
	Thailand, 1986[a]	130	21.5
Chicken	Cuba, 1990[b]	200	62.5
	Denmark, 1995[b]	4,099	45.7
	France, 1994[a]	616	19.8
	Germany, 1994[a]	630	28.6
	United States, 1995[b]	1,297	20.0
	Mexico, 1993[a]	70	68.6
Sheep/lamb	Germany, 1987[b]	264	9.8
	Iran, 1992[b]	500	2.6
	Portugal, 1987[b]	300	7.3

[a] Retail samples.

[b] Postslaughter carcasses.

Source: Adapted from ref. 29.

workers with poor personal hygiene, washing of produce in contaminated water basins, and packaging in ice made from nonpotable water could adversely affect food quality and safety. Consumption of these raw products further heighthens public health concerns and reinforces the need to thoroughly wash fresh fruits and vegetables before serving. Increasing reports of human salmonellosis from the consumption of fresh cantaloupes, watermelons, alfalfa and other vegetable sprouts, tomatoes, and orange juice underline the expansive ecological boundaries of salmonellae.

The occurrence of *Salmonella* in $\leq$9.0% of raw milk samples confirms the importance of this food as a potential vehicle of human infection. Pasteurization effectively eliminates milkborne *Salmonella*. Isolated reports of human illness from the consumption of pasteurized dairy products were associated with human error in industrial processing or to post-pasteurization contamination of finished products. The continued preferences of some consumers for fresh and putatively more nutritious raw milk and for cheeses made from raw or thermized (unpasteurized) milk contribute regularly to national statistics on foodborne bacterial diseases in humans.

The importance of the aquaculture fish and shellfish industries as sources of foodborne *Salmonella* should not be underestimated. The rapid depletion of feral stocks of marine fish and shellfish and the growing consumer demand for these delicate foods in recent decades favored a rapid development of aquaculture where more than half of the global production originates from Asian countries. The feeding of raw meat scraps, human excreta, and rendered animal proteins to fish and shellfish in aquaculture ponds and exposure of estuarine rearing sites to agricultural wastes and contaminated surface waters can only aggravate the *Salmonella* problem within this industry. Moreover, favorable water temperatures and abundant nutrients in these microenvironments favor the proliferation and maintenance of large populations of bacterial pathogens. *Salmonella* contamination rates of 16.0–37.0% in raw aquacultural shrimps from India (1995) and the Philippines (1990) underline the health hazards associated with the consumption of raw or lightly cooked fish and shellfish.

Antibiotics are used increasingly in the agriculture and aquaculture industries as therapeutic, prophylactic, and growth-promoting agents to protect the vigor of reared animal species. These well-entrenched practices carry potentially serious human health implications in the form of selection and persistence of foodborne bacterial pathogens that are resistant to medically important drugs (Table 11.4). Therapeutic doses of antibiotics are administered to diseased animals according to specific treatment regimens for the abatement of clinical symptoms and elimination of infectious agents. The propensity for cross-contamination of farm animals in barns and feedlots and of aquacultural species in earthen ponds and basins with antibiotic-resistant salmonellae is significant. Interestingly, several therapeutic drugs used in veterinary medicine are also used for the treatment of human bacterial infections. In contrast, growth-promoting drugs such as apramycin, virginiamycin, avoparcin, and tylosin are administered to meat animals at subtherapeutic levels to enhance feed conversion and rapidly increase body weight. It is most disturbing that half of the estimated 20×10^6 kg of antibiotics produced in the United States in 1986 were used in animal husbandry, of which approximately 90% were administered at subtherapeutic levels. New evidence indicates that growth-promoting drugs such as apramycin, avoparcin, and tylosin can engender bacterial resistance to gentamicin, vancomycin, and erythromycin, respectively. The prophylactic use and abuse of sulfonamides, tetracycline, and streptomycin in animal husbandry has dramatically influenced the prevalence of resistant phenotypes in meat animals (Table 11.4). The use of streptomycin to treat blights and black rot disease in fruit and vegetable crops also threathens the safety of fresh produce. The magnitude and dynamics of the rising tide of antibiotic resistance in foodborne bacterial pathogens and attendant human health significance are highly controversial issues currently under review by many government agencies and consumer groups. An alarming development in recent years relates to the introduction of fluoroquinolones in agricultural and aquacultural operations. This concern stems from the medical importance of these drugs for the treatment of systemic salmonellosis and other deeply seated human bacterial infections pursuant to the worldwide increase in bacterial resistance to ampicillin, chloramphenicol, and trimethoprim sulfamethoxazole. The approved veterinary use of enrofloxacin in several European countries (1987–1994) and sarafloxacin in the United States (1995) was most unfortunate because it led to the emer-

TABLE 11.4. Antibiotic Resistance in Nonhuman Isolates of *Salmonella* spp.

Country	Period	Sample	Number of Strains Tested	Antibiotic Resistance (%)[a]							Reference
				A	C	TP[b]/SXT[c]	S	G	Cip	Multiple[d]	
Argentina	1993–1995	Chicken carcasses	93	21.5	8.6	9.7[c]	37.6	0.0	NL	52.7	54
Canada	1996	Raw poultry	239	13.4	4.2	1.3[c]	56.5	15.9	0.0	49.3	30
		Animal feed	134	2.2	1.5	0.7[c]	30.6	3.0	0.0	17.1	
		Raw pork	63	17.5	15.9	0.0[c]	50.8	0.0	0.0	47.6	
		Raw beef	19	10.5	5.3	5.3[c]	31.6	0.0	0.0	21.2	
Denmark	1993	Cattle	48	4.0	4.0	0.0[c]	6.0	0.0	0.0	8.3	53
		Swine	99	0.0	0.0	1.0[c]	6.0	0.0	0.0	8.1	
		Poultry	98	7.0	0.0	0.0[c]	0.0	0.0	0.0	0.0	
France	1994	Farm animals	100	100.0	91.0	10.0[c]	88.0	2.0	NL	100.0	19
Thailand	1994	Chicken meat	199	6.0	8.5	13.6[c]	NL	5.0	NL	NL	16
United States	1990	Pork carcasses	121	80.1	11.6	80.2[b]	41.3	0.0	NL	93.3	33
	1990–1992	Eggs, layer flocks	467	83.0	4.0	2.0[c]	NL	5.0	1.0	87.0	41

[a] Ampicillin (A); chloramphenicol (C); streptomycin (S); gentamicin (G); ciprofloxacin (Cip). NL = not listed.

[b] Trimethoprim.

[c] Trimethoprim sulfamethoxazole.

[d] Resistant to ≥ 2 antibiotics that may not be listed above.

gence and spread of fluoroquinolone-resistant *Salmonella*, *Campylobacter*, and other bacterial pathogens in foods and in consumers. Knowledge that the acquisition of resistance from exposure to a single fluoroquinolone may endow carrier strains with cross-resistance to other fluoroquinolones is disquieting. The foregoing concerns on the selection of antibiotic-resistant salmonellae in the global food chain and the propensity for the genetic exchange of resistance determinants between related bacterial species sharing a common ecological niche place new emphasis on the need for stringent controls on the use of antibiotics in food production.

In many countries, the number of foodborne incidents attributed to *Salmonella* spp. greatly exceeds that reported for *Staphylococcus aureus*, *Campylobacter* spp., *Clostridium perfringens*, *Escherichia coli*, and *Listeria* spp. The annual number of foodborne *Salmonella* outbreaks also varies widely between countries. For example, Spain (1992) and Canada (1991), with populations of approximately 40 and 30 million inhabitants, respectively, reported 482 and 28 outbreaks of foodborne salmonellosis. Poultry, eggs, and egg products figured prominently as vehicles of infection in these episodes. Several major outbreaks of human salmonellosis associated with commercially distributed foods are of historical interest. In 1984, cheddar cheese manufactured and distributed in Canada was implicated in 2700 confirmed cases of *S.* Typhimurium PT 10 infections. The cause of the outbreak was ascribed to human error that allowed the mixing of raw and heat-treated milk in the cheese plant. The largest outbreak in the United States (1985) involving 16,284 confirmed cases of *S.* Typhimurium infections and seven fatalities was associated with pasteurized fluid milk. In-depth investigations of the incriminated plant identified cross-connections between raw and pasteurized milk lines as the likely source of product contamination. In France (1993), cooked tuna fish mixed with raw egg mayonnaise served in a school cafeteria resulted in 751 cases of *S.* Enteritidis infection where many cases required hospitalization. In the same year, potato chips that had been sprinkled with paprika contaminated with *S.* Saintpaul, *S.* Javiana, and *S.* Rubislaw infected approximately 1000 consumers in Germany. More recent episodes of foodborne salmonellosis involving different foods and serovars are of interest (Table 11.5). Alfalfa sprouts germinated from seeds imported from Australia were identified as the source of a major outbreak of *S.* Bovismorbificans in several Scandinavian countries. The consumption of alfalfa sprouts contaminated with *S.* Stanley and *S.* Newport led to three additional outbreaks (1995–1996) in Canada, the United States, and Finland and Denmark. In 1994, consumption of a nationally distributed brand of ice cream resulted in 740 cases of *S.* Enteritidis infection. Cross-contamination of pasteurized ice cream mix during transportation in a truck that had recently hauled raw liquid eggs was identified as the cause of this outbreak. In 1995, consumption of fresh unpasteurized orange juice at a world-renowned theme park in Orlando, Florida, resulted in 62 cases of salmonellosis. The original source of contamination could not be ascertained. Fresh orange juice contaminated with *S.* Typhimurium PT 135A infected more than 427 consumers in Australia. In June 1999, un-

TABLE 11.5. Recent Outbreaks of Foodborne Salmonellosis

			Number[a]		
Country	Vehicle	Serovar	Cases	Deaths	Reference
Finland/Sweden, 1994	Alfalfa sprouts	*S.* Bovismorbificans	492	0	44
United States, 1994	Ice cream	*S.* Enteritidis	740	0	36
Spain, 1994	Infant formula	*S.* Virchow	48	0	56
United Kingdom, 1994–1995	Peanut snacks	*S.* Agona PT 15	41	0	3
Italy, 1995	Salami	*S.* Typhimurium PT 193	83	0	45
United States, 1995	Orange juice	*S.* Hartford	62	0	20
		S. Gaminara			
		S. Rubislaw			
	Chicken/restaurant	*S.* Newport	>850	NS	4
Australia, 1996	Peanut butter	*S.* Mbandaka	>100	1	5
United States, 1997	Stuffed ham	*S.* Heidelberg	746	1	15
United Kingdom, 1997	Infant milk formula	*S.* Anatum	12	0	6
Canada, 1998	Cheddar cheese	*S.* Enteritidis PT 8	700	0	47
United States, 1998	Toasted oat cereal	*S.* Agona	209	0	7
Majorca/Spain, 1998	Mousse/tomato sauce	*S.* Enteritidis PT 1	284	0	8
Australia, 1998	Gelati ice cream	*S.* Oranienburg	102	NS	9

[a] NS = not stated.

pasteurized orange juice distributed within the United States and to Canada from a single plant in Arizona was implicated in more than 200 cases of infections with *S.* Muenchen and other serovars. In 1996, peanut butter contaminated with *S.* Mbandaka was associated with more than 100 cases of illness and one fatality in Australia. A total of 746 cases of *S.* Heidelberg infection were reported following a community dinner in Maryland involving 1400 guests. The episode was linked to bulk cooking of stuffed hams under inadequate time/temperature conditions. In 1998, cheddar cheese in prepackaged luncheon meals was incriminated in a Canadian outbreak of *S.* Enteritidis PT 8. School children were primarily affected in this incident involving more than 700 infected individuals. In the same year, a multistate outbreak of 209 cases of *S.* Agona infection and hospitalization of 47 patients was traced to toasted oats cereals manufactured at a single plant in the United States. These incidents underline the significant human health impact of foodborne salmonellae and reiterate the need for continued vigilance in all aspects of food production and food service.

11.5. DETECTION

11.5.1. Culture Methods

The reliable detection of foodborne salmonellae hinges on the use of statistically significant sampling plans and sensitive analytical methods. The probable isolation of salmonellae varies directly with the number of random sample units withdrawn from a lot and with the total quantity of product analyzed. Sampling intensity is product dependent and varies from 5 to 60 replicate sample units per lot according to the perceived human health hazard from product abuse during processing and consumer handling. Although standard culture methods are based on 25-g analytical units, testing of larger amounts of product increases method sensitivity. Standard culture methods for the detection of foodborne *Salmonella* spp. generally consist of five distinct steps: preenrichment, selective enrichment, plating on differential agar, biochemical screening, and serological confirmation. Preenrichment in a nonselective broth medium for 18–24 h at 35–37°C allows salmonellae that may be stressed or injured to resuscitate and proliferate to detectable levels. The use of short (6- to 8-h) preenrichment for greater method brevity is not recommended because it fails to provide salmonellae with sufficient time to adapt to its new environment, repair cellular damage, and actively grow to high numbers. Traditional preenrichment media include nonfat dry milk with added brilliant green dye for the preeenrichment of cocoa and chocolate products, brilliant green water for milk powder, trypticase soy broth supplemented with potassium sulfite to neutralize spice-dependent bacteriostasis, and buffered peptone water, nutrient, or lactose broths for other foods and agricultural products. Selective enrichment of preenrichment cultures in tetrathionate brilliant green (TBG), selenite cystine

(SC), Rappaport–Vassiliadis (RV), or other less favored enrichment broth media facilitates the recovery of salmonellae on plating media through repression of competitive microflora. The concurrent use of two enrichment media where one medium is incubated at 41–43°C and the other at a more permissive temperature (35–37°C) maximizes detection of the target microorganism. These conditions accommodate fastidious *Salmonella* species that grow poorly or are inhibited at elevated temperatures and whose growth is hampered in highly selective enrichment media such as TBG and RV. It is likely that the productivity of TBG_{43} and SC_{35} would be superior to TBG_{43} and TBG_{35} or TBG_{43} and RV_{42} because the former combination includes two different enrichment media incubated at two different temperatures. Numerous studies on the reliability of enrichment conditions support the following decreasing order of effectiveness: $TBG_{43} \geq RV_{43} > TBG_{35} > SC_{35}$. Although short (6- to 7-h) enrichment is advocated in the analytical protocol of several commercial systems for the detection of foodborne *Salmonella*, short enrichment adversely affects the sensitivity of standard culture methods. Enrichment cultures are then plated onto selective agar media for the presumptive identification of *Salmonella* colonies on the basis of discriminating biochemical reactions. Standard plating media include brilliant green (BGA), brilliant green sulfa (BGS), xylose–lysine–desoxycholate (XLD), and Hektoen enteric (Hek) agars that report on acid production from lactose and/or sucrose utilization through determinant color changes in the media. The increasing occurrence of salmonellae that utilize one or both disaccharides tends to undermine the diagnostic value of these plating media. Bismuth sulfite (BSA) is a highly selective, saccharide-independent medium that is commonly used in standard culture methods. *Salmonella* species that typically produce hydrogen sulfide appear as charcoal black colonies with or without a black halo that produce a metallic sheen under reflected light. The Rambach and SM-ID plating media produce discriminating color reactions in the presence of isolates that are ß-galactosidase positive and that produce acid from propylene glycol (Rambach) and glucuronate (SM-ID). Plating media generally yield suspect colonies within 18–24 h incubation at 35–37°C, except BSA, which may require 48 h incubation for the development of presumptive *Salmonella* colonies. Comparative studies support the following ranking in decreasing order of effectiveness: BSA > BGS > BGA ≥ Rambach = SM-ID > XLD > Hek. The biochemical screening of suspect isolates using appropriate tube media or commercially available diagnostic kits is based primarily on the production of hydrogen sulfide gas, the presence of lysine decarboxylase and absence of urease activities, and the inability to produce acid from lactose and sucrose. Biochemically suspect isolates are then confirmed serologically using somatic (O) and flagellar (H) polyvalent and single-grouping antisera.

11.5.2. Rapid Methods

Several cultural approaches have been proposed for the rapid detection of foodborne salmonellae. In the Salmosyst system (E. Merck, Darmstadt,

Germany), preenrichment and selective enrichment are combined into a single analytical step. Following preenrichment for 6–8 h at 35°C, a tablet containing selective agents is added to a portion (10 mL) of preenrichment culture, and the resulting preparation is reincubated for an additional 18 h. The rapid dissolution of the tablet converts the preenrichment broth into a selective enrichment medium. Reported levels of agreement between the Salmosyst and standard culture methods vary widely. The Simple Preenrichment and Rapid Isolation Novel Technology (SPRINT) marketed by Oxoid Limited (Basingstoke, United Kingdom) consists of a soya peptone preenrichment medium supplemented with oxyrase and a timed release of RV selective agents from insoluble capsules sealed with hydrogel plugs. Although preliminary validation data generated by the manufacturer are favorable, independent studies have yet to confirm the reliability of this technique. The modified semisolid Rappaport–Vassiliadis (MSRV) medium provides for selective enrichment in an agar matrix. The MSRV is point inoculated with an overnight preenrichment culture and incubated at 42–43°C for 24 h. Growth from the peripheral zone of radial migration is then streaked onto differential plating media, and suspect colonies are screened biochemically using conventional tests. Numerous studies have underlined the high sensitivity of the MSRV medium notwithstanding its inability to detect nonmotile *Salmonella* strains. The refrigeration of preenrichment or enrichment cultures for 72 h (i.e., over the weekend) can expedite the detection of foodborne *Salmonella* and greatly increase laboratory productivity. The benefits of this novel technique stem from the restricted initiation of sample analyses by standard culture methods to Mondays and Tuesdays if weekend work is to be avoided and if analyses are to be interrupted only by the refrigeration of incubated plating media. Extensive validation studies on the recovery of salmonellae from refrigerated preenrichment and enrichment cultures reported sensitivity levels of $\geq$88% and $\geq$99% for high- and low-moisture products, respectively.

In recent years, commercial interests in the development of methodologies for the rapid detection of *Salmonella* in foods and agricultural products have generated an inventory of no fewer than 35 diagnostic systems. The underlying scientific principles for these novel approaches include colorimetric and fluorometric enzyme-linked immunosorbent assays (ELISA), DNA probe hybridization, immunoimmobilization of motile *Salmonella*, polymerase chain reaction (PCR), *Salmonella*-specific phage transduction of an ice nucleation gene, *Salmonella*-induced conductance changes in liquid media, and systems for the identification of *Salmonella* based on generic biochemical reactions. Although a discussion of the operating characteristics, sensitivity, and specificity of each technique is beyond the scope of this chapter, salient features of selected systems will be reviewed. In direct ELISA procedures, *Salmonella* antigens in a test sample specifically react with polyclonal or monoclonal antibodies previously adsorbed to a solid matrix. The resulting antigen–antibody complex is then reacted with a conjugate consisting of a *Salmonella*-specific antibody coupled to an enyme protein. The subsequent addition of enzyme substrate yields a colored or fluorescent reaction product that presumptively identifies the presence of salmonellae in the test sample. Threshold sensitivities for ELISA systems generally range from 10^5 to 10^6 salmonellae per milliliter of test culture. Sensitivity of ≥94.1% and specificity of ≥79.4% were compiled from a total of 15 independent studies involving the following ELISA systems: Salmonella-Tek (Organon Teknika, Durham, NC), Tecra microplate (Bioenterprises PTY, Roseville, NSW, Australia), Locate (Rhone-Poulenc, Glasgow, Scotland), Unique (Bioenterprises PTY, Roseville, NSW, Australia), Equate (Binax, Portland, ME), Assurance (BioControl Systems, Bothell, WA), and VIDAS (bioMérieux, Marcy-l'Etoile, France). It is notable that the results of two major Canadian studies comparing the VIDAS fluorometric assay to a standard culture method showed complete method agreement for the detection of salmonellae in naturally contaminated foods and agricultural products (D'Aoust, unpublished). The 1-2 Test (BioControl Systems, Bothell, WA) is a self-contained vial of semisolid agar that provides for the immunoimmobilization of motile salmonellae by specific antisera. Polyvalent antiserum is first dispensed onto the top surface of the agar followed by the inoculation of a lower side chamber with a portion of selective enrichment culture. During the overnight incubation of the inoculated vial at 35°C, an inverted-bell-shaped line of precipitation forms at the interface between the downward-diffusing antisera and upward-migrating *Salmonella*. Performance assessment results from six independent studies showed levels of sensitivity and specificity of ≥89.6% and ≥94.9%, respectively.

The Oxoid Rapid Test (Unipath, Basingstoke, Hampshire) also detects motile salmonellae in preenrichment cultures using generic biochemical reactions. The system consists of two tubes each separated into an upper and lower chamber by a porous partition. The lower chambers of both tubes provide for the selective enrichment of *Salmonella*, which then migrates into the upper chambers to produce discriminating color reactions. Several validation studies have reported levels of sensitivity and specificity of ≥77.7% and ≥93.6%, respectively. The GENE-TRAK colorimetric probe assay (GENE-TRAK Systems, Framingham, MA) is based on two DNA probes that hybridize with homologous nucleotide sequences in ribosomal ribonucleic acid (rRNA) released from enzymically lysed *Salmonella* cells previously grown in gram-negative (GN) broth. The capture probe contains a polydeoxyadenylic (poly-dA) tail whereas the detector probe is directly labeled with peroxidase. In the assay, both probes are permitted to react concurrently with the lysed cell preparation. The detector probe/peroxidase–rRNA–capture probe/poly-dA complexes are retrieved from the mixture by reaction of the poly-dA tail with polydeoxythymidylic nucleotides (poly-dT) on the surface of a dipstick. The dipstick with the attached complex is then transferred to a separate tube containing peroxidase substrate and appropriate chromogen. A discriminating color reaction is observed if the lysed GN broth culture contained *Salmonella* rRNA. The results of eight independent validation studies showed levels of sensitivity and specificity of ≥89.3% and ≥95.2%, respectively.

Sustained research interest in nucleic acid–based technologies has given prominence to PCR for the detection of foodborne *Salmonella* spp. The strength of the PCR technique stems from its ability to produce multiple copies of a single targeted DNA sequence, thereby facilitating the confirmation of *Salmonella*-specific nucleotides by conventional DNA hybridization techniques. Three commercial PCR diagnostic systems that use overnight preenrichment cultures as test material are currently available. The BAX system (DuPont, Wilmington, DE) generates amplified DNA products (amplicons) that are subsequently separated by gel electrophoresis and identified using positive control bands. Preliminary results with artificially and naturally contaminated foods have been favorable, and a threshold sensitivity of 10^2–10^3 salmonellae per milliliter has been determined using pure cultures. The Probelia system (Sanofi Diagnostics Pasteur, Marnes La Coquette, France) is a colorimetric PCR assay where amplicons are immobilized in microplate wells and then reacted with a *Salmonella*-specific probe previously conjugated with peroxidase. The reaction complex is then visualized by the addition of enzyme substrate to individual microwells. Preliminary validation results from the manufacturer indicate that the technique is highly sensitive and specific. The automated TaqMan system (Perkin-Elmer Biosystems, Foster City, CA) is based on the hybridization of a fluorogenic *Salmonella*-specific probe within the chromosomal region targeted for PCR amplification and emission of a discriminant fluorescent signal when the targeted chromosomal region is fragmented during PCR amplification. Preliminary results from the manufacturer indicate high levels of sensitivity (99.7%) and specificity (100%) with pure cultures.

Salmonella-specific phages have been used for the detection of foodborne salmonellae. The Bacterial Ice Nucleation Detection (BIND) system (Idetek, Sunnyvale, CA) is a microplate assay that is applied to preenrichment or direct enrichment cultures. A mixture of phages transduce an ice nucleation gene into viable salmonellae, thereby enabling the transduced organism to synthesize a template protein for ice crystal formation. The microwell freezing (orange-red) of a fluorescent dye at −9.3°C presumptively identifies the presence of salmonellae in a test sample whereas no color change (green) coresponds to a negative result. Studies by the manufacturer suggest a threshold sensitivity of $\geq 10^2$ salmonellae per milliliter of preenrichment culture. The automated Malthus conductimetric system (IDG Limited, Bury, United Kingdom) presumptively detects the presence of salmonellae in preerichment cultures. The assay is based on the ability of *Salmonella* to characteristically alter the conductivity of liquid media through the breakdown of substrates into charged molecular species. Replicate portions of an overnight preenrichment culture are inoculated into two conductance cells each containing distinct differentiating media. Both conductance cells are incubated for $\leq$30 h at 35°C during which conductance changes are automatically monitored at regular intervals. Comparison of conductance changes with internal reference values enables the instrument to flag presumptive-positive samples. Two evaluations of the Malthus system using naturally and artificially contaminated foods and agricultural products showed good sensitivity ($\geq$88.9%) and specificity ($\geq$86.1%) whereas less favorable results were obtained in a third validation study. Recent years have witnessed major progress in the development of sensitive methods that are 24–36 h more rapid than standard culture procedures for the presumptive identification of foodborne *Salmonella*.

11.6. PHYSICAL METHODS FOR DESTRUCTION

Thermal processes are widely used in the food industry to enhance product quality and safety. Bacterial lethality is commonly expressed as decimal reduction time (D), the amount of time required to effect a one $\log_{10}$ reduction in the number of viable microorganisms at a given temperature. A $D_{72°C} = 0.2$ min for salmonellae signifies that the number of viable cells in the test product will decrease by one $\log_{10}$ (90% inactivation) for every 0.2 min the product is held at 72°C. The heat resistance of *Salmonella* is markedly affected by the a_w, the pH and composition of foods, the serovar, and the physiological state of contaminating microorganisms. The considerably larger D values for salmonellae in dry than in high-moisture foods clearly underline the importance of a_w on bacterial heat resistance. For example, the addition of 1–3% water to solid milk chocolate or cocoa liquor markedly decreased the D values of *S.* Anatum and *S.* Eastbourne at 71.0 and 100.0°C (Table 11.6). The nature of dissolved solutes also impacts heat resistance, as evidenced by the mean $D_{57.2°C}$ values of 62.5 and 2.3 min for *S.* Anatum heated, respectively, in aqueous solutions of sucrose and glycerol adjusted to $a_w = 0.90$. Similarly, greater heat resistance was observed with pooled *Salmonella* strains heated at 63.3°C in egg yolk supplemented with 10% NaCl than with 10% sucrose (Table 11.6). Heat resistance increased upon the acidification of liquid whole egg from pH 8.0 to 5.5 and egg albumen from pH 9.3 to 7.8 (Table 11.6). The ability of salmonellae to withstand potentially thermolytic conditions is strain dependent, as indicated by the response of *S.* Senftenberg 775W in environments of high water activity (Table 11.6). Heat resistance also increases as cells enter the stationary phase of growth and when salmonellae are stressed as a result of prolonged storage of dry foods at greater than ambient temperatures. Exposure to sublethal temperatures ($\leq$50.0°C) increases heat resistance through a rapid chloramphenicol-sensitive synthesis of protective heat-shock proteins and a physical reconfiguration of the bacterial cell membrane that is more resilient to heat damage.

Several nonthermal processes such as irradiation and high hydrostatic pressure are currently used commercially to extend product shelf life and/or inactivate bacterial pathogens. The assessment of other technologies such as oscillating magnetic fields, high-intensity pulsed electric fields, and intense light pulses is currently the focus of major research interest under laboratory and pilot plant conditions. In contrast to the adverse effects of bactericidal heat treatments on the organoleptic and physical properties of foods, nonthermal methods of preservation are generally innocuous. Three types of ionizing radiation including gamma rays, x-rays, and accelerated electrons are deemed suitable for the treatment of foods. The reduced penetration of electrons and the

TABLE 11.6. Thermal Resistance of Foodborne *Salmonella* spp.

Product	Serovar	Thermal Resistance		Reference
		Temperature (°C)	*D* (min)	
	High Moisture			
Liquid whole egg (pH 8.0/5.5)	*S.* Typhimurium	60.0	0.55/2.2	1
	S. Senftenberg 775W	60.0	1.5/9.5	
Liquid whole egg	*S.* Enteritidis PT 4	60.0	0.44	38
	S. Enteritidis PT 13a	60.0	0.22	
	S. Typhimurium PT 110	60.0	0.26	
	S. Senftenberg 775W	60.0	5.60	
Egg yolk	Mixture (6 strains)	63.3	0.20	42
Egg yolk + 10% sucrose	Mixture (6 strains)	63.3	0.72	
Egg yolk + 10% NaCl	Mixture (6 strains)	63.3	11.50	
Egg albumen				
pH 9.3/7.8	Mixture (6 strains)	56.6	1.08/3.60	43
pH 8.2	Mixture (2 strains)	56.7	2.96	52
Whole milk (raw)	*S.* Typhimurium	62.8	0.11	17
		68.3	0.02	
		71.7	0.004	
	S. Muenster	60.0	0.06–0.09	31
		63.0	0.04	
	S. Senftenberg 775W	60.0	0.10–0.19	
		67.3	0.04	
	S. Anatum	62.8	0.09	48
		68.3	0.008	
	S. Senftenberg 775W	65.6	0.57	
		68.3	0.17	
		71.7	0.02	

Skim milk (pasteurized)	*S.* Senftenberg 775W	60.0	3.60	39
		68.0	0.31	
		71.0	0.07	
Shrimp (homogenate)	*S.* Weltevreden	60.0	0.68	49
		62.8	0.38	
	Low Moisture			
Wheat flour ($a_w = 0.4$)	*S.* Weltevreden	60.0–62.0	825.0	10
		69.0–71.0	50.0	
Chocolate (solid milk)	*S.* Anatum	71.0	1200.0	14
Chocolate (solid milk) + 2% H_2O		71.0	240.0	
Chocolate (solid milk)	*S.* Typhimurium	70.0	678.0–1050.0	35
		80.0	222.0	
		90.0	72.0–78.0	
	S. Senftenberg 775W	70.0	360.0–480.0	
		80.0	96.0–144.0	
		90.0	30.0–42.0	
Cocoa liquor + 1% moisture	*S.* Eastbourne	100.0	16.9	32
Cocoa liquor + 2% moisture		100.0	10.6	
Cocoa liquor + 3% moisture		100.0	1.5	

high cost of x-ray emitters have favored the use of gamma-ray emission from ^{60}Co and ^{137}Cs for the reduction or elimination of foodborne bacterial pathogens and the increased shelf-life of perishable products through inactivation of spoilage microorganisms. From the early application of gamma irradiation to disinfest spices, fruits, and cereals, inhibit the sprouting of vegetable crops, and delay fruit ripening, studies in the last few decades have confirmed the efficacy of radiation doses of $\geq$3.0 kGy to significantly reduce or eliminate *Salmonella* in foods and in agricultural products. Salmonellae are more susceptible to irradiation in the logarithmic than in the stationary phase of growth, whereas lethality increases with increasing food temperature and moisture content. Benefits of this technology also include marked reductions in the functional requirement of nitrites in irradiated meat products, replacement of fumigation technologies involving potentially toxic chemical agents such as ethylene oxide and methyl bromide. Irradiation of packaged products also minimizes post-process contamination. However, ionizing radiation is not without limitations as evidenced by its inability to eliminate preformed bacterial toxins, its degradation of certain amino acids and vitamins lowering the nutritive quality of foods, its softening of some fruits and vegetables from an induced breakdown of cell wall cellulose, and the creation of free radicals that oxidize lipids in fatty foods and engender rancidity. Irradiation of *Salmonella* in shell eggs and in dairy products is not feasible because the process adversely affects the viscosity of egg albumen and the flavor of milk and milk products. Moreover, irradiation of raw meats and other food products that can sustain bacterial growth could increase human health risks because of the facilitated growth of *Salmonella* in the absence of large populations of competing microflora. The World Health Organization has recognized ionizing radiation at doses of $\leq$10.0 kGy (1.0 Mrad) as an acceptable process to enhance the bacterial safety and preservation of foods. The on-going lack of consensus among regulatory agencies on whether irradiation is a food process or a food additive has hampered the international harmonization of processing standards and labeling requirements for irradiated foods. The need to closely control the use of irradiation in the global food chain has prompted major research efforts in the development of methods for the rapid identification of irradiated foods and estimation of applied radiation doses. Electron spin resonance spectroscopy, x-ray scattering, and detection of specific radiolytic products are currently favored for the identification of irradiated foods. Although general consumer acceptance of irradiated foods remains a significant challenge to proponents of this technology, increasing international approval of the process by government agencies as a valid alternative to the traditional pasteurization of foods may lead to a general public endorsement of this new food technology.

High hydrostatic pressure in the range of $4.0\text{–}9.0 \times 10^3$ atm is used commercially to enhance product safety and extend the shelf life of perishable foods. Lethality increases with increasing hydrostatic pressure where levels of *S.* Typhimurium and *S.* Senftenberg inoculated into pork slurries decreased 3.0–5.0 $\log_{10}$ under pressures of $2.3\text{–}3.4 \times 10^3$ atm. The susceptibility of *Salmonella* to hyperbaric conditions is also strain dependent and is higher in nutrient-limited than in complete media. Bacterial inactivation arises from the pressure-dependent disruption of bacterial cell membranes. The antibacterial potential of pulsed electric fields has generated considerable research interest in recent years. This technology is based on the passage of liquid food between electrodes that transmit a pulsating electric field to the food matrix. The resulting potential difference in the bacterial cell membrane induces electroporation, the formation of pores in the cell membrane that increase cell permeability, ultimately resulting in cell death. Lethality of the process increases with increasing strength and pulsation of applied electric fields, increasing temperature, decreasing ionic strength, and in the absence of divalent cations. Cells in the logarithmic phase of growth are particularly susceptible to inactivation. The elimination of *S.* Dublin inoculated at 3.8×10^3 mL^{-1} of fluid milk following exposure to 40 pulses at an energy level of 36.7 kV/cm underlines the potential of this technology in commercial food processing.

11.7. PREVENTION AND CONTROL

The effective control of *Salmonella* in fresh and processed foods constitutes a formidable challenge because of the complexity and sectorial dependence within the meat, dairy, fresh produce, and other food industries. The lack of coordinated control efforts in the last few decades has favored the continued prominence of salmonellae in the global food chain and in human bacterial foodborne illnesses. Government regulations on the mandatory pasteurization of foods, including fluid milk and powdered and liquid eggs, have contributed significantly to product safety. However, the ubiquity of *Salmonella* in the natural environment, the increasing number of food industries commensurate with the demands of a growing world population, and the massive trade between western and third world countries are public health concerns that continue to strain limited government human resources dedicated to food control. Nonetheless, the major impact of industry commitment to the quality and microbial safety of foods needs to be recognized. In addition to good manufacturing practices (GMPs), the expansive application of hazard analysis and critical control point (HACCP) measures within the food industry will further strengthen consumer confidence in the safety of the global food supply.

Intense husbandry practices in the bovine, porcine, and poultry industries and exposure of reared species to *Salmonella*-contaminated drinking waters and grazing pastures synergistically promote intestinal carriage and surface contamination of live animals. The asymptomatic shedding of salmonellae in animals stressed from water deprivation, induced molting, prolonged holding in lairage pens, and transport to slaughterhouses further potentiate the surface contamination of meat animals. The introduction of infected replacement stocks and provision of *Salmonella*-contaminated feeds remain primary sources of contamination in rearing facilities. Although feed pelleting can effectively reduce or destroy salmonellae, postprocess contamination of pelleted feeds during bulk storage in rendering and feed mill plants, transportation, and on-farm storage in unprotected bins undermine the benefits of the pelleting process. The addition of commercial bactericidal mixtures of formic and propionic acids such as Bio-Add (Trouw UK, Norwich, United Kingdom) and Sal Curb (Kemin Europa N. V., Herentals, Belgium) to prevent the recontamination of finished feeds is gaining worldwide endorsement. Parenteral and oral vaccines protect meat animals against *Salmonella* colonization and infection with host-specific serovars. Commercial vaccines consisting of *Salmonella* antigens or viable but avirulent salmonellae afford protection against *S.* Gallinarum (poultry), *S.* Choleraesuis (porcine), *S.* Dublin (bovine), and *S.* Typhimurium (meat animals). The efficacy of oral vaccination of poultry chicks with bacterial mixtures prepared from the ceca of *Salmonella*-free birds has been extensively researched. In this approach, the early establishment of adult bird microflora in chicks competitively excludes salmonellae from colonizing their gastrointestinal tract. Commercial oral vaccines such as Broilact (Orion Corporation Farmos, Turku, Finland) and Aviguard (Life-Care Products, Malvern Link, United Kingdom) contain bacterial mixtures of unknown composition, whereas Preempt (MS BioScience, Madison, WI) is a mixture of 29 characterized strains of aerobic and anaerobic bacteria. Preparations of known composition circumvent industry concerns on the possible presence of poultry disease agents in undefined mixtures. Unfortunately, performance assessment studies have generally found greater levels of protection in birds treated with undefined mixtures.

Meat slaughtering practices such as the mechanical or manual removal of animal hides, fleece, hair, and feathers soiled with fecal material contribute to carcass contamination. The treatment of meat surfaces by immersion or spray washing with hot chlorinated water, hot dilute organic acid, and highly alkaline trisodium phosphate (AvGuard) is rapidly gaining favor over the more traditional water sanitizing of raw meat carcasses. The high incidence of *Salmonella* in raw poultry derives mainly from cross-contamination during the immersion of carcasses in a common scald tank of water held at 52°C (soft scald) or 60°C (hard scald), and from the mechanical action of defeathering machines that successively spread salmonellae on the surface of scalded birds on conveyor lines. Alternate approaches, including the use of highly alkaline or acidified scald waters, steam scalding, and simultaneous scalding and plucking under a hot water (62°C) spray, have been proposed for a more effective reduction of salmonellae on poultry carcasses. The immersion and chilling of eviscerated birds in large tanks of cold water (spin chillers) further increase cross-contamination of carcasses. Improved counterflow dynamics within spin chillers, greater replacement of chill tank water with fresh or recycled (decontaminated) chill water, and the use of antibacterial agents that are more potent than the inlet chlorine levels (20 ppm) commonly used in spin chillers would enhance the safety of raw poultry.

The prominence of intact shell eggs and egg products as vehicles of human salmonellosis in the ongoing pandemic of *S.* Enteritidis (SE) has led to a worldwide review of production and processing practices within the table egg industry. The internal contamination of intact shell eggs results from the ability of SE to infect the ovaries of layer birds and migrate into the ovules of layer hens prior to shell deposition. The localization of SE within shell eggs preempts decontamination by the surface sanitizing practices currently used in commercial egg-grading operations. National control programs are contributing to an abatement of salmonellae in the egg industry. Control measures include the destruction of primary, multiplier breeder and layer flocks infected with SE, close monitoring of replacement stocks for the presence of SE in hatching eggs and day-old chicks, stringent surface decontamination of hatching eggs, enhanced biosecurity in barns to prevent exposure of breeder and layer birds to potentially infected rodents and insects, environmental monitoring of layer barns and diversion of shell eggs from SE-infected layer flocks to pasteurization plants, extensive sanitizing of environmentally contaminated premises, exclusion of animal proteins in poultry feeds, increased monitoring of domestic and imported feeds, refrigeration ($\leq 7.0°C$) of sanitized shell eggs at grading stations and during their distribution to and display in retail outlets, and coding of retail egg cartons to facilitate trace-back to producer farms. Recent decreases in the number of human outbreaks of eggborne SE in the United States and the United Kingdom are noteworthy. Shell eggs, nevertheless, should be cooked until the yolk has congealed and is no longer fluid. The use of raw eggs in the home preparation of ice cream, mayonnaise, and Cesar salads and other foods containing lightly cooked eggs should be avoided.

Fresh fruits and vegetables have gained notoriety in recent years as vehicles of human salmonellosis (Table 11.5). The exposure of field crops to polluted irrigation waters and untreated animal waste fertilizers together with the repeated manual handling of these foods during harvesting, processing, and distribution could result in the surface contamination of ready-to-eat produce. The use of effective concentrations of disinfectants in processing washwaters, final product rinsing in chlorinated or potable water, and other physicochemical interventions for the abatement of salmonellae in produce need to be promoted. The rapid distribution of fresh fruits and vegetables to retail outlets preempts their premarket testing for *Salmonella* spp. because detection methods require 3–4 days to ascertain the absence or presumptive presence of the pathogen. Although the pasteurization of fresh fruits and vegetables with low doses of gamma irradiation is generally feasible, consumer acceptance of irradiated foods remains a major hurdle to the food industry.

The widespread occurrence of *Salmonella* in the global food chain and the prominence of this pathogen as the leading cause of human foodborne bacterial disease reiterate the importance of effective and coordinated control measures at the levels of food production, harvesting, processing, and distribution. Recent events such as the human pandemic of *S.* Enteritidis from the consumption of poultry meat and internally contaminated shell eggs, the emergence and global spread of multiple antibiotic-resistant *S.* Typhimurium DT4 strains in meat animals and human populations, the identification of vegetable sprouts as vehicles of human salmonellosis, and the resurgence of fresh fruits and fruit juices as sources of *Salmonella* infections emphasize the need for greater vigilance and commitment of the food industry and government agencies to the safety of the global food supply. The dramatic increases in the resistance of salmonellae to medically important antibiotics in the last decade is disconcerting and immediately requires a more judicious use of antimicrobials in the agricultural and aquacultural sectors and in human medicine. Sadly, attempts at consumer education on the health hazards associated with the consumption of raw fluid milk, raw milk cheese, sushi, steak tartare, raw fish and shellfish, mayonnaise and ice cream made with raw eggs, and undercooked meats, fish, and shellfish have met with mitigated success. The historical prominence of *Salmonella* in foods and its major impact on human health are a painful reality that will likely prevail in the next century.

REFERENCES

1. A. Anellis, J. Lubas, and M. M. Rayman. Heat resistance in liquid eggs of some strains of the genus *Salmonella*. *Food Res.* **19**, 377–395 (1954).
2. F. J. Angulo. Epidemiology of antimicrobial resistance in *Salmonella* isolates from humans. Paper presented at annual meeting of the American Society of Microbiology (ASM), New Orleans (1996).
3. Anonymous. An outbreak of *Salmonella agona* due to contaminated snacks. *Comm. Dis. Rep.* **5**, 29, 32 (1995).
4. Anonymous. Florida *Salmonella* outbreak caused changes in state inspections. *Food Chem. News* **37**, 27 (1995).
5. Anonymous. Salmonellosis and peanut butter. *Int. Food Safety News* **5**(6), 11 (1996).
6. Anonymous. *Salmonella anatum* infection in infants linked to dried milk. *Comm. Dis. Rep.* **7**, 33, 36 (1997).
7. Anonymous. Multistate outbreak of *Salmonella* serotype Agona infections linked to toasted oats cereal—United States, April–May, 1998. *MMWR* **47**, 462–464 (1998).
8. Anonymous. *Salmonella enteritidis* PT 1 outbreak in a hotel in Majorca. *Eurosurveillance Wkly*, July 2 (1998).
9. Anonymous. *Salmonella oranienburg* in Australia. *Comm. Dis. Intelligence* **22**(8), (1998).
10. J. Archer, E. T. Jervis, J. Bird, and J. E. Gaze. Heat resistance of *Salmonella weltevreden* in low-moisture environments. *J. Food Prot.* **61**, 969–973 (1998).
11. A. R. Bahrmand and A. A. Velayati. Antimicrobial resistance pattern and plasmid profile of *Salmonella typhi* isolated from an outbreak in Tehran province. *Scand. J. Infect. Dis.* **29**, 265–269 (1997).
12. L. Bakir, O. Chourou, and N. Ben Salah. Epidémiologie des salmonelles au CHU de La Marsa (Tunisie). *Méd. Mal. Infect.* **27**, 838–841 (1997).
13. G. V. Barbosa-Canovas, U. R. Pothakamury, E. Palou, and B. G. Swanson, *Nonthermal Preservation of Foods*. New York: Marcel Dekker, 1997.

14. J. C. Barrile and J. F. Cone. Effect of added moisture on the heat resistance of *Salmonella anatum* in milk chocolate. *Appl. Microbiol.* **19**, 177–178 (1970).
15. A. Beers. Maryland church dinner *Salmonella* outbreak blamed on faulty food preparation. *Food Chem. News* **39**, 19–20 (1997).
16. S. Boonmar, A. Bangtrakulnonth, S. Pornruangwong, S. Samosornsuk, K. Kaneko, and M. Ogawa. Significant increase in antibiotic resistance of *Salmonella* isolates from human beings and chicken meat in Thailand. *Vet. Microbiol.* **62**, 73–80 (1998).
17. J. G. Bradshaw, J. T. Peeler, J. J. Corwin, J. E. Barnett, and R. M. Twedt. Thermal resistance of disease-associated *Salmonella typhimurium* in milk. *J. Food Prot.* **50**, 95–96 (1987).
18. J. Breuil, L. Armand-Lefevre, I. Casin, A. Dublanchet, and E. Collatz. Surveillance de la sensibilité aux antibiotiques des salmonelles et shigelles isolées dans 77 hôpitaux français. *Bull. Epidemiol. Hebdo.* **51**, 219–221 (1998).
19. I. Casin, A. Brisabois, N. Berger, J. Breuil, and E. Collatz. Phénotypes et génotypes de résistance de 182 souches de *Salmonella* sérotype Typhimurium résistantes à l'ampicilline d'origine humaine et animale. *Méd. Mal. Infect.* **26**, 426–430 (1996).
20. K. A. Cook, T. E. Dobbs, W. G. Hlady, J. G. Wells, T. J. Barrett, N. D. Puhr, G. A. Lancette, D. W. Bodager, B. L. Toth, C. A. Genese, A. K. Highsmith, K. E. Pilot, L. Finelli, and D. L. Swerdlow. Outbreak of *Salmonella* serotype Hartford infections associated with unpasteurized orange juice. *J. Am. Med. Assoc.* **280**, 1504–1509 (1998).
21. M. Cormican, C. Butler, D. Morris, G. Corbett-Feeney, and J. Flynn. Antibiotic resistance amongst *Salmonella enterica* species isolated in the Republic of Ireland. *J. Antimicrob. Chemother.* **42**, 116–118 (1998).
22. J.-Y. D'Aoust. *Salmonella*. In: M. P. Doyle (Ed.), *Foodborne Bacterial Pathogens*. New York: Marcel Dekker, 1989, pp. 327–445.
23. J.-Y. D'Aoust. Manufacture of dairy products from unpasteurized milk: A safety assessment. *J. Food Prot.* **52**, 906–914 (1989).
24. J.-Y. D'Aoust. Pathogenicity of foodborne *Salmonella*. *Int. J. Food Microbiol.* **12**, 17–40 (1991).
25. J.-Y. D'Aoust. Psychrotrophy and foodborne *Salmonella*. *Int. J. Food Microbiol.* **13**, 207–216 (1991).
26. J.-Y. D'Aoust. *Salmonella* and the international food trade. *Int. J. Food Microbiol.* **24**, 11–31 (1994).
27. J.-Y. D'Aoust. Methods for the detection of foodborne *Salmonella* spp. *Southeast Asian J. Trop. Med. Publ. Hlth.* **26**, 195–208 (1995).
28. J.-Y. D'Aoust. *Salmonella* species. In: M. P. Doyle, L. R. Beuchat, and T. J. Montville (Eds.), *Food Microbiology: Fundamentals and Frontiers*. Washington, DC: ASM Press, 1997, pp. 129–158.
29. J.-Y. D'Aoust. *Salmonella*. In: B. M. Lund, A. C. Baird-Parker, and G. W. Gould (Eds.), *The Microbiological Quality and Safety of Food*. London: Chapman and Hall, 1999, pp. 261–327.
30. J.-Y. D'Aoust. unpublished results.
31. J.-Y. D'Aoust, D. B. Emmons, R. McKellar, G. E. Timbers, E. C. D. Todd, A. M. Sewell, and D. Warburton. Thermal inactivation of *Salmonella* species in fluid milk. *J. Food Prot.* **50**, 494–501 (1987).

32. A. R. Davies, R. M. Blood, and P. A. Gibbs. Effect of moisture level on the heat resistance of salmonellae in cocoa liquor. *Leatherhead Food R. A.,* Report No. 666, p. 31 (1990).
33. L. K. Epling and J. A. Carpenter. Antibiotic resistance of *Salmonella* isolated from pork carcasses in northeast Georgia. *J. Food Prot.* **53**, 253–254 (1990).
34. S. M. Glidewell, N. Deighton, B. A. Goodman, and J. R. Hillman. Detection of irradiated food: A review. *J. Sci. Food Agric.* **61**, 281–300 (1993).
35. J. M. Goepfert and R. A. Biggie. Heat resistance of *Salmonella typhimurium* and *Salmonella senftenberg* 775W in milk chocolate. *Appl. Microbiol.* **16**, 1939–1940 (1968).
36. J. Henkel. Ice cream firm linked to *Salmonella* outbreak. *FDA Consumer* **29**, 30–31 (1995).
37. N. T. T. Hoa, T. S. Diep, J. Wain, C. M. Parry, T. T. Hien, M. D. Smith, A. L. Walsh, and N. J. White. Community-acquired septicaemia in southern Viet Nam: The importance of multidrug-resistant *Salmonella typhi. Trans. Roy. Soc. Trop. Med. Hyg.* **92**, 503–508 (1998).
38. T. J. Humphrey, P. A. Chapman, B. Rowe, and R. J. Gilbert. A comparative study of the heat resistance of salmonellae in homogenized whole egg, egg yolk or albumin. *Epidemiol. Infect.* **104**, 237–241 (1990).
39. J. L. Kornacki and E. H. Marth. Thermal inactivation of *Salmonella senftenberg* and *Micrococcus freudenreichii* in retentates from ultrafiltered milks. *Lebensmitt. Wiss. U-Technol.* **26**, 21–27 (1993).
40. C. Lecos. Of microbes and milk: Probing America's worst *Salmonella* outbreak. *Dairy Food Sanit.* **6**, 136–140 (1986).
41. U. S. Nair, A. M. Saeed, P. M. Muriana, R. A. Kreisle, B. Barrett, C. L. Sinclair, and M. L. Fleissner. Plasmid profiles and resistance to antimicrobial agents among *Salmonella enteritidis* isolates from human being and poultry in the midwestern United States. *J. Am. Vet. Med. Assoc.* **206**, 1339–1344 (1995).
42. M. S. Palumbo, S. M. Beers, S. Bhaduri, and S. A. Palumbo. Thermal resistance of *Salmonella* spp. and *Listeria monocytogenes* in liquid egg yolk and egg yolk products. *J. Food Prot.* **58**, 960–966 (1995).
43. M. S. Palumbo, S. M. Beers, S. Bhaduri, and S. A. Palumbo. Thermal resistance of *Listeria monocytogenes* and *Salmonella* spp. in liquid egg white. *J. Food Prot.* **59**, 1182–1186 (1996).
44. A. Pönkä, Y. Andersson, A. Siitonen, B. de Jong, M. Jahkola, O. Haikala, A. Kuhmonen, and P. Pakkala. *Salmonella* in alfalfa sprouts. *Lancet* **345**, 462–463 (1992).
45. M. Pontello, L. Sodano, A. Nastasi, C. Mammina, M. Astuti, M. Domenichini, G. Belluzzi, E. Soccini, M. G. Silvestri, M. Gatti, E. Gerosa, and A. Montagna. A community outbreak of *Salmonella enterica* serotype Typhimurium associated with salami consumption in northern Italy. *Epidemiol. Infect.* **120**, 209–214 (1998).
46. M. Y. Popoff, J. Bockemühl, and F. W. Brenner. Supplement 1997 (no. 41) to the Kauffmann-White scheme. *Res. Microbiol.* **149**, 601–604 (1998).
47. S. Ratman, F. Stratton, C. O'Keefe, A. Roberts, R. Coates, M. Yetman, S. Squires, R. Khakhria, and J. Hockin. *Salmonella enteritidis* outbreak due to contaminated cheese—Newfoundland. *Can. Comm. Dis. Rep.* **25**, 17–21 (1999).

48. R. B. Read, J. G. Bradshaw, R. W. Dickerson, and J. T. Peeler. Thermal resistance of salmonellae isolated from dry milk. *Appl. Microbiol.* **16**, 998–1001 (1968).
49. P. J. A. Reilly and D. R. Twiddy. *Salmonella* and *Vibrio cholerae* in brackishwater cultured tropical prawns. *Int. J. Food Microbiol.* **16**, 293–301 (1992).
50. S. Rubino, E. Muresu, M. Solinas, M. Santona, B. Paglietti, A. Azara, A. Schiaffino, A. Santona, A. Maida, and P. Cappuccinelli. IS200 fingerprint of *Salmonella enterica* serotype Typhimurium human strains isolated in Sardinia. *Epidemiol. Infect.* **120**, 215–222 (1998).
51. G. Samonis, S. Maraki, A. Christidou, A. Georgiladakis, and Y. Tselentis. Bacterial pathogens associated with diarrhoea on the island of Crete. *Eur. J. Epidemiol.* **13**, 831–836 (1997).
52. J. D. Schuman and B. W. Sheldon. Thermal resistance of *Salmonella* spp. and *Listeria monocytogenes* in liquid egg yolk and egg white. *J. Food Prot.* **60**, 634–638 (1997).
53. A. M. Seyfarth, H. C. Wegener, and N. Frimodt-Møller. Antimicrobial resistance in *Salmonella enterica* subsp. *enterica* serovar Typhimurium from humans and production animals. *J. Antimicrob. Chemother.* **40**, 67–75 (1997).
54. M. A. Tessi, M. S. Salsi, M. I. Caffer, and M. A. Moguilevsky. Drug resistance of *Enterobacteriaceae* isolated from chicken carcasses. *J. Food Prot.* **60**, 1001–1005 (1997).
55. E. J. Threlfall, J. A. Frost, L. R. Ward, and B. Rowe. Increasing spectrum of resistance in multiresistant *Salmonella typhimurium*. *Lancet* **347**, 1053–1054 (1996).
56. M. A. Usera, A. Echeita, A. Aladueña, M. C. Blanco, R. Reymundo, M. I. Prieto, O. Tello, R. Cano, D. Herrera, and F. Martinez-Navarro. Interregional foodborne salmonellosis outbreak due to powdered infant formula contaminated with lactose-fermenting *Salmonella virchow*. *Eur. J. Epidemiol.* **12**, 377–381 (1996).
57. D. Wolday. Increase in the incidence of multidrug-resistant salmonellae in Ethiopia. *J. Antimicrob. Chemother.* **41**, 421–423 (1998).

CHAPTER 12

Shigella

DAVID W. K. ACHESON

12.1. INTRODUCTION

Shigella species are not typically high on the list of organisms when one thinks of foodborne pathogens. Yet *Shigella* can be and is transmitted in food and water and may result in a very unpleasant episode of gastroenteritis, and even major and life-threatening complications, in those who are exposed. According to the U.S. Food and Drug Administration, *Shigella* is considered to be the cause of around 10% of outbreaks of foodborne illness in the United States. *Shigella* is unusual in that it is not typically present in fecal material from animals such as poultry, beef, and pork that we usually associate with foodborne illness. Therefore it is not transmitted in food and water in the same way as many other foodborne pathogens such as *Salmonella, Campylobacter*, or *Escherichia coli. Shigella* is highly host adapted and infects only humans and some nonhuman primates; therefore transmission via food can usually be traced back to human fecal contamination of the food product. There are four different species of *Shigella* (*S. dysenteriae, S. flexneri, S. sonnei*, and *S. boydii*) and all cause human disease, but some species are much more prevalent in developed countries than others. Also, the diseases associated with the various *Shigella* species are clinically different yet all have the common theme that they are transmitted via the fecal–oral route. During the course of this chapter we will discuss the various microbiological and clinical differences between the different types of *Shigella* and address some of the issues in relation to transmission of these organisms in food.

12.2. NATURE OF ILLNESS

Of the four species of *Shigella* the diseases associated with them do have some common features and yet may have varying degrees of clinical severity and

Guide to Foodborne Pathogens, Edited by Ronald G. Labbé and Santos García
ISBN 0-471-35034-6

TABLE 12.1. Geographic Distribution, Clinical Manifestations, and Outcome Following Infection with Various *Shigella* Species

Species	Geographic Distribution	Clinical Manifestations	Outcome
S. sonnei	Principal cause of disease in the United States	Watery diarrhea	Usually self-limiting
S. boydii	Rare in developed countries, found in Indian subcontinent	Some dysentery	Requires antibiotic therapy
S. flexneri	Seen in many parts of the world, including developed countries	Usually causes dysentery	Requires antibiotic therapy
S. dysenteriae	Mainly a problem in developing countries	Causes dysentery and major complications such as hemolytic uremic syndrome (HUS)	Requires antibiotic therapy

complications (Table 12.1). The condition caused by *Shigella* is known as shigellosis and usually begins with generalized constitutional symptoms that include fever, aches, fatigue, and loss of appetite. These symptoms will typically develop around 24 h after exposure to the organisms, but there is a range of around 12–48 h following exposure in which symptoms often develop. Following these initial symptoms, patients usually develop diarrhea. The diarrhea will typically begin as a watery stool but may then progress to bloody stools and dysentery. The development of bloody stools may happen in a few hours or may take several days. This progression is not inevitable and is more likely in patients infected with *S. dysenteriae* and *S. flexneri* than those infected with *S. sonnei.* The classic dysenteric stool consists of small amounts of blood and mucus and is sometimes grossly purulent; more often in shigellosis the blood and mucus is mixed in with small amounts of fecal material. Typically, patients with shigellosis do not lose large volumes of fluid, but the passage of each stool is often accompanied by severe abdominal cramps and tenesmus. Stool frequency may be variable but is usually at least 8–10 movements per day but can on occasion be much more than that. While severe dehydration is not typical in shigellosis, anorexia is often a major feature and can continue well into the recovery phase of the disease. This is of particular importance in developing countries, where patients may be malnourished to begin with and the combination of a highly catabolic illness such as shigellosis with anorexia may result in long-term problems in relation to malnutrition that increase the likelihood of other enteric infections or diseases becoming manifest.

In the United States and other developed countries infection with *Shigella* is usually due to *S. sonnei* or *S. flexneri*. *Shigella sonnei* is the most likely cause of an outbreak of shigellosis in the United States, and there have been several examples of this (see below). *Shigella flexneri* has been more often associated with person-to-person transmission in the United States, especially amongst homosexual men. *Shigella sonnei* produces a relatively mild illness that is unlikely to progress to dysentery and is usually self-limiting. Progression to dysentery occurs more often with *S. boydii* than *S. sonnei*, even more often with *S. flexneri*, and in most patients with *S. dysenteriae.* Some of the complications following shigellosis can be life threatening. Situations such as toxic megacolon and intestinal obstruction are associated with a high mortality but are rare in developed countries. Bacteremia during shigellosis is not routinely looked for but is considered to be unusual. Neurological complications, especially seizures, are well documented as are longer term complications such as reactive arthritis and Reiter's syndrome. Complications following infection with *S. dysenteriae* are much more common and serious than with the other *Shigella* species. However, *S. dysenteriae* is thankfully very unusual in the United States and is really only a significant problem in developing countries. *Shigella dysenteriae* type 1 is the only member of the various *Shigella* species that make Shiga toxin. This is a chromosomally encoded toxin that is homologous to the Shiga toxins made by Shiga toxin–producing *E. coli* (STEC) such as serotype O157:H7. As with STEC, patients infected with *S. dysenteriae* type 1 are at risk of developing hemolytic uremic syndrome (a triad of renal failure, thrombocytopenia, and hemolytic anemia), which carries a mortality of around 10%.

12.3. CHARACTERISTICS OF AGENT

Shigella belongs to the family Enterobacteriacae and closely resembles *E. coli* at the genetic level. The four species of *Shigella* are differentiated by group-specific polysaccharide antigens of lipopolysaccharide (LPS). They are nonmotile and do not possess H antigens. *Shigella dysenteriae* consists of 10 antigenic types of which type 1 is the only one to produce Shiga toxins. *Shigella flexneri* is divided into 6 types and 14 subtypes. *Shigella boydii* includes 18 serological types, and while there is only 1 *S. sonnei* serotype, there are at least 20 colicin types. *Shigella* species are gram-negative, nonencapsulated bacilli and are usually non–lactose fermenting (*S. sonnei* ferments lactose slowly), non–gas producing, and lysine decarboxylase–acetate and mucate negative.

Once ingested, *Shigella* colonizes the lower portions of the intestinal tract, adheres to intestinal epithelial cells, and subsequently invades the mucosal surface. All *Shigella* species have a large number of virulence genes on both the chromosome and on plasmids that are involved in the invasion process. It is thought that the M cells in the intestine are the initial site of entry for *Shigella* and that the organisms then gain access to adjacent epithelial cells through their basolateral surface. *Shigella* also induces a local inflammatory response that results in the infiltration and subsequent transmigration of neutrophils across the intestinal epithelial cell barrier. This process results in an opening of the tight junctions between cells, which may allow access of *Shigella* on the apical surface to the basolateral surface of the epithelial cells. Much is known about the mechanisms of pathogenesis and the genes involved in the various invasion and intra- and intercellular spread of *Shigella*. However it is beyond the scope of this chapter to discuss these issues, and interested readers are referred to the bibliography.

12.4. EPIDEMIOLOGY

As mentioned previously, *Shigella* occurs throughout the world, but different species cause different problems in various geographic areas (Table 12.1). One of the most striking features of shigellosis is the very small inoculum of organisms required to cause disease. As few as 10–100 of the most virulent species, *S. dysenteriae*, are sufficient to cause clinical dysentery in healthy adult volunteers. This low infectious dose leads to a significant level of person-to-person spread, and approximately 20% of persons in a household acquire infection when an index case is identified in a family. This has also resulted in significant spread in residential institutions, day care centers, and other gatherings leading to large outbreaks of *S. sonnei* in the United States and other developed countries. In developing countries shigellosis is prominently associated with poverty, overcrowding, poor personal hygiene, inadequate water supplies, and malnutrition. In countries where these problems exist, the estimated incidence of shigellosis is 750–2000 per 1000 children per year, compared with 0.22 cases per 1000 children per year in the United States.

The question therefore arises as to how much of the spread of *Shigella* is due to food and water and how much is person to person. Given that these organisms are not typically present on food other than via human contamination—either directly during food preparation or indirectly from contamination with human fecal material—all shigellosis could be considered to be due to person-to-person spread in one way or another. Food has been a significant vehicle for the spread of *Shigella* in a number of documented instances. An example of this is a recent report of *Shigella* following a wake in Adelaide, Australia. In this outbreak 13 out of 32 people who ate lunch at the wake developed diarrhea. The majority became symptomatic within 3 days with diarrhea, abdominal pain, vomiting, and bloody stools in 3 of the 13. Most of the food was purchased in the 24 h prior to the lunch and refrigerated overnight. During the night some of the sliced ham that had been bought for the lunch was used and then returned to the refrigerator. The person who handled the ham had recovered earlier that week from a diarrheal illness of unknown type. Following the outbreak in which *S. sonnei* was found in four people and *S. dysenteriae* type 2 in one person, a cohort study implicated only the sliced ham as a possible source. The attack rate in those who consumed ham was considered to be 64%. A subsequent examination of the establishment from where the ham was bought revealed no public health violations, and in view of the lack of other reported cases, it was not considered to be the source of the contamination. The suspicion was that the person recently recovered from the diarrhea who handled the ham had contaminated it. The person who had the *S. dysenteriae* type 2 did not eat the ham and the assumption is that the illness in this case was unrelated to the outbreak.

A variety of foods have been implicated in the spread of *Shigella*, including salads (potato, tuna, shrimp, macaroni, and chicken), raw vegetables, milk and dairy products, and poultry, as well as common-source water supplies. In 1994, an outbreak of shigellosis on a cruise ship resulted in diarrhea and vomiting in 586 passengers out of 1589 and 24 out of 594 crew. *Shigella flexneri* 2a was isolated from at least 12 of the sick passengers. While the source of this outbreak has not been identified, it is highly likely it was due to contaminated food or water on board the ship. An outbreak of *S. sonnei* was associated with drinking water from ground wells in Idaho in 1995. *Shigella sonnei* was also incriminated in an outbreak in several European countries, including Sweden, Norway, and the United Kingdom, due to contamination of iceberg lettuce imported from Spain. A number of other documented food-related outbreaks have occurred. In 1985, nearly 5000 people developed shigellosis in Texas with chopped, bagged lettuce as the implicated source. Interestingly, subsequent testing demonstrated that the *S. sonnei* isolate from the lettuce could survive in chopped lettuce at refrigeration temperatures without compromising the look of the produce. Contamination of salad bars has been implicated in a number of outbreaks on college campuses that occurred in the mid-1980s. Often in these situations it was demonstrated that a sick food handler was the likely source of the contamination. In 1988 there were a number of people who developed shigellosis following consumption of food on a commercial airline. The food was prepared at a central facility, and although various sandwiches were suspected, no specific food was incriminated. In an outbreak of *S. dysenteriae* in Thailand coconut milk desert prepared at a local school was considered to be the source of infection. In 1991 an outbreak of *S. sonnei* in Alaska was associated with the consumption of moose soup. It later transpired that the soup was prepared in private homes and in at least five situations in which women had prepared the soup in their homes there was at least one person who had a gastrointestinal illness at the time the soup was prepared or shortly before. In 1988 over 3000 people became ill with shigellosis while attending a music festival in Michigan. An uncooked tofu salad was implicated as the outbreak vehicle. The same strain of *S. sonnei* had caused a smaller outbreak in some of the food handlers just prior to the festival and was believed to be the source of the infection in the larger outbreak situation. Oysters have also been incriminated as a source of shigellosis in an outbreak involving 24 people in southeastern Texas. All had eaten raw oysters within 5 days at various restaurants. The oysters in all implicated restaurants were from the same supplier, and the suspect oysters were traced to a single boat. One of the crew was found to be an asymptomatic carrier of *S. sonnei* that was very similar to the strain that infected the 24 people. Subsequent investigation found that the toilets on the boat were in fact 5-gal pails that were tipped overboard in the oyster harvesting area after they had been used. These situations add further evidence that humans are often the original source in food-related outbreaks of shigellosis. Other implicated foods in the transmission of *Shigella* include watermelon and other fresh vegetables.

12.5. DETECTION OF ORGANISM

In an infected person shigellosis may be diagnosed clinically, microbiologically, or serologically. Clinically, the classical presentation of dysentery in a febrile patient with abdominal pain and sheets of leukocytes in the stool should make one think of shigellosis. Microbiologically, *Shigella* species are fastidious and tend to die off between sampling from the patient and arrival in the laboratory. Stool samples are better than swabs and rapid transport to the laboratory is essential. A number of selective media are used for *Shigella* isolation, including MacConkey agar, deoxycholate, and eosin–methylene blue agar. Highly selective media such as Hektoen-enteric, *Salmonella–Shigella*, and xylose–lysine–deoxycholate agars can also be used. Serologically, antibodies to *Shigella* lipopolysaccharide have been used for the diagnosis of shigellosis. However, these methods are cumbersome and require a variety of reagents and are not routinely used in microbiology laboratories.

Detection of *Shigella* species in food may be difficult, and the growth and survival characteristics of the various *Shigella* species in different types of food have received little attention. A recent study by Zaika et al. (1998) examined the growth of *S. flexneri* inoculated into various commercially available sterile foods such as canned broths, meat, fish, UHT milk, and baby foods. They found that the organisms did not grow well at temperatures less than 19°C. Others have shown that *S. flexneri* will survive on certain foods such as boiled rice, lentil soup, and milk at 5°C and that keeping the food at 25–37°C resulted in multiplication of the organisms by several logs. A detection method for *Shigella* that involved selective enrichment in GN broth supplemented with novobiocin (10 μg/mL) followed by subculture to *Salmonella–Shigella* agar was tested in a variety of foods. Test samples of minced beef, fresh cut vegetables, and cooked shrimp were contaminated with *Shigella*. The *Shigella* could be detected at a contamination level of 1 organism per 25 g in the shrimp in 20 out of 20 samples tested and at 25 organisms per 25 g in 13 out of 18 cut vegetable samples but not in the minced beef even at levels of 100 organisms per 25 g.

Non-culture-based methods such as polymerase chain reaction (PCR) have also been used for the detection of *Shigella* in food. These have been based on amplification of specific virulence genes that are unique to *Shigella* and entero-invasive *E. coli*. These genetic methods have been combined with novel methods of sample preparation such as buoyant density centrifugation. This method involves the layering of the suspect food sample onto Percoll. Centrifugation results in the bacteria moving to the layer below the Percoll and the food residue remaining on the top. This has the advantage of removing some of the inhibitors of PCR that may be present in food.

12.6. PREVENTION AND CONTROL

Shigella organisms are highly infectious as defined by a low inoculation dose required to produce illness in humans. As already discussed, *Shigella* only colonizes humans and some nonhuman primates. Therefore, someone who becomes sick from *Shigella* has almost certainly ingested human fecal material either directly from another person or indirectly via contamination of food or water. Prevention of shigellosis basically comes down to preventing the transmission of the bacteria from one person to another either in the food or water or by direct person-to-person spread. One of the principal areas in our community where the risks of spread are high are in day care centers. To decrease the likelihood of transmission of *Shigella* and other enteric pathogens in day care centers, facility operators should ensure that staff and children be instructed in rigorous and consistent hand washing, including the use of soap and running water. Hands should be washed after toilet use and changing diapers and before handling, preparing, serving, and eating food. It is even better if those individuals who are preparing and handling the food are not the same ones that are changing the diapers. If there is an outbreak of shigellosis, it is important to decontaminate surfaces regularly and to exclude children with diarrhea from a day care facility until they have recovered. Convalescing children should be placed in a separate room with separate staff and a separate bathroom until they have two stool cultures that are negative for *Shigella* collected 48 h or more after completing a 5-day course of antibiotics. Unlike many other enteric pathogens such as *Salmonella* and *Campylobacter*, the risks of cross-contamination with *Shigella* from uncooked to cooked food is remote. This is simply because the majority of food that arrives in the household is not contaminated with *Shigella* when it arrives.

It has been shown in Bangladesh, where *Shigella* is endemic, that hand washing with soap prevents transmission of shigellosis. The main problem is how to motivate people to change age-old customs and wash hands adequately after defecation. Flies are also a well-recognized transmission route of *Shigella* between human fecal material and food. In developing countries the use of baited fly traps has been shown to reduce the rates of shigellosis.

In terms of treating patients with shigellosis the disease is not usually severely dehydrating; nevertheless maintaining adequate hydration is an important part of patient management. It is beyond the scope of this chapter to discuss treatment of specific complications, but it is important to remember that major complications such as hemolytic uremic syndrome following infection with *S. dysenteriae* can occur and require specialized treatment. Generally, patients with *S. sonnei* infection will recover without antibiotic therapy. Infection with other *Shigella* species usually requires antibiotic treatment. There are increasing problems with antibiotic resistance in *Shigella* species. Virtually all strains carry resistance for streptomycin, tetracycline, and chloramphenicol, and many are now resistant to ampicillin and trimethoprim-sulfamethoxazole.

Because of this increasing resistance, the use of a third-generation cephalosporin in children or a 4-fluoroquinolone in adults is recommended in the United States. In developing countries naladixic acid is frequently used and is generally effective, but some resistance has been reported.

REFERENCE

Zaika, L. L., J. G. Phillips, J. S. Fanelli, and O. J. Scullen. 1998. Revised model for aerobic growth of *Shigella flexneri* to extend the validity of predictions at temperatures between 10 and 19 degrees C. *Int. J. Food Microbiol.* **41**, 9–19.

BIBLIOGRAPHY

Acheson, D. W. K., and G. T. Keusch. 1995. *Infections of the Gastrointestinal Tract.* New York: Raven, pp. 763–784.

Anonymous. 1999. Shigella at a wake in Adelaide. *Dairy, Food Environ. Sanit.* Feb.

Hale, T. L. 1991. Genetic basis of virulence in *Shigella* species. *Micriobiol. Rev.* **55**, 206–224.

Hedberg, C. W., W. C. Levine, K. E. White, R. H. Carlson, D. K. Winsor, D. N. Cameron, K. L. MacDonald, and M. T. Osterholm. 1992. An international foodborne outbreak of shigellosis associated with a commercial airline. *JAMA*. **268**, 3208–3212.

Kapperud, G., L. M. Rorvik, V. Hasseltvedt, B. G. Iversen, K. Staveland, G. Johnsen, J. Leitao, Y. Herikstad, Y. Andersson, G. Langeland, B. Gondrosen, and J. Lassen. 1995. Outbreak of *Shigella sonnei* infection traced to imported lettuce. *J. Clin. Microbiol.* **33**, 609–614.

Lampel, K. A., J. A. Jagow, M. Trucksess, and W. E. Hill. 1990. Polymersase chain reaction for detection of invasive *Shigella flexneri* in food. *Appl. Environ. Micriobiol.* **56**, 1536–1540.

Reeve, G., D. L. Martin, J. Pappas, R. E. Thompson, and K. D. Greene. 1989. An outbreak of shigellosis associated with the consumption of raw oysters. *New Engl. J. Med.* **321**, 224–227.

CHAPTER 13

Staphylococcus aureus

R. W. BENNETT

13.1. INTRODUCTION

The first association of *Staphylococcus* with foodborne illness dates back as early as 1884. Spherical organisms in cheese caused a large food-poisoning outbreak in the United States, other earlier outbreaks now attributed to the consumption of staphylococcal contaminated foods occurred in France in 1894, in the United States in Michigan in 1907, and in the Phillippines in 1914. In 1930, Gail Dack and his colleagues at the University of Chicago were able to demonstrate that the cause of a staphylococcal food poisoning that occurred by the consumption of a contaminated Christmas sponge cake with cream filling was a toxin produced by the isolated staphylococci.

The growth and proliferation of *Staphylococcus aureus* in foods presents a potential hazard to consumer health since many strains of *S. aureus* produce enterotoxins. The primary reasons for examining foods for *S. aureus* and/or their toxins are to confirm that this organism is the causative agent in a specific food-poisoning episode, determine whether a food or ingredient is a source of enterotoxigenic staphylococci, and demonstrate postprocessing contamination. The latter is usually due to human contact with processed food or exposure of food to inadequately sanitized food-processing surfaces. Foods subjected to postprocess contamination with enterotoxigenic staphylococci also represent a potential hazard because of the absence of competitive organisms that might restrict the growth of *S. aureus* and subsequent production of enterotoxins.

Of the various metabolites produced by the staphylococci, the enterotoxins pose the greatest risk to consumer health. Enterotoxins are proteins produced by some strains of staphylococci, which, if allowed to grow in foods, may produce enough enterotoxin to cause illness when the contaminated food is consumed. These structurally related, toxicologically similar proteins are produced primarily by *S. aureus*, although *Staphylococus intermedius* and *Staphylococcus hyicus* also have been shown to be enterotoxigenic. Normally considered a vet-

Guide to Foodborne Pathogens, Edited by Ronald G. Labbé and Santos García
ISBN 0-471-35034-6

erinary pathogen, *S. intermedius* was isolated from butter blend and margarine in a food-poisoning outbreak. A coagulase-negative *Staphylococcus epidermidis* was reported to have caused at least one outbreak. These incidents support testing staphylococci other than *S. aureus* for enterotoxigenicity, if they are present in large numbers in a food suspected of causing a food-poisoning outbreak.

Foods commonly associated with staphylococcal food poisoning fall into general categories such as meat and meat products, salads, cream-filled bakery products, and dairy products. Many of these items are contaminated during preparation in homes or food service establishments and subsequently mishandled prior to consumption. In processed foods, contamination may result from human, animal, or environmental sources. Therefore, the potential for enterotoxin development is greater in foods that are exposed to temperatures that permit the growth of *S. aureus*. This is especially true for fermented meat and dairy products. Though the potential is there, it is only when improper fermentation takes place that the development of staphylococcal enterotoxin occurs.

In processed foods in which *S. aureus* is destroyed by processing, the presence of *S. aureus* usually indicates contamination from the skin, mouth, or nose of food handlers. This contamination may be introduced directly into foods by process line workers with hand or arm lesions caused by *S. aureus* coming into contact with the food or by coughing and sneezing, which is common during respiratory infections. Contamination of processed foods also may occur when deposits of contaminated food collect on or adjacent to processing surfaces to which food products are exposed. When large numbers of *S. aureus* are encountered in processed food, it may be inferred that sanitation, temperature control, or both were inadequate.

In raw food, especially animal products, the presence of *S. aureus* is common and may not be related to human contamination. Staphylococcal contamination of animal hides, feathers, and skins is common and may or may not result from lesions or bruised tissue. Contamination of dressed animal carcasses by *S. aureus* is common and often unavoidable. Raw milk and unpasteurized dairy products may contain large numbers of *S. aureus*, usually a result of staphylococcal mastitis.

The significance of the presence of *S. aureus* in foods should be interpreted with caution. The presence of large numbers of the organism in food is not sufficient cause to incriminate a food as the vehicle of food poisoning. Not all *S. aureus* strains produce enterotoxins. The potential for staphylococcal intoxication cannot be ascertained without testing the enterotoxigenicity of the *S. aureus* isolate and/or demonstrating the presence of staphylococcal enterotoxin in food. Neither the absence of *S. aureus* nor the presence of small numbers is complete assurance that a food is safe. Conditions inimical to the survival of *S. aureus* may result in a diminished population or death of viable microbial cells, while sufficient toxin remains to elicit symptoms of staphylococcal food poisoning.

The method to be used for the detection and enumeration of *S. aureus* depends, to some extent, on the reason for conducting the test. Foods suspected to be vectors of staphylococcal food poisoning frequently contain a large population of *S. aureus*, in which case a highly sensitive method will not be required. A more sensitive method may be required to demonstrate an unsanitary process or postprocess contamination, since small populations of *S. aureus* may be expected. Usually, *S. aureus* may not be the predominant species present in the food, and therefore, selective inhibitory media are generally employed for isolation and enumeration.

The methods for identifying enterotoxins involve the use of specific antibodies. The fact that there are several antigenically different enterotoxins complicates their identification because each one must be assayed separately. Another problem is that unidentified enterotoxins exist for which antibodies are not available for in vitro serology. These unidentified toxins, however, appear to be responsible for only a very small percentage of food-poisoning outbreaks.

13.2. NATURE OF ILLNESS

13.2.1. Symptoms

The onset of symptoms in staphylococcal food poisoning is usually rapid (2–6 h) and in many cases acute, depending on individual susceptibility to the toxin, the amount of contaminated food eaten, the amount of toxin in the food ingested, and the general health of the victim. The most common symptoms are nausea, vomiting, retching, abdominal cramping, and prostration. Some individuals may not always demonstrate all the symptoms associated with the illness. In more severe cases, headache, muscle cramping, and transient changes in blood pressure and pulse rate may occur. Recovery generally takes 2 days; however, it is not unusual for complete recovery to take 3 days and sometimes longer in severe cases. Death from staphylococcal food poisoning is very rare, although such cases have occurred among the elderly, infants, and severely debilitated persons.

13.2.2. Dose

A toxin dose of less than 1.0 μg in contaminated food will produce symptoms of staphylococcal intoxication. This toxin level is reached when *S. aureus* populations exceed 10^5 g^{-1}. However, in highly sensitive people, a dose of 100–200 ng is sufficient to cause illness.

13.3. CHARACTERISTICS OF AGENT

13.3.1. Organism

Staphylococcus aureus is a spherical bacterium (coccus) that on microscopic examination appears in pairs, short chains, or bunched, grapelike clusters. The organisms are gram positive. Some strains can produce a highly heat stable protein toxin capable of causing illness in humans. Other salient characteristics are that they are nonmotile and asporogenous; capsules may be present in young cultures but are generally absent in stationary-phase cells. The *Staphylococcus* species are aerobes or facultative anaerobes and have both respiratory and fermentative metabolism. They are catalase positive and utilize a wide variety of carbohydrates. Amino acids are required as nitrogen sources, and thiamine and nicotinic acid are required among the B vitamins. When grown anaerobically, they appear to require uracil.

Although the staphylococci are mesophilic, some strains of *S. aureus* grow at a temperature as low as 6.7°C. In general, growth of *S. aureus* range from 7 to 47.8°C with an optimum temperature for growth at 35°C. The pH range for growth is between 4.5 and 9.3, with the optimum between pH 7.0 and 7.5. As is true with other parameters, the exact minimum growth pH is also dependent on the degree to which all other parameters are at optimal conditions. With regard to water activity (A_w), the staphylococci are unique in being able to grow lower than other nonhalophilic bacteria. Growth has been demonstrated as low as 0.83 under ideal conditions. These low-A_w conditions are too low for the growth of many competing organisms. Most strains of *S. aureus* are highly tolerant to the presence of salts and sugars and can grow over an A_w range of 0.83 to >0.99. *Staphylococcus aureus* grows best at an A_w of >0.99 and growth at low-A_w values depends on other growth conditions being optimal.

Staphylococcus aureus is capable of producing a large number of extracellular enzymes, toxins, and other chemical components. It has been shown that *S. aureus* is capable of producing at least 34 different extracellular proteins, although no one strain of the organism is capable of producing all of these proteins. Some of these extracellular metabolites have been useful in the identification of *S. aureus* and differentiation from other commonly encountered staphylococcal species. The salient characteristics of *S. aureus* and some other staphylococcal species are presented in Table 13.1. The two most common metabolites that have been the most useful in the identification of *S. aureus* are coagulase, a soluble enzyme that coagulates plasma, and thermonuclease (TNase), a heat-stable phosphodiesterase that can cleave either TNase deoxyribonucleic acid (DNA) or ribonucleic acid (RNA) to produce 3′-phosphomononucleosides. Thermonuclease is much more heat stable than ribonuclease and is useful in speciating staphylococci.

13.3.2. Enterotoxins

The staphylococcal enterotoxins are single-chain proteins that are antigenic with molecular weights of 26,000–29,000. They are basic proteins with isoelectric points of 7.0–8.6. They are resistant to proteolytic enzymes such as trypsin and pepsin, which makes it possible for them to travel through the digestive tract to the site of action. The toxins are highly stable to heat, which makes these toxins remain a potential health hazard when they appear in

TABLE 13.1. Characteristics of *Staphylococcus* Species

	Species[a]			
Property	*S. aureus*	*S. intermedius*	*S. hyicus*	*S. epidermidis*
Pigment	+[b]	–	–	–
Coagulase	+	+	+/–	–
DNase	+	+	+/–	–
Hemolysis	+	+	–	+/–
Mannitol (anerobic conditions)	+	–	–	–
Acetoin	+	–	–	+
Clumping	+	+	+/–	–
Hyaluronidase	+	–	+	–
Lysostaphin	HS[3]	HS	HS	SS[4]

[a] Abbreviations: HS, high sensitivity; SS, slight sensitivity.
[b] Over 90%.

canned foods. Their other general properties such as amino acid composition and immunological characteristics have been readily described elsewhere.

13.4. EPIDEMIOLOGY

The epidemiology of foodborne disease is changing in intensity and concept to better trace established organisms as well as newly recognized or emerging pathogens as etiological agents of foodborne illnesses. Many of the pathogens have reservoirs in healthy food animals, from which they spread to a wide variety of foods. These pathogens, including staphylococcal species other than *S. aureus*, cause millions of sporadic illnesses and chronic complications as well as massive and challenging outbreaks around the world. Recently developed technologies and commercially available rapid methods have allowed for improved surveillance of such outbreaks. An outbreak investigation or epidemiological study should go beyond identifying a suspected food and removing it from the shelf to defining the chain of events that allowed contamination with an organism in large enough numbers to cause illness. This approach provides strategies for preventing similar occurrences in the future.

13.4.1. Frequency of Illness

The true incidence of staphylococcal food poisoning is unknown for a number of reasons, including poor responses from victims during interviews with health officials; misdiagnosis of the illness, which may be symptomatically similar to other types of food poisoning (such as vomiting caused by *Bacillus cereus* emetic toxin); inadequate collection of samples for laboratory analyses; improper laboratory examination; and, in many countries, unreported cases.

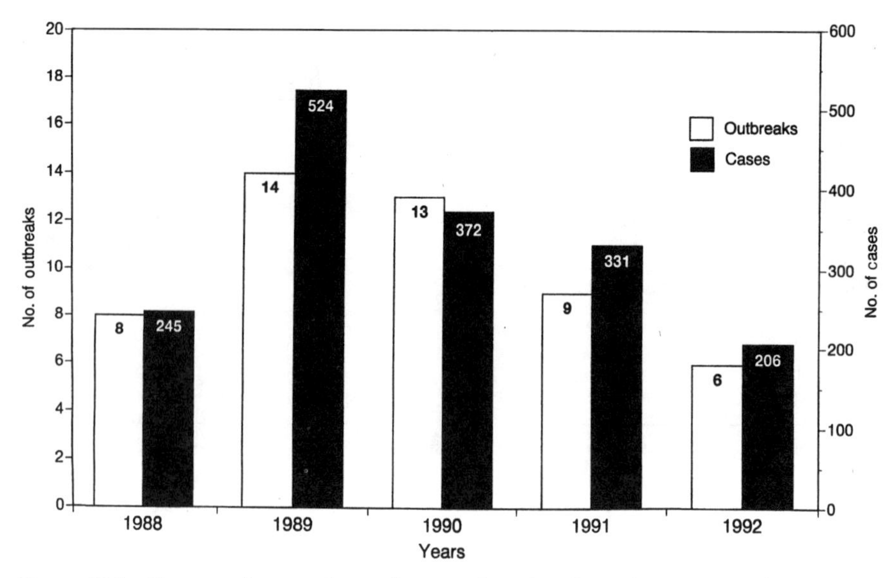

Figure 13.1. Comparative numbers of reported outbreaks and cases caused by staphylococcal species in the United States, 1988–1992.

During a 5-year period (1988–1992) in the United States, bacterial pathogens accounted for 789 outbreaks involving 33,206 cases. Of those outbreaks, 50 outbreaks involving 1678 cases were attributed to staphylococcal intoxication. Figure 13.1 shows the comparative numbers of outbreaks and cases caused by staphylococcal foodborne illnesses from 1988 to 1992 in the United States.

13.4.2. Diagnosis of Human Illness

In the diagnosis of staphylococcal foodborne illness, proper interviews with the victims and gathering and analyzing epidemiological data are essential. Incriminated foods should be collected and examined for staphylococci. The presence of relatively large numbers of enterotoxigenic staphylococci is good circumstantial evidence that the food contains toxin. The most conclusive test is the linking of an illness with a specific food or, in cases where multiple vehicles exist, the detection of the toxin in the food sample(s). In cases where the food may have been treated to kill the staphylococci, as in pasteurization or heating, direct microscopic observation of the food may be an aid in the diagnosis. A number of serological methods for determining the enterotoxigenicity of *S. aureus* isolated from foods as well as methods for the separation and detection of toxins in foods have been developed and used successfully to aid in the diagnosis of the illness. Phage typing may also be useful when viable staphylococci can be isolated from the incriminated food, from victims, and from suspected carriers such as food handlers. However, this approach can be limiting because there are strains of *S. aureus* that are not typable by this system. More recently, genetic fingerprinting techniques are being applied to characterize strains of staphylococci. Two of these approaches are pulse-field gel electrophoresis (PFGE) of DNA restriction fragment profiles and DNA restriction fragment polymorphism of ribosomal RNA (rRA) genes (ribotyping).

13.4.3. Vectors of Transmission

Staphylococci exist in air, dust, sewage, water, milk, and food or on food equipment, environmental surfaces, humans, and animals. Humans and animals are the primary reservoirs. Staphylococci are present in the nasal passages and throats and on the hair and skin of 50% or more of healthy individuals. This incidence is even higher for those who associate or come in contact with sick individuals and hospital environments. Although food handlers are usually the main source of food contamination in food-poisoning outbreaks, equipment and environmental surfaces can also be sources of contamination with *S. aureus*. Human intoxication is caused by ingesting enterotoxins produced in food by some strains of *S. aureus*, usually because the food has not been kept hot enough [60°C (140°F) or above] or cold enough [7.2°C (45°F) or below].

13.4.4. Foods Incriminated

Foods that are frequently incriminated in staphylococcal food poisoning include meat and meat products; poultry and egg products; salads such as egg, tuna, chicken, potato, and macaroni; bakery products such as cream-filled pastries, cream pies, and chocolate eclairs; sandwich fillings; and milk and dairy products. Foods that require considerable handling during preparation and that are kept at slightly elevated temperatures after preparation are frequently involved in staphylococcal food poisoning. The types of food incriminated in food-poisoning outbreaks in the United States from 1988 to 1992 are listed in Table 13.2.

13.4.5. Typical Food-Poisoning Outbreak

Recently, 1364 children became ill out of a total of 5824 who had eaten lunch served at 16 elementary schools in Texas. The lunches were prepared in a central kitchen and transported to the schools by truck. Epidemiological studies revealed that 95% of the children who became ill had eaten a chicken salad. The afternoon of the day preceding the lunch, frozen chickens were boiled for 3 h. After cooking, the chickens were deboned, cooled to room temperature with a fan, ground into small pieces, placed into 12-inch-deep aluminum pans, and stored overnight in a walk-in refrigerator. The following morning, the remaining ingredients of the salad were added and the mixture was blended with an electric mixer. The food was placed in thermal containers and transported to the various schools at 9:30 AM to 10:30 AM, where it was kept at room temper-

TABLE 13.2. Reported Vehicles of Transmission Causing Staphylococcal Foodborne Outbreaks in the United States, 1988–1992

	Number of Outbreaks				
Food Type	1988	1989	1990	1991	1992
Baked food			1	2	
Beef	1		2		2
Chicken			1		
Chinese food		1	1	1	
Ham	3	3	3		1
Mexican food		1	1	1	
Multiple-dairy				1	
Multiple	3	7	1	2	3
Mushrooms		1			
Other fish		1			
Other salad			1	4	
Pork			1	1	
Potato salad			1		
Salad (poultry, fish, egg)					
Turkey	1				
Unknown			1		

Source: Compiled from *CDC Surveillance Summaries*, Vol. 45, No. SS5, 1996.

ature until served between 11:30 AM and noon. Bacteriological examination of the chicken salad revealed the presence of large numbers of *S. aureus*.

Contamination of the chicken probably occurred when it was deboned. The chicken was not cooled rapidly enough because it was stored in 12-inch-deep layers. Growth of the staphylococcus probably occurred also during the period when the food was kept in the warm classrooms. Prevention of this incident would have entailed screening the individuals who deboned the chicken for carriers of the *Staphylococcus*, more rapid cooling of the chicken, and adequate refrigeration of the salad from the time of preparation to its consumption.

13.4.6. Atypical Food-Poisoning Outbreaks (Thermally Processed Food)

In 1989, multiple staphylococcal foodborne diseases were associated with the consumption of canned mushrooms.

Starkville, Mississippi. On February 13, 22 people became ill with gastroenteritis several hours after eating at a university cafeteria. Symptoms included nausea, vomiting, diarrhea, and abdominal cramps. Nine people were hospitalized. Canned mushrooms served with omelets and hamburgers were associated with the illness. No deficiencies in food handling were found. Staphylococcal enterotoxin type A was identified in a sample of implicated mushrooms from the omelet bar and in unopened cans from the same lot.

Queens, New York. On February 28, 48 people became ill a median of 3 h after eating lunch in a hospital employee cafeteria. One person was hospitalized. Canned mushrooms served at the salad bar were epidemiologically implicated. Two unopened cans of mushrooms from the same lot as the implicated can contained staphylococcal enterotoxin A.

McKeesport, Pennsylvania. On April 17, 12 people became ill with gastroenteritis a median of 2 h after eating lunch or dinner at a restaurant. Two people were hospitalized. Canned mushrooms, consumed on pizza or with a parmigiana sauce, were associated with the illness. No deficiencies were found in food preparation or storage. Staphylococcal enterotoxin was found in samples of remaining mushrooms and in unopened cans from the same lot.

13.5. DETECTION AND IDENTIFICATION

13.5.1. Tests Used for Identification

Sometimes additional diagnostic features may be required to confirm *S. aureus* colonies because the inhibitors used may not completely prevent growth of other organisms, such as bacilli, micrococci, streptococci, and some yeasts. Microscopic morphology helps to differentiate bacilli, streptococci, and yeasts from staphylococci, which form irregular or grapelike clusters of cocci. Staphylococci may be further differentiated from streptococci on the basis of the catalase test, with the former being positive. Additional features are needed to differentiate staphylococci further from micrococci. Usually, staphylococci are lysed by lysostaphin but not by lysozyme, and they can grow in the presence of 0.4 μg/mL of erythromycin. Micrococci are not lysed by lysostaphin, may be lysed by lysozyme, and will not grow in the presence of erythromycin. In a deep stab culture, micrococci will grow at the surface, whereas most staphylococci grow throughout the agar. Staphylococci will grow and produce acid from glucose and mannitol anaerobically, whereas micrococci do not. Staphylococcal cells contain teichoic acids in the cell wall and do not contain aliphatic hydrocarbons in the cell membrane, whereas the reverse is true with micrococci. Further, the G + C content (mole percentage) of staphylococci is 30–40 and 66–75 for micrococci. Testing for some of these features is difficult, time consuming, and expensive and usually not required for routine detection and enumeration procedures. Several commercially available miniaturized systems have been developed to speciate staphylococci. A number of commercially available nucleic acid and serological based assays for the detection and confirmation of *S. aureus* are listed in Table 13.3.

13.5.2. Diagnostic Features

The principal diagnostic features of contemporary media include (a) the ability of *S. aureus* to grow in the presence of 7.5 or 10% NaCl; (b) the ability to grow in the presence of 0.01–0.05% lithium chloride and from 0.12–1.26% glycine or

TABLE 13.3. Commercially Available Nucleic Acid and Serological Based Assays for Detection and Confirmation of *S. Aureus*

Trade Name	Assay Format	Company
AccuProbe	Probe[a]	Gen-Probe, San Diego, CA
GENE-TRAK	Probe	Gene-Trak Hopkinton, MA
Staphylo-slide	Latex agglutination	Becton Dickinson, Cockeysville, MD
Aureus Test	Latex agglutination	Trisum, Taipei, Taiwan
Staphlococcus aureus Visual Immunoassay	ELISA	TECRA Diagnostics, Roseville, NSW, Australia
Staphaurex	Latex agglutination	Rhone Poulenc, Glasgow, United Kingdom
Staphylase	Latex agglutination	Unipath—Oxoid Division, Ogdensburg, NY
Slidex[b]	Latex/RBC agglutination	BioMérieux, Marcy-l'Etoile, France
Mastastaph	Latex agglutination	Mast Laboratories
Staphytest	Latex agglutination	Unipath—Oxoid Division, Ogdensburg, NY
Avistaph	Latex agglutination	Omega Diagnostics
Staph Latex Test	Latex agglutination	Difco Laboratories, Detroit, MI
Bactident Staph	Latex	Merck

[a] Used for identification of pure culture isolates.

[b] A combined latex and hemagglutination test. RBCs sensitized with fibrinogen; latex particles sensitized with monoclonal antibodies to protein A or immunogens on the surface of *S. aureus* strains.

Source: Updated, in part, from P. Feng, 1998, *FDA Bacteriological Analytical Manual*, 8th ed. (rev. A/1998), AOAC International, Gaithersburg, MD.

40 ng/mL polymyxin; (c) the ability of *S. aureus* to reduce potassium tellurite, producing black colonies, aerobically and anaerobically; (d) the colonial form, appearance, and size; (3) the pigmentations of colonies; (f) coagulase activity and acid production in a solid medium; (g) the ability of *S. aureus* to hydrolyze egg yolk; (h) the production of phosphase; (i) the production of thermonuclease; and (j) growth at 42–43°C on selective agar. Media used in the detection and enumeration of *S. aureus* may employ one or more of these diagnostic features.

13.5.3. Media Selection

Enrichment isolation and direct plating are the most commonly used approaches for detecting and enumerating *S. aureus* in foods. Enrichment procedures may be selective or nonselective. Nonselective enrichment is useful for demonstrating the presence of injured cells, whose growth is inhibited by toxic components of selective enrichment media. Enumeration by enrichment isolation, or selective enrichment isolation, may be achieved by determining either an indicated number or the most probable number (MPN) of *S. aureus* present. Common MPN procedures use three or five tubes for each dilution.

For enumeration, samples may be applied to a variety of selective media in two main ways: surface spreading, and pour plates used in direct plating procedures. Surface spreading is advantageous in that the form and appearance of surface colonies are somewhat more characteristic than the subsurface colonies encountered with pour plates. The principal advantage of pour plates is that greater sample volumes can be used.

Selective media employ various toxic chemicals, which are inhibitory for *S. aureus* to a varying extent as well as to competitive species. The adverse effect of selective agents is more acute in processed foods containing injured cells of *S. aureus*. A toxic medium may help prevent overgrowth of *S. aureus* by competing species.

13.5.4. Direct-Plating Method

This method is suitable for the analysis of foods in which more than 100 *S. aureus* cells/g may be expected. The basic equipment, media, reagents, preparation of sample, and procedures for the isolation and enumeration of staphylococci are described in the Food and Drug Administration's (FDA) *Bacteriological Analytical Manual.*

13.5.5. Enrichment Isolation Method

The MPN method is recommended for routine surveillance of products in which small numbers of *S. aureus* are expected and in foods expected to contain a large population of competing species.

13.5.6. Differential Characteristics

Staphylococcus aureus is differentiated from the other staphylococcal species by a combination of the following features: colonial morphology and pigmentation; production of coagulase, thermonuclease, acetone, β-galactosidase, phosphatase, and alpha toxin (hemolysis); acid from mannitol, maltose, xylose, sucrose, and trehalose; novobiocin resistance; presence of ribitol teichoic acid; protein A; and clumping factor in the cell wall. The ultimate species identification may be established by DNA–DNA hybridization with reference strains. A nonisotopic DNA hybridization assay and a polymerase chain reaction (PCR) procedure have been used to successfully identify *S. aureus*.

13.5.7. Coagulase

The confirmation procedure most frequently used to establish the identity of *S. aureus* is the coagulase test. Coagulase is a substance that clots plasma of humans and other animal species. Differences in suitability among plasmas from various animal species have been demonstrated. Human or rabbit plasma

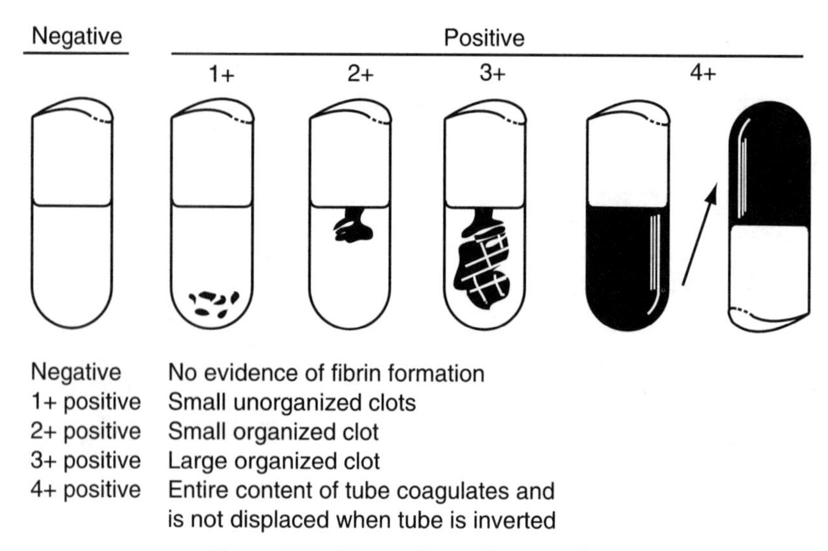

Figure 13.2. Types of coagulase reactions.

is most frequently used for coagulase testing and is available commercially. The use of pig plasma sometimes has been found advantageous, but it is not widely available. Coagulase production by *S. aureus* may be affected adversely by physical factors, such as culture storage condition, pH of the medium, and denigration. The extent to which the production of coagulase may be impaired by the toxic components of selective isolation media has not been demonstrated clearly.

Presence of clumping factor in cells is another unique feature of *S. aureus*. It can be used to distinguish tube-coagulase-positive *S. aureus* from other tube-coagulase-positive species such as *S. hyicus*. Clumping factor present in *S. aureus* cells binds to fibrinogen or fibrin present in human or rabbit plasma, resulting in agglutination of cells. This is referred to as slide coagulase, bound coagulation, or agglutination. Clumping of cells in this test is very rapid (less than 2 min), and the results are more clearcut than 1+ or 2+ clotting in the tube coagulase test. Clumping factor can be detected using commercially available latex agglutination reagents. Anti–protein A immunoglobulin G (IgG) and fibrinogen are used to coat polystyrene latex beads to simultaneously bind protein A and coagulase, both of which are specific cell surface components of *S. aureus*. One latex kit was collaboratively studied by comparing a latex agglutination method to the coagulase test. The types of coagulase test reactions are shown in Figure 13.2.

13.5.8. Thermonuclease

Thermonuclease is also frequently used as a simple, rapid, and practical test for routine identification of *S. aureus*. Coagulase and heat-stable nuclease tests are very efficient for the identification of foodborne *S. aureus* strains isolated on Baird Parker agar. However, the use of the coagulase and/or thermonuclease test may result in erroneous species designation from a taxonomic standpoint. Two species, *S. intermedius* and *S. hyicus* sp. *hyicus* are both coagulase and thermonuclease positive. However, the latter species can easily be differentiated from *S. aureus* on the basis of the clumping factor test. Coagulase and/or thermonuclease negative staphylococci are being reported to be enterotoxigenic.

13.5.9. Ancillary Tests

Additional tests in the identification of *S. aureus* include catalase anaerobic utilization of glucose and mannitol and lysostaphin sensitivity.

13.6. DETECTION OF ENTEROTOXINS

The need to identify enterotoxins in foods encompasses basically two areas: (1) foods that have been incriminated in food-poisoning outbreaks and (2) foods that are suspect of containing enterotoxin. In the former case, the identification of enterotoxin in foods is confirmation to support a staphylococcal food-poisoning outbreak or episode. In the latter case, the presence or absence of toxin will determine the marketability of the product. The latter cannot be overemphasized because it is difficult to prevent the presence of staphylococci in some types of foods. The isolation and determination of enterotoxigenicity of staphylococcal isolates in foods can serve as a signal of potential toxin formation if the food is time–temperature abused, which would allow for the proliferation of the organism. The two most common approaches are biological and serological testing.

13.6.1. Methods for Toxin Identification

Biological Assays. Prior to the advent of serological identification of toxins, all the toxins were identified by emetic responses in a monkey feeding assay. However, such assays had to be limited in quantity and possessed variable sensitivity, making interpretation sometimes difficult. In this method, the test sample is injected by catheter into the stomach of a young monkey; the animal is observed for 5 h, and if vomiting occurs during the observation period, the sample is judged to contain toxin. While this animal assay is considered specific, a number of disadvantages exist. An alternative bioassay is through the IV injection of cats or kittens. However, other bacterial metabolites have been found to cause nonspecific emetic responses, although these nonspecific components can be neutralized or inactivated.

Serological Methods. Most laboratory methods for the identification of the enterotoxins are based on the use of specific antibodies to each of the various toxin serotypes. While all of the enterotoxins are similar in composition and biological activity, they can be differentiated based on serology. Seven serological distinct types of enterotoxin have been characterized and designated as SEA, SEB, SEC (subtypes C_1, C_2, C_3), SED, and SEE. Approximately 5.0% of the staphylococcal foodborne outbreaks are caused by unidentified toxins. Their existence can be demonstrated by biological tests and are not serologically related to the previously established toxin serotypes. More recently, a new enterotoxin, SEH, was identified and partially characterized and a rapid method developed for its identification. A number of methods employing polyclonal or monoclonal antibodies have been used to identify and measure enterotoxins. Earlier developed methods utilized precipitation and agglutination approaches, while more recently developed methods employ tracer-labeled or tagging methods to increase assay sensitivity. Systems based on serological assays can, in general, be divided into a number of antigen–antibody reaction types: (1) gel immunodiffusion by direct precipitation or precipitation inhibition assays, (2) agglutination assays, and (3) tracer-labeled or tagged immunoassays. The most commonly used earlier developed methods have been previously described in a general way, in reviews, and in stepwise procedural detail by a number of investigators. Earlier developed classical methods such as the microslide gel double-diffusion test have been described, in detail, in the FDA *Bacteriological Analytical Manual* and the American Public Health Association (APHA) *Compedium of Methods for the Microbiological Examination of Foods* (Vanderzant and Splittstoesser, 1992). Some commercially available rapid methods for the identification of the enterotoxins are presented in Table 13.4.

Several enzyme-linked immunosorbent assay (ELISA) methods have been proposed for the identification of the staphylococcal enterotoxins and are currently the most commonly used methods. Of the competitive and noncompetitive ELISA-based methods, the noncompetitive, double-antibody Sandwich-ELISA appears to be the most popular for routine toxin identification. With the noncompetitive ELISA, specific antibodies (polyclonal or monoclonal) are absorbed onto a solid support such as paper disks, polyvinyl, or polystyrene in the form of polystyrene balls, plastic microtiter wells, plastic tubes, or other solid-phase supports. The antibody absorbed onto the solid-phase support is the capture antibody. The enterotoxin in samples is bound to the capture antibodies and subsequently detected by the addition of an enzyme-labeled secondary antibody whose enzyme acts on a suitable substrate producing a color reaction. The intensity of the color reaction is proportional to the amount of toxin in the assay food extract or culture fluid. The advantages and limitations of some of the commercially available rapid methods have been reviewed in detail by Su and Wong (1997). Of the rapid methods proposed for the identification of staphylococcal enterotoxins in foods, only the microtiter plate polyvalent ELISA method has been studied exhaustively and is approved by the Association of Official Analytical Chemists (AOAC) International. As a consequence, these methods should always be used with the recommended controls. A polyvalently configured automated enzyme-linked fluorescent im-

TABLE 13.4. Commercially Available Agglutination and ELISA-Based Methods for Identification of Staphylococcal Enterotoxins

Kit Name	Assay Format	Company (Distributor)[b]
TECRA-SET[c]	ELISA, polyvalent[d] (A–E)[e]	TECRA Diagnostics, Roseville, NSW, Australia (International BioProducts, Redmond, WA)
TECRA-SET	ELISA, monovalent[f] (A–E)	TECRA Diagnostics, Roseville, NSW, Australia (International BioProducts, Redmond, WA)
SET-RPLA	Latex agglutination monovalent (A–D)	Unipath—Oxoid Division, Ogdensburg, NY
VIDAS-SET	ELFA, polyvalent (A–E)	BioMérieux-Vitek, Hazelwood, MO
Transia Tube-SET	ELISA, polyvalent (A–E)	TRANSIA Diffchamb, SA, Lyon, France, (Gene-Trak Systems, Hopkinton, MA)
Transia Plate-SET	ELISA, polyvalent (A–E)	TRANSIA Diffchamb, SA, Lyon, France (Gene-Trak Systems, Hopkinton, MA)
SET-EIA	ELISA, polystyrene ball, monovalent (A–D)	Diagnostische Laboratorien, Bern, Switzerland (Toxin Technology, Sarasota, FL)
Microtiter Plate-SET	ELISA, polyvalent (A–E)	W. Brommeli, Bern, Switzerland
Ridascreen SET	ELISA, monovalent (A–E)	R. Biopharm GmbH, Darnstadt, Germany (BioTeck Instruments, Winooski, VT)

[a] Abbreviations: SET, staphylococcal enterotoxin; reversed passive latex agglutination, RPLA; enzyme-linked fluorescent immunoassay, ELFA.

[b] U.S. distributor.

[c] AOAC International status-adopted "Official First" action.

[d] Does not distinguish between toxin serotypes.

[e] Identifiable toxin serotype.

[f] Distinguishes between toxin serotypes.

Source: Updated, in part, from P. Feng, 1998, *FDA Bacteriological Analytical Manual*, 8th Edition (rev. A/1998), AOAC International, Gaithersburg, MD.

munoassay has also been developed and is commercially available. This multiparametric immunoanalyzer is highly sensitive because of the fluorescent tag. This is a labor-saving approach in that only the sample is added and the analyzer automatically completes the ELISA steps—providing print-out data in approximately 80 min. Preliminary evaluation of this system shows that it is highly sensitive ($\leq$1 ng/g) and, generally, specific. Other ELISA-based methods are dedicated to determining specific serotypes of staphylococcal enterotoxins. Some of these monovalently configured assays are also listed in Table 13.4. One such method, TECRA, utilizes a single-specific serotype antibody as the capture antibody with polyvalent antibodies conjugated to the enzyme instead of each secondary antibody conjugated separately to the enzyme.

13.6.2. Toxin Production by Staphylococci

Determining the enterotoxigenicity of staphylococci by examining staphylococcal isolates for toxin production is helpful for identifying enterotoxin in foods and is desirable for examining strains isolated from various sources. A number of methods for the laboratory production of enterotoxins have been developed. Of the methods developed elsewhere for the laboratory production of enterotoxin, only the semisolid agar method is an AOAC International approved method. It is simple to perform and requires a minimum of items found in the routine analytical laboratory. To determine the presence of enterotoxin in culture fluid, any of the classical as well as rapid methods can be utilized to determine the enterotoxigencity of suspect staphylococcal isolates from foods or other sources. Although commercial kits generally recommend broth media, these media are comparable in enterotoxin production to semisolid agar. The potential enterotoxigencity of staphylococcal strains may also be determined by DNA hybridization techniques in cases where the nuclotide sequence has been determined. As a consequence, oligonucleotides can be synthesized and used as DNA probes to demonstrate that the enterotoxin gene exist in an isolate, although laboratory demonstration of toxin production is more direct and serves a better signal for the possible presence of toxin in a suspect food. Determining the enterotoxigenicity of staphylococcal isolates is particularly important in determining prevalence of the various serotypes of toxin producers in foods and other sources.

13.6.3. Toxin Identification in Foods

The major problem of identifying enterotoxin in foods is the small amount that may be present in foods incriminated in foodborne outbreaks. The amount of enterotoxin that may be present in foods involved in food-poisoning outbreaks may be as little as 50 ng/g of food, although the normal amount of toxin in foods involved in food-poisoning outbreaks is easy to detect since frequently the amounts may be larger. Toxin can be readily identified if the counts are or were at some time $\geq 10^6$ enterotoxigenic staphylococci/g of food. Such high counts are not acceptable. Marketable foods should contain no enterotoxin and should be readily demonstrated by rapid as well as classical methods.

Most outbreaks of staphylococcal intoxication are caused by foods that do not receive a high thermal treatment; staphylococci survive in sufficient numbers in these foods to form the toxin before the food is consumed. However, some heated foods have also been incriminated in illnesses that display the typical symptoms of intoxication. Foods that receive enough heat to render the bacterium nonviable and yet cause food poisoning have included boiled goat's milk, spray-dried milk, cooked sausage, and canned lobster bisque.

In two instances, the FDA has taken regulatory action on staphylococcal enterotoxin contaminated thermally processed foods. In 1982, thermally processed infant formula was incriminated in foodborne illness, and in 1989 mushrooms that received a higher than normal thermal treatment as a means of product preservation were implicated in staphylococcal foodborne illnesses.

In the food-poisoning episodes involving canned mushrooms, analysis of the product initially proved serologically negative, although there was a retention in toxicological activity in individuals who consumed the product as indicated by symptoms that were consistent with staphylococcal intoxication. To determine the disparity between serological inactivity and human intoxication, studies were conducted to better understand the kinetics of thermal stress on the enterotoxin protein. In these experimental studies, it was determined that the enterotoxin underwent conformational changes, thus preventing antibody recognition because of toxin denaturation. Methods were developed to renature (reactivate) the heat-altered toxin utilizing urea or urea combined with zinc acetate. This renatured toxin could then be identified serologically. The utilization of urea or urea–zinc acetate to restore serological activity to heat-denatured enterotoxin has been confirmed by other investigators. The only practical way to eliminate future staphylococcal food-poisoning outbreaks in thermally processed foods is to prevent the contamination and proliferation of enterotoxigenic staphylococci in foods before processing.

13.7. PHYSICAL METHODS FOR DESTRUCTION

The total destruction or significant reduction in the bacterial load in foods during growth, harvesting, processing, packaging, and storage prior to consumption has always been a general goal. However, the wide array of parameters for proliferation of foodborne pathogens is staggering. Some of the same methods for the control of organisms in the food supply are used separately or in combination in the preservation of foods. Staphylococci may be totally destroyed or injured when subjected to lethal or sublethal stresses respectively by heat, cold, drying, irradiation, or chemicals. While total destruction of these organisms might be ideal, sublethal injury may occur, thus providing the organism an opportunity to recover and proliferate, if conditions are conducive, and continue to be a potential hazard to consumer health. The kinetics of sublethal injury and stressed cell rejuvenation has been reviewed and examined by a number of investigators. Conditions that have been studied include heating, freezing and freeze drying, irradiation, reduced water activity, and exposure to various chemicals such as acids and salts. The effects of various nutritional and environmental factors on the growth of *S. aureus* with major emphasis on enterotoxin synthesis in foods and model systems have been reviewed previously. Occurrences of food-poisoning outbreaks have demonstrated that growth of staphylococcal species and subsequent enterotoxin synthesis are determined by a variety of nutritional and environmental factors including temperature, pH,

TABLE 13.5. Leading Factors Attributed to Staphylococcal Foodborne Outbreaks in the United States, 1988–1992

Causes	Number of Outbreaks
Improper holding temperatures	37
Poor personal hygiene	18
Inadequate cooking	15
Contaminated equipment	4
Food from unsafe source	4
Other	4

Source: CDC Surveillance Summaries, Vol. 45, No. SS5, 1996.

water activity, salt and sugar concentrations, bacterial load, bacterial competition, and atmospheric conditions.

To better understand the behavior of staphylococcal growth and toxin production in foods, a greater emphasis must be placed on the multiplicity of interactive factors involved in the proliferation of staphylococci in food matrices. Such characterization is necessary to make predictive microbiology a reality.

13.8. PREVENTION AND CONTROL

Staphylococci are ubiquitous and impossible to eliminate from the environment. The most frequent source of contamination of food is the food handler involved in preparing food for serving. Whenever a food is exposed to human handling, there is always a potential that it will be contaminated. Not all strains of *S. aureus* and other species are enterotoxigenic, although a large percent (50–70%) may be. To prevent food-poisoning outbreaks, it is necessary to keep foods either refrigerated (below 10°C) or hot (above 45°C) to prevent proliferation of the organism to such numbers (10^5 cells/g) that are necessary for detectable toxin formation. Additionally, foods to be refrigerated should be placed in shallow layers or small portions to facilitate quick cooling. While the bacterial inoculum size is important, time–temperature abuse and the nutritional composition of the contaminated food are of utmost importance. The leading factors that were attributed to staphylococcal food poisoning outbreaks in the United States during 1988–1992 are listed in Table 13.5.

REFERENCES

Dack, G. M., W. E. Cary, O. Woolpert, and H. Wiggers. 1930. An outbreak of food poisoning proved to be due to a yellow homolytic *Staphylococcus*. *J. Prev. Med.* **4**, 167–175.

Food and Drug Administration. *Bacteriological Analytical Manual.* 8th ed. (rev. A). 1998. Gaithersburg, MD: AOAC International.

Su, Y-C., and A. C. L. Wong. 1997. Current perspectives on detection of staphylococcal enterotoxins. *J. Food Prot.* **60**, 195–202.

Vanderzant, C., and D. Splittstoesser. (Eds.). 1992. *Compendium of Methods for the Microbiological Examination of Foods.* Washington, DC: American Public Health Association.

BIBLIOGRAPHY

Bennett, R. W. 1994. Urea renaturation and identification of staphylococcal enterotoxin. In: R. C. Spencer, E. P. Wrights, and S. W. B. Newson (Eds.), *Rapid Methods and Automation in Microbiology and Immunology*, RAMI-93. Andover, Hampshire, England: Intercept Limited, pp. 193–207.

Bennett, R. W. 1998. Current concepts in the rapid identification of staphylococcal enterotoxin in foods. *Food Test. Anal.* **4**, 16–18, 31.

Bennett, R. W., et al. 1992. Staphylococcal enterotoxins. In: C. Vanderzant and D. Splittstoesser (Eds.), *Compendium of Methods for the Microbiological Examination of Foods.* Washington, DC: American Public Health Association, pp. 551–592.

Bergdoll, M. S. 1989. *Staphylococcus aureus.* In: M. P. Doyle (Ed.), *Foodborne Bacterial Pathogens.* New York: Marcel Dekker, pp. 463–523.

Bergdoll, M. S. 1990. Staphylococcal food poisoning. In: D. O. Cliver (Ed.), *Foodborne Diseases.* San Diego: Academic, pp. 85–106.

Bergdoll, M. S., and R. W. Bennett. 1984. Staphylococcal enterotoxins. In: M. L. Speck (Ed.), *Compendium of Methods for the Microbiological Examination of Foods.* Washington, DC: American Public Health Association, pp. 428–456.

Centers for Disease Control Surveillance Summaries (1988–1992). 1996. Atlanta, GA: U.S. DHHS, Centers for Disease Control and Prevention, 45, #SS-5.

FDA Bacteriological Analytical Manual, 8th ed. 1995. Gaithersburg, MD: AOAC International.

Halpin-Dohnalek. M. I., and E. H. Marth. 1989. *Staphylococcus aureus*: Production of extracellular compounds and behavior in foods—review. *J. Food Prot.* **52**, 267–282.

Jay, J. M. 1992. Staphylococcal gastroenteritis. In: J. M. Jay (Ed.), *Modern Food Microbiology.* New York: Van Nostrand Reinhold, pp. 455–478.

Khambaty, F. M., et al. 1994. Application of pulsed field gel electrophoresis to the epidemiological characterization of *Staphylococcus intermedius* implicated in a food-related outbreak. *Epidemiol. Infect.* **113**, 75–81.

Lancette, G., and R. W. Bennett. 2001. *Staphylococcus aureus* and *Staphylococcal* enterotoxins. In: K. Ito (Ed.), *Compendium of Methods for the Microbiological Examination of Foods.* Washington, DC: American Public Health Association, pp. 387–403.

Martin, S. E., and E. R. Myers. 1994. *Staphylococcus aureus.* In: Y. H. Hui, J. R. Gorham, K. D. Murrell, and D. O. Cliver (Eds.), *Foodborne Disease Handbook, Diseases Caused by Bacteria.* New York: Marcel Dehker, pp. 345–394.

Official Methods of Analysis, 16th ed. 1997. Gaithersburg, MD: AOAC International.

Smith, J. L., et al. 1983. Effect of food environment on staphylococcal enterotoxin synthesis: A review. *J. Food Prot.* **46**, 545–555.

Tatini, S. R., and R. W. Bennett. 1999. *Staphylococcus*/detection by cultural and modern techniques. In: *Encyclopedia of Food Microbiology*. London: Academic, pp. 2071–2076.

Tauxe, R. V. 1997. Emerging foodborne diseases: An evolving public health challenge. *Dairy Food Environ. Sanit.* **17**, 788–795.

CHAPTER 14

Vibrio vulnificus, *Vibrio parahaemolyticus*, and *Vibrio cholerae*

MARK L. TAMPLIN

Bacteria of the genus *Vibrio* are indigenous to fresh, estuarine, and marine environments. Worldwide, the most notable species that are pathogenic to humans include *Vibrio cholerae*, *Vibrio parahaemolyticus*, and *Vibrio vulnificus*. In recent years, there have been fundamental discoveries about the genomics of *Vibrio* spp. and how environmental signals modulate the expression of specific genes involved in virulence and survival. This chapter provides an overview of the diseases caused by *V. vulnificus*, *V. parahaemolyticus*, and *V. cholerae*, the epidemiology of infections, purported virulence factors, standard methods of detection, control interventions, as well as selected characteristics of the organisms.

14.1. *Vibrio vulnificus*

14.1.1. Introduction

In the United States, *V. vulnificus* is the leading reported cause of death attributed to the consumption of seafoods, and it has one of the highest mortality rates (~50%) of all foodborne pathogens. Its has had a profound effect on public health policies, including those in Florida, California, and Louisiana, where consumer advisories must be posted at points of retail sale for raw molluscan shellfish. In addition, the state government of Hawaii has gone further by banning the sale of raw shellfish that contain *V. vulnificus* and are destined for human consumption.

Guide to Foodborne Pathogens, Edited by Ronald G. Labbé and Santos García
ISBN 0-471-35034-6

14.1.2. Nature of Illness Caused

Vibrio vulnificus causes fulminant and life-threatening infections in humans who consume raw oysters from the Gulf of Mexico and in persons who suffer skin wounds in estuarine environments. The consumption of molluscan shellfish presents an increased risk to human health for two primary reasons. First, shellfish filter and concentrate particles from overlying seawater by approximately 100-fold. Second, shellfish are frequently eaten raw or undercooked, thus allowing live infectious microbes access to internal human tissues. The fact that *V. vulnificus* can cause wound infections following skin punctures indicates that the infectious dose may be relatively small, since *V. vulnificus* levels in seawater range from 100 to 1000 colony-forming units (CFU)/mL.

The typical disease caused by *V. vulnificus* is a fulminating infection that results in extensive invasion of host tissues. A primary predisposing host condition is high levels of tissue iron, which is classified as hemochromatosis, a result of gene mutations and liver disease, such as cirrhosis from alcoholism or hepatitis. Additionally, people with diabetes mellitus and immunodeficiency are at risk for *V. vulnificus* disease. Disease and death can occur as soon as 24 h after consumption of oysters containing *V. vulnificus* or after bringing a skin wound in contact with seawater.

The opportunistic nature of *V. vulnificus* disease is illustrated by the large quantity of organisms typically consumed in a meal of oysters without clinical illness. For instance, it is not unusual for one dozen raw summer Gulf of Mexico oysters to contain more than 10 million *V. vulnificus*. Currently, the infective dose of *V. vulnificus* for humans is not well understood, but evidence indicates that it is associated with concentrations of *V. vulnificus* strains greater than 1000 CFU/g of oyster meat, or approximately 10^5 CFU in one dozen oysters.

14.1.3. Characteristics of Agent

Vibrio vulnificus is an autochthonous estuarine organism found in temperate and tropical waters throughout the world. It is isolated commonly from seawater, sediment, and various marine life forms. As with many *Vibrio* spp., it exists in seawater and on a variety of surfaces including plankton, sediment, and fish as well as within tissues of filter-feeding molluscan shellfish. Due to the filter-feeding nature of molluscan shellfish, *V. vulnificus* numbers in shellfish tissues are many times higher than the levels in surrounding seawater.

The ecology of *V. vulnificus* in seawater is markedly affected by temperature and enhanced by salinities of 5–20 parts per thousand (ppt). In a study involving collaborators from health agencies in 15 coastal states, our laboratory observed that *V. vulnificus* can be found in many coastal U.S. waters, ranging from Maine to Washington. When salinity is adequate for *V. vulnificus* survival, such as in Gulf of Mexico estuaries, temperature has a strong influence on *V. vulnificus* growth and survival. In estuarine environments with temperatures below 20°C, *V. vulnificus* numbers decline in seawater, but persist for some time period in sediment and oysters. When water temperatures remain below 15°C, *V. vulnificus* cells die and numbers decline to undetectable levels in seawater, sediment, and shellfish. In a seasonal context, marine entities are likely re-colonized by *V. vulnificus* cells that are introduced by water currents that carry cells from warm environments.

The effect of salinity on *V. vulnificus* survival has been observed in environments where temperature is relatively constant year-round, such as Hawaiian coastal waters. Along the shores of these islands, *V. vulnificus* is not isolated from near-shore high-salinity seawater or in freshwater beach upwellings, yet it can be found in remarkably high concentrations at salinity–fresh water interfaces. These observations have also been confirmed in laboratory microcosms of defined salinity, showing that *V. vulnificus* survival is enhanced at 5–10 ppt, that it declines to nondetectable levels at salinities greater than 20 ppt, and that it lyses in fresh water.

Vibrio vulnificus exists in biofilms on sediments and on various other physical forms in the marine environment. Consequently, as shown for other aquatic bacteria, *V. vulnificus* may utilize quorum sensing mechanisms to modulate between attached and free-swimming life forms. In this regard, *V. vulnificus* possesses a gene that is homologous to the transcriptional regulator, *lux*R, of the *lux* operon found in *Vibrio harveyi.* A similar finding has been reported for *V. cholerae* and *V. parahaemolyticus*, and it indicates that the *lux* operon may have been inherited from a common ancestor, resulting in widespread appearance in marine *Vibrio* spp.

The highest environmental levels of *V. vulnificus* occur in warm months of the year, and these levels parallel the incidence of human infections. Health risks do not occur when the organism cannot be isolated from the environment. In the past decade, there has been significant controversy concerning a purported "viable but nonculturable" (VBNC) state of *V. vulnificus*. In brief, researchers report that following prolonged exposure to low temperature, *V. vulnificus* cells cannot be cultured using conventional microbiological techniques. However, some degree of metabolic activity can be detected that researchers associate with viability. Furthermore, reports state that when the temperature is increased to more favorable conditions, the VBNC forms become culturable on conventional media. Importantly, some laboratories show that this resuscitation is due to the presence of low numbers of culturable cells in test flasks, while others contend that VBNC cells are truly "resuscitated."

It can be expected that many cellular enzymes of *V. vulnificus* would show some degree of activity even after cellular deoxyribonucleic acid (DNA) has degraded and replication is not possible. For example, enzymes are routinely purified and refrigerated and still retain activity; the same could occur with enzymes of dying cells. It is very likely that the term VBNC will not be a static one and its definition will reflect the current knowledge of optimum cultivation techniques needed to isolate and grow an organism in broth or on agar.

Currently, there is no microbiological indicator for *Vibrio* spp., such as one that is analogous to fecal coliforms. However, there has been good progress in the development of predictive models based on quantitative measurements of seawater temperature and salinity. Two groups have conducted extensive environmental research to define environmental factors associated with the incidence and levels of *V. vulnificus* in molluscan shellfish. One group investigated temperature and salinity parameters associated with U.S. waters and oysters linked to foodborne *V. vulnificus* infections. At sites along the Gulf of Mexico, *V. vulnificus* numbers increased with water temperature up to 26°C and were constant at higher temperatures. High (>10,000 g^{-1}) *V. vulnificus* levels were found in oysters from salinities of 5–25 ppt. Smaller *V. vulnificus* numbers (<100 g^{-1}) were found at salinities greater than 28 ppt, typical of Atlantic Coast sites.

A second group studied levels of *V. vulnificus* in seawater, sediment and oysters, collected at monthly test intervals from commercial shellfish harvesting sites in 15 coastal states: Maine, Connecticut, Rhode Island, Massachusetts, New Jersey, Virginia, South Carolina, Florida, Mississippi, Louisiana, Texas, California, Oregon, Washington, and Hawaii. The results of this study showed that of the two predictors (temperature and salinity), temperature had the highest correlation with *V. vulnificus* levels in oysters. A linear regression formula was derived that predicted the levels of *V. vulnificus* in Gulf of Mexico shellfish based on seawater temperature and salinity:

$$\log_{10} \text{CFU/g} = -6.32 + (0.23S) + (0.347T) - (0.0056ST) - (0.0039SS)$$

where S is salinity (ppt) and T is temperature (°C).

14.1.4. Epidemiology

Vibrio vulnificus infections have been reported throughout the world, indicating the ubiquitous nature of the organism and a wide distribution of virulent strains. The vast majority of *V. vulnificus* foodborne infections are restricted to seafoods harvested from temperate and tropical waters where high levels of the organism occur in warm months. Wound infections occur in the same geographical regions as foodborne infections. However they are also reported in areas where foodborne infections are rare and where seawater salinity and temperature are suboptimal for high environmental levels of *V. vulnificus*.

In a recent overview of *V. vulnificus* disease in the United States, 422 infections were reported between 1988 and 1996. Of these, 45% were wound infections, 43% primary septicemia, and 5% gastroenteritis. Of those who developed primary septicemia, 61% died. All successful trace-backs of implicated oyster product showed that the oysters were harvested from the Gulf of Mexico when the water temperature was >22°C. In Florida, between 1981 and 1993, there were 71 cases of *V. vulnificus* oyster-associated septicemia that required hospitalization; of these cases, 77% resulted in death. In addition, there were 34 cases of wound infection with a mortality rate of 15%. A recent report of food-related illnesses by the U.S. Centers for Disease Control and Prevention (CDC) estimates that there are 97 *V. vulnificus* infections and 48 associated deaths each year in the United States.

Prior to 1996, epidemiological reports showed that cases of human *V. vulnificus* disease were sporadic. However, in the summer and autumn of 1996 and 1997, an outbreak of invasive *V. vulnificus* infection occurred in Israel in people who had recently handled fresh, whole fish purchased from inland artificial fish ponds. Overall, 62 cases of wound infection and bacteremia were reported. The isolates were indistinguishable by restriction fragment length polymorphism–polymerase chain reaction (RFLP-PCR) and could not be typed by pulse-field gel electrophoresis (PFGE). Based on phenotype and RFLP tests, the authors concluded that the outbreak was caused by a new strain of *V. vulnificus*, which they classified as biogroup 3. A detailed understanding of its ecology and virulence properties has yet to be reported.

***Vibrio vulnificus* Levels Associated with Human Infections.** Individual oysters can contain more than 100 *V. vulnificus* strains per oyster, as evidenced by PFGE. This finding demonstrates the challenge for retrospective studies of *V. vulnificus* infections, since the pathogenic strain(s) must be enumerated apart from the total *V. vulnificus* flora. This approach has been supported by observations that only one PFGE profile is found in the blood of an infected individual, even though multiple strains were ingested.

To gather evidence about environmental levels associated with human disease, total *V. vulnificus* levels in oysters were measured following harvest from a commercial site in Apalachicola Bay, Florida, over a 3-year period. During this study, the incidence of human *V. vulnificus* infections reported in Florida and associated with oysters harvested from the Apalachicola Bay sampling location were recorded. Environmental *V. vulnificus* levels were considered relevant if the case-associated oysters were harvested within 72 h of the Apalachicola Bay sampling time. Eight *V. vulnificus* infections met these criteria, and results showed that *V. vulnificus* disease occurred when environmental *V. vulnificus* levels exceeded 1000 CFU/g of oyster meat (range 10^3–10^5 g^{-1}). This concentration translated to the ingestion of approximately 3×10^5 *V. vulnificus* in a meal of one dozen raw oysters.

In a separate series of studies, *V. vulnificus* was enumerated in batches of oysters directly implicated in human infections. Logistical constraints included identifying cases, obtaining implicated oysters from homes and retail outlets, and collecting clinical samples from patients. In a total of four cases where clinical specimens were obtained, each patient displayed a single strain, as defined by PFGE. This finding has been substantiated in other reports. In one human infection, oysters obtained from the implicated restaurant were analyzed to determine both the concentration and genetic diversity of *V. vulnificus* isolates. The oysters contained 9.6×10^3/g *V. vulnificus*, indicating that the patient ingested approximately 6×10^5 CFU. The PFGE tests of oyster *V. vulnificus* isolates revealed eight unique DNA profiles among the ingested strains. The strain isolated from the patient's blood was present in the implicated oysters at 2×10^3 CFU/g meat, indicating that approximately 1×10^5 CFU of the clinical strain were consumed.

Virulence. The virulence traits of *V. vulnificus* observed in humans are believed to have first evolved in the marine environment as a mechanism to enhance its replication. For example, it has been shown that extracellular products of *V. vulnificus* that damage mammalian cells cause pathogenic changes in certain fish and algal species and also enhance its replication.

Since the original description of *V. vulnificus*, intraspecies diversity has been measured primarily by phenotype variation. Such studies have investigated metabolizable substrates, colony morphotype, and presence of membrane proteins and lipopolysaccharide. *Vibrio vulnificus* virulence has been measured by its median lethal dose (LD_{50}) in mice, opaque versus translucent colony morphotype, the presence of specific membrane proteins and extracellular enzymes, resistance to animal host defense systems, the presence of specific genes, and DNA profiles using various RFLP techniques (discussed below).

Role of Iron. *Vibrio vulnificus* causes septicemia and serious wound infections in patients with iron overload and preexisting liver disease. Indeed, saturation of transferrin appears to be an important prerequisite for a successful infection. The ability of *V. vulnificus* to acquire iron from the host has been shown to correlate with virulence. Siderophores produced by *V. vulnificus* assist it in acquiring iron from transferrin. The ability of *V. vulnificus* to acquire host iron is believed to directly result in growth of the pathogen to potentially high levels. This has been shown in studies of the survival of *V. vulnificus* in whole blood from healthy volunteers and patients with chronic hepatitis, liver cirrhosis, and hepatoma. Researchers have demonstrated that bacterial numbers in human blood increased with the severity of a patient's liver disease.

Extracellular Toxins. Research shows that nearly all environmental and clinical isolates of *V. vulnificus* produce some form(s) of extracellular enzymes with cytolytic and/or cytotoxic effects. A long list of exoenzymes have been described including collagenase, elastase, hemolysin, DNase, lipase, mucinase, and phospholipase. More recently, there has been considerable research of new hemolysins, their biological properties, and relevance to virulence in humans and other animals. A new hemolysin and its gene (*vllY*) has been cloned and sequenced. The *vllY* gene expresses a unique peptide, and it shows high similarity to the sequence of legiolysin, responsible for hemolysis, pigment production, and fluorescence in *Legionella pneumophila*.

Capsule. Significant attention has been given to the role of capsular polysaccharide in virulence. For example, it is reported that the quantity of capsular polysaccharide (CPS) is positively associated with *V. vulnificus* virulence in mice. Investigators report that large capsules are associated with low LD_{50}, resistance to the bactericidal action of human serum, antiphagocytic activity, and tissue invasion.

Although multiple CPS types have been described in the literature, virulence does not appear to correlate with a specific CPS type. Numerous reports confirm that the quantity of CPS expressed by *V. vulnificus* can result in reversible opaque and translucent colony morphologies. In contrast to translucent colony morphotypes, opaque morphotypes have been associated with virulence and resistance to phagocytosis by human and oyster phagocytes. Translucent variants express intermediate quantities of CPS and expression of CPS varies with growth phase. Temperature also has an effect on CPS expression, whereby more CPS is produced at 30°C than at 37°C. Following *V. vulnificus* infection, IgG antibodies can be demonstrated in patient serum that reacts with the capsular polysaccharide of the infecting strain. Anticapsular polysaccharides antibodies have also been demonstrated in persons without a history of *V. vulnificus* infection, thus indicating the occurrence of cross-reacting antibodies.

14.1.5. Detection of Organism

A well-established method for the isolation, detection, and enumeration of *V. vulnificus* is found in the eighth edition of the U.S. Food and Drug Administration (FDA) *Bacteriological Analytical Manual* (Elliott et al., 1998). The methods for preparing and enriching samples for *V. vulnificus* are derived from methods developed for *V. parahaemolyticus* and *V. cholerae*, with some modification.

Isolation begins by diluting the sample 10- and 100-fold (and further dilutions for enumeration) in alkaline peptone water (APW) and blending for 2 min at high speed. The APW is incubated at 35–37°C for 18–24 h. After incubation and without shaking the flask, a loopful of surface growth is transferred to a plate of mCPC agar. The mCPC agar is incubated for 18–24 h at 39–40°C. *Vibrio vulnificus* produces flattened translucent yellow colonies with opaque centers. For further tests, colonies are streaked for isolation on a nonselective media such as tryptic soy agar containing 1–2% NaCl.

Enzyme immunoassay (EIA) and gene probe methods are available as confirmatory tests for *V. vulnificus*, and therefore there is less demand to perform a battery of biochemical tests to speciate suspected colonies. When these latter methods are used, *V. vulnificus* can be differentiated from non-*Vibrio* using TSI, KIA, and arginine glucose slants, oxidation fermentation tests, the oxidase test, and various other methods.

These characteristics are presumptive of *V. vulnificus*:

1. Gram negative
2. Growth at 42°C, positive
3. Arginine dihydrolase test, negative
4. Lysine decarboxylase test, positive
5. Voges–Proskauer test, negative
6. NaCl test: 0% NaCl, negative; 3, 6% NaCl, positive; 8% NaCl, negative
7. Sucrose fermentation, negative
8. ONPG test, positive

9. Arabinose fermentation, negative
10. Sensitivity to O/129, sensitive to 10 and 150 μg
11. TSI reaction: alkaline slant (rarely acidic)/acid butt; gas production, negative; H_2S, negative
12. Hugh–Leifson test: glucose oxidation and fermentation, positive
13. Cytochrome oxidase, positive

An EIA can be used to confirm colonies as *V. vulnificus*. In this method, isolates from mCPC agar are transferred to wells of a 96-well plate containing APW. These cultures are incubated at 35–37°C, and a sample of the APW is coated onto the wells and then reacted with monoclonal antibody specific for an intracellular antigen of *V. vulnificus*. A gene probe to the *V. vulnificus* cytotoxin–hemolysin gene is also available as a confirmatory test.

14.1.6. Prevention and Control

Persons with immunocompromising conditions need to take precautions to reduce the risk of *V. vulnificus* infections. These safeguards include not eating raw shellfish, cooking shellfish thoroughly, avoiding cross-contaminating other foods with raw seafood, avoiding exposure of open wounds or broken skin to brackish water, avoiding raw clams and oysters harvested from warm brackish waters, and wearing protective gloves when handling raw shellfish.

14.2. *Vibrio parahaemolyticus*

14.2.1. Introduction

Vibrio parahaemolyticus is one of the most widely recognized pathogenic *Vibrio* spp. due to its etiology in numerous outbreaks and its wide occurrence in marine environments. Like *V. cholerae*, it has been the subject of extensive study, particularly since its initial discovery as the causative agent of numerous seafood-borne outbreaks in Japan and other Asian countries. It has gained renewed attention in the United States following three recent outbreaks involving the consumption of raw molluscan shellfish.

Vibrio parahaemolyticus can be isolated from a variety of entities in the marine environment, including shellfish, finfish, plankton, sediment, and seawater. Typically, it can be recovered from most biological and physical structures in marine environments where environmental temperatures and salinities are compatible with its growth or survival.

14.2.2. Nature of Disease Caused

Human illness is normally associated with cooked seafood that has been cross-contaminated with seawater or with uncooked seafood. The disease presents as a self-limiting form of mild to severe gastroenteritis. The associated symptoms include gastroenteritis (59%), wound infections (34%), septicemia (5%), and other consequences (2%). Approximately 45% of patients are hospitalized. Eighty-eight percent of the individuals who report acute gastroenteritis claim to have eaten raw oysters during the week before their illness.

14.2.3. Characteristics of Agent

Vibrio parahaemolyticus inhabits many salt water environments. Even though its numbers increase with increasing seawater temperature, it is less restricted to warm, low-salinity coastal waters as observed for *V. cholerae* and *V. vulnificus*.

In a large survey, *V. parahaemolyticus* was recovered from 45.9% of 686 samples of seafood imported from southeast Asian countries. The incidence rates in shrimp, crab, snail, lobster, sand crab, fish, and crawfish were 75.8, 73.3, 44.3, 44.1, 32.5, 29.3, and 21.1%, respectively. None of the isolates possessed the thermal direct hemolysin (*tdh*) or thermal related hemolysin (*trh*) genes.

It is well known that all *Vibrio* spp. produce chitinase, an enzyme capable of hydrolyzing chitin. Chitin is the most abundant carbohydrate-based polymer in the marine environment and is found in the exoskeletons of various forms of marine life, including plankton. Various reports show that *Vibrio* spp. colonize chitinous surfaces of plankton and are likely involved in remineralization of molted plankton exoskeletons. Interestingly, unlike *V. cholerae* non-O1 and *Aeromonas hydrophila*, *V. parahaemolyticus* fails to colonize either live or dead copepods.

Another factor that may contribute to the persistence of *V. parahaemolyticus* in the environment is its ability to resist phagocytic cells of marine invertebrates and vertebrates, such as those of molluscan hemocytes. Using a tetrazolium dye reduction assay to study factors governing the killing of bacteria by oyster hemocytes, researchers have shown that the opaque *V. parahaemolyticus* phenotype resists oyster cellular host defenses better than the translucent phenotype.

Vibrio parahaemolyticus has been shown to produce molecules capable of inducing luminescence in *V. harveyi*. Consequently, quorum signaling may assist it in adapting to communal life on surfaces in the marine environment. For example, it has been shown that when grown on a surface or in a viscous layer, *V. parahaemolyticus* differentiates to swarmer cells capable of movement over and colonization of surfaces. *Vibrio parahaemolyticus* can also switch between translucent and opaque colony types.

With the recent global spread of the O3:K6 strain, there has been speculation that this strain may possess characteristics that distinguish it from other serotypes. Investigators have found that all O3:K6 strains possess a common plasmid having 8782 base pairs and 10 open reading frames. Filamentous phage has been isolated from various strains of *V. parahaemolyticus*, and they have been suspected of being vectors for moving genetic material among *V. parahaemolyticus* and other *Vibrio* spp.

14.2.4. Epidemiology

Vibrio parahaemolyticus infections are associated with the consumption of raw shellfish, cross-contaminated food, and the exposure of wounds to seawater containing *V. parahaemolyticus*. It is the cause of numerous outbreaks of foodborne illnesses where raw or undercooked seafoods are commonly consumed. In Japan, *V. parahaemolyticus* is one of the most frequent causes of foodborne disease.

In the United States, *V. parahaemolyticus* is the most common reported cause of bacterial seafood-borne disease. Forty outbreaks, constituting over 1000 cases of *V. parahaemolyticus* infection, were reported to the CDC by four Gulf Coast states between 1973 and 1998. Between May and September of 1997, over 250 human *V. parahaemolyticus* infections occurred in the Pacific Northwest of the United States and were associated with raw oyster consumption. All illnesses were caused by a strain with the O4 serotype, which also possessed the virulence-associated *tdh* gene. At the time of the outbreak, seawater temperatures were 1–5°C higher than normal in the implicated estuaries, and it was suggested that El Niño–induced weather changes may have produced higher levels (i.e. >11,000 CFU/g) of *V. parahaemolyticus* in shellfish.

In the following year, more than 300 *V. parahaemolyticus* infections were linked to consumption of oysters harvested from Galveston Bay, Texas. All clinical specimens displayed one PFGE profile belonging to the O3 serotype. Interestingly, this serotype was also associated with other recent outbreaks in various geographical regions of the world. In the same year, oysters and clams harvested from the Long Island Sound in New York were implicated in 23 culture-confirmed cases of *V. parahaemolyticus* serotype O3 gastroenteritis. This was the first reported outbreak of *V. parahaemolyticus* linked to consumption of shellfish harvested from New York waters. Water temperature was reportedly 8°F higher than in the previous 2-year period.

In other epidemiological research, researchers conducted a 1-year case–control study of sporadic *Vibrio* infections in two coastal areas of Louisiana and Texas to identify risk factors related to the consumption of seafoods. They reported an association between consumption of cooked crayfish and *V. parahaemolyticus* infection. In addition, *V. parahaemolyticus* infections have been associated with wounds (e.g., shark bites) that occur in marine environments. *Vibrio parahaemolyticus* disease can also be manifested as septicemia and ocular infections. Its ability to invade tissues has been supported by research showing that *V. parahaemolyticus* isolates invade Caco-2 cells, a human colon carcinoma-derived cell line.

In a survey of travelers quarantined at the Osaka and Kansai Airport Quarantine Station during 1979–1995, investigators analyzed the stool of 29,587 persons who reported that they suffered from diarrhea. Various enteropathogenic bacteria were isolated from 9766 (33.0%) patients of the stools examined. *Vibrio parahaemolyticus* ranked third among the isolated pathogens (1959 cases).

A detailed examination of 1319 *V. parahaemolyticus* strains isolated from traveler's diarrhea showed that the majority of the strains (87.3%) were positive by the Kanagawa phenomenon (KP) test on Wagatsuma agar, of which 91% and 9% of the strains were only *tdh* and both *tdh* and *trh* positive, respectively. However, approximately 13% of the total 1319 strains were KP negative, of which 94 and 24 strains were only *trh* positive and both *tdh* and *trh* positive, respectively.

Emergence of O139 Strain. Prior to 1995, there was little evidence that specific serotypes of *V. parahaemolyticus* were selectively spreading in geographical regions. In analyses of hundreds of strains collected over time, a variety of K serotypes have been reported. Compelling evidence has been presented that the recent pandemic of *V. parahaemolyticus* has resulted from the spread of a single clone, O3:K6. *Vibrio parahaemolyticus* O3:K6 strains isolated from cases of diarrhea in Calcutta, India, beginning in February 1996 and strains isolated from southeast Asian travelers since 1995 belong to a unique clone that possesses the *tdh* gene but not the *trh* gene. These data agree with a previous report of clonality using arbitrarily primed PCR. Human illness resulting from this clone has increased in Bangladesh since 1997 and also in Taiwan, Laos, Japan, Thailand, Korea, and the United States between 1997 and 1998. Researchers have provided further evidence for this clonality by analysis of the *tox*RS sequence. The *tox*RS sequences of the representative strains of the new O3:K6 clone differ from those of the O3:K6 strains isolated before 1995 by at least seven base positions within a 1346-bp region. Furthermore, a PCR method has been developed that targets two of the base positions unique to the new O3:K6 clone. This method differentiated 172 strains belonging to the new O3:K6 clone from O3:K6 strains isolated from earlier time periods and only showed cross-reactions with *tdh*-positive and *trh*-lacking strains that belonged to the O4:K68 and O1:K untypeable serovars, which had been isolated from travelers beginning in 1997. It is proposed that these strains may have diverged from the new O3:K6 clone by alteration of the O:K antigens.

Hemolysin. The 46-kDa homodimer secreted thermal direct hemolysin (TDH) has been proposed to be a major virulence factor of *V. parahaemolyticus*. Thermal direct hemolysin is a pore-forming toxin and has been extensively studied for its ability to hemolyze erythrocytes of various mammalian species. Moreover, TDH induces a variety of cytotoxic effects that often lead to cell degeneration and loss of viability. In addition, TDH can cause intestinal fluid secretion. It induces a rapid transient increase in intracellular calcium that leads to a decreased rate of advancement through the cell cycle. The resulting morphological alterations appear to depend on the structure of the microtubular network.

It has been shown that KP-positive strains possess both *tdh*1 and *tdh*2 genes. Only the *tdh*2 gene has been shown to be responsible for the hemolytic activity of a KP-positive strain, although both genes may be relevant to pathogenicity.

In this regard, KP-negative strains of *V. parahaemolyticus* have been isolated from clinical cases. Investigators have demonstrated that two bases within the *tdh*2 promoter sequence are largely responsible for the difference in the promoter strength between the *tdh*2 and *tdh*1 genes in *V. parahaemolyticus*. They show that a base substitution of the *tdh* promoters of KP-negative strains only at one position is sufficient to increase the expression of the *tdh* genes to the KP-positive level. Consequently, *tdh* genes of KP-negative strains are significant since they can produce a KP-positive subclone by a single point mutation in the promoters.

Studies have shown a coincidence between urease-producing strains of *V. parahaemolyticus* and possession of the *trh* gene. These genes lie in close proximity with the *tdh* gene on the chromosome. Pulse-field gel electrophoresis of *Not*1-digested DNA from clinical isolates followed by Southern hybridization with probes to *tdh*, or *trh*, or *ure* genes displays a single fragment. The majority of strains (81%) have two copies of *tdh* on the chromosome and no copies of *trh* or *ure*. Seven percent of the strains have the *tdh*, *trh*, and *ure* genes on chromosomal DNA. Of these latter strains, all three genes were also detected on a single *Not*I fragment. Using long-polymerase-chain reactions, the distance between *trh* and *ure* was determined to be less than 8.5 kb.

Thermal direct hemolysin induces the phosphorylation of two proteins on membranes of human erythrocytes. Research indicates that phosphorylation of a 25-kDa protein is essential for the hemolysis by TDH after it binds to erythrocyte membranes. Interestingly, researchers report that TDH causes a dose-dependent increase in intracellular free calcium in both human and rat cell monolayers.

Iron. The influence of the regulation of iron on the pathogenesis of *V. parahaemolyticus* has not been well characterized compared to other proposed virulence factors. It has been shown that under iron-restricted conditions, *V. parahaemolyticus* produces a siderophore, vibrioferrin, along with the production of two outer membrane proteins. *Vibrio parahaemolyticus* utilizes heme and hemoglobin as iron sources. Some suggest that the heme iron utilization systems of *V. parahaemolyticus* are similar at the DNA level, the functional level, and the amino acid sequence or protein level to that of *V. cholerae*.

Effects of Bile. Several investigators have proposed that the presence of bile in the human intestine stimulates the production of TDH by *V. parahaemolyticus* strains. It was found that either glycocholic acid or taurocholic acid stimulated 4- to 16-fold greater production of TDH into cell medium than compared to medium without the bile acids. Also, investigators have shown that the addition of bile or the bile acid deoxycholic acid to estuarine water-cultured bacteria led to an increase in the direct viable count and colony counts among the virulent strains. Furthermore, it has been proposed that ToxR is a conserved protein involved in the modulation of outer membrane proteins and bile resistance of pathogenic *Vibrio* species and that ToxR-mediated bile resistance was an early step in the evolution of *Vibrio* spp. as intestinal pathogens.

Methods to Measure Strain Diversity. A fundamental discovery about the structure of the *V. parahaemolyticus* genome has been recently reported. Researchers found that *V. parahaemolyticus* strains, similar to most *Vibrio* spp., possess two chromosomes. Specifically, a physical map of the genomic DNA (5.1 Mb) for *V. parahaemolyticus* strain AQ4673 showed two circular replicons of 3.2 and 1.9 Mb. The presence of chromosomes rather than large plasmids was indicated by the observation of 16S ribosomal ribonucleic acid (rRNA) genes on both the replicons. Various laboratories have evaluated different methods to characterize (type) *V. parahaemolyticus* strains of clinical and environmental origin. Researchers have analyzed isolates by enterobacterial repetitive intergenic consensus (ERIC) sequence PCR, RFLP in rRNA genes (ribotyping), PFGE, and RFLP analysis of the genetic locus (*Fla*) encoding the polar flagellum. They reported that ERIC sequence PCR and ribotyping were the most useful typing methods when used in combination; while *Fla* locus RFLP was the least discriminatory. A high level of DNA degradation was observed when applying PFGE tests with various restriction endonucleases.

14.2.5. Detection of Organism

A standard method for the isolation, detection, and enumeration of *V. parahaemolyticus* is found in the FDA *Bacteriological Analytical Manual* (Elliott et al., 1998). Sample preparation and enrichment are similar to that described above for *V. vulnificus*. On TCBS agar, *V. parahaemolyticus* colonies are green or blue-green, round, and 2–3 mm in diameter. *Vibrio parahaemolyticus* can be differentiated from non-*Vibrio* using TSI, KIA, and arginine glucose slants, oxidation fermentation tests, the oxidase test, and various other methods. *Vibrio parahaemolyticus* will not grow in 0% NaCl but does grow in 3–8% NaCl.

Traits that are presumptive of *V. parahaemolyticus* include:

14. Growth at 42°C, positive
15. NaCl test: 0% NaCl, negative; 3, 6, and 8% NaCl, positive; 10% NaCl, negative or poor
16. Sucrose fermentation, negative
17. ONPG test, negative
18. Arabinose fermentation, usually positive
19. Gram-negative rods
20. TSI appearance: alkaline slant/acid butt; gas production, negative; H_2S, negative
21. Hugh–Leifson test: glucose oxidation and fermentation, positive
22. Cytochrome oxidase, positive
23. Arginine dihydrolase test, negative
24. Lysine decarboxylase test, positive
25. Voges–Proskauer test, negative
26. Sensitivity to O/129: sensitive to 150 μg, resistant to 10 μg

Vibrio parahaemolyticus is typed by capsular (K) and lipopolysaccharide (O) antigens. Researchers have conducted extensive serological analyses of *V. parahaemolyticus* isolates that form the basis of a common serological typing scheme. Currently, there are approximately 59 K and 12 O serotypes.

The presence of the thermal direct hemolysin (*tdh*) and the TDH-related hemolysin (*trh*) genes and their gene products (TDH and TRH) serves as a useful tool to identify virulent *V. parahaemolyticus* strains. The Kanagawa reaction demonstrates the presence of TDH on Wagatsuma agar. The PCR and gene probe methods are also available to detect the *tdh* and *trh* genes.

14.2.6. Prevention and Control

Most *V. parahaemolyticus* infections can be prevented by properly cooking seafood and avoiding the contamination of cooked seafood with seawater and raw seafood. The wound infection syndrome can be prevented by avoiding exposure of open wounds to warm, brackish seawater.

14.3. *Vibrio cholerae*

14.3.1. Introduction

Vibrio cholerae is the etiological agent of pandemic cholera, a human gastrointestinal disease manifested by severe diarrhea, dehydration, and electrolyte imbalance. The endemic nature of cholera in various areas of the world illustrates that it is one of various transmittable diseases that occur when untreated human waste and domestic water mix. Of the various human gastrointestinal pathogens, *V. cholerae* has been the subject of numerous studies to elucidate its disease processes and how it survives and replicates in the environment. Today, there is a wealth of new information that illustrates the complex bacterial and host processes that produce cholera as well as sustain the existence of *V. cholerae* in the aquatic environment.

Historically, a small subset of *V. cholerae* strains were considered to be capable of causing epidemic cholera. These strains produced cholera toxin (CT) and expressed the O1 antigen. Other *V. cholerae* strains were not pathogenic or they only caused mild diarrhea and were referred to as "non-O1" strains. However, this dogma was quickly dispelled in the early 1990s when another serovar, *V. cholerae* O139, was defined as the cause of a new *V. cholerae* epidemic in India. It is believed that the *V. cholerae* O139 epidemic strain likely emerged from the seventh pandemic O1 El Tor strain through a genetic rearrangement involving the horizontal transfer of exogenous O antigen and capsule genes. It has been suggested that strains of *V. cholerae* O22 from the environment might have been the source of the exogenous DNA resulting in the emergence of epidemic *V. cholerae* O139.

In the last decade, very profound discoveries have been made about the virulence of *V. cholerae*, including the finding of two replicons, the virulence cassette, and how environmental signals control and orchestrate the expression of specific genes. These findings have shown that the processes that lead to cholera are much more dynamic than originally thought. In addition, they offer new insights into potential virulence properties of other bacterial pathogens.

This section is a general overview of *V. cholerae* and discusses selected virulence factors, methods of detection, and other characteristics. For a more in-depth discussion of the pathogenicity, virulence, and ecology of *V. cholerae*, the reader is directed to other publications, including those by Barua and Greenough (1992), Colwell et al. (2000), Kaper et al. (1994), and Wachsmuth et al. (1994).

14.3.2. Nature of Illness Caused

Human volunteer experiments show that high levels ($\sim 10^{11}$ CFU) of *V. cholerae* are needed to survive passage through the stomach, colonize the small intestine, and cause acute diarrhea. Only 0.001% of this level is needed to cause acute diarrhea when stomach acidity is neutralized with sodium bicarbonate or with nonacidic food.

Once in the small intestine, the flagella of *V. cholerae* move it to mucosal enterocytes of the small bowel, where it attaches and proliferates to concentrations as high as 10^7–10^8 CFU/g of small intestine. While actively colonizing the mucosa and closely approximating themselves to enterocytes, *V. cholerae* produces CT, which is responsible for changes in fluid transport across epithelial surfaces of the small bowel.

Adherence to intestinal mucosa may involve a variety of factors. To date, the most well characterized colonization factor is the toxin coregulated pili (TCP). It consists of long filaments that are attached to the lateral surface of *V. cholerae* O1 and O139 cells. When the *tcpA* gene is mutated, human volunteers do not experience diarrheal illness. Other factors which have a potential role in *V. cholerae* colonization of the small intestine include accessory colonization factor (ACF), core-encoded pilus, LPS, mannose-sensitive hemagglutinin, mannose-fucose-resistant hemagglutinin, outer membrane protein U, and the polysaccharide capsule of O139 strains.

Following attachment and colonization of the small intestine, CT is produced and affects normal ion transport across epithelial cell surfaces. This leads to chloride secretion by the cell, water loss, and severe diarrhea. Cholera toxin is comprised of two subunits, A and B. Subunit A (CT-A) is present as a single subunit and possesses toxic activity that results in activation of membrane-bound adenylate cyclases. There are five B (CT-B) subunits that are receptors for epithelial GM_1 ganglioside, and they bring the holotoxin in proximity to the epithelial cell surface. At least three immunological forms of CT-B have been described.

Cholera toxin catalyzes the transfer of adenosine diphosphate (ADP) ribose of nicotinamide adenine dinucleotide (NAD) to the G_s protein, a regulator of adenylate cyclase. This action leads to increased Cl^- secretion by intestinal crypt cells and decreased NaCl-coupled absorption by villus cells. The net movement of water from the cell into the intestinal lumen results in the watery diarrhea manifested in cholera.

Lysis of sheep erythrocytes is one test used to differentiate the El Tor biotype of *V. cholerae* from the classical biotype, although more recent El Tor strains have less hemolytic activity. The role of El Tor cytolysin/hemolysin in human infections has been assessed, and it was shown that the recombinant strain still caused diarrhea in 33% of the volunteers. Therefore, its importance as a virulence factor remains unresolved. Interestingly, it has been shown that a single base-pair difference in classical and El Tor promoters is responsible for the differential regulation of virulence gene expression in these two biotypes.

Vibrio cholerae also produces a toxin that changes the permeability of the small intestine by affecting the cytoskeletal structure of the intercellular tight junction (zonula occludens). Normally, the tight junction prevents the movement of compounds through the intercellular space. This toxin (Zot), in conjunction with other virulence factors, may also contribute to the diarrheal symptoms of cholera. The *zot* gene has been demonstrated in various O1 and non-O1 strains and is coregulated with the *ctx* gene. Another *ctx-zot* coregulated factor is the accessory cholera enterotoxin (Ace).

14.3.3. Characteristics of Agent

Vibrio cholerae replicates in various environmental niches, including fresh and estuarine waters, in association with plants and animals, and in foods of plant and animal origin. The primary environmental factors that affect growth are temperature, nutrient level, acidity, and water activity (a_W).

In the aquatic environment, the organism is part of the endogenous (autochthonous) microflora, where growth is enhanced by moderate to high environmental temperatures and low to moderate salinities. The ecology of *V. cholerae* has been extensively studied in cholera-endemic regions of the world, where researchers have sought to determine the environmental factors that cause the seasonal periodicity of cholera. In this regard, it has been demonstrated that plankton support the growth and survival of *V. cholerae* and that *V. cholerae* 01 colonizes the exoskeletons of certain zooplankton and phytoplankton species. In recent studies, satellite imagery has been used as an indirect means to assess the risk of *Vibrio* spp. infections by monitoring global surface water temperature and associated plankton levels.

14.3.4. Epidemiology

The seventh pandemic of cholera is still in progress and is caused by *V. cholerae* O1, biotype El Tor. This pandemic began in 1961 with the report of epidemic cholera in Celebes (Sulawesi), Indonesia, and rapidly spread to other countries of eastern Asia. It was reported in Bangladesh in 1963, India in 1964, and Iran,

Iraq, and the USSR in 1965–1966. Cholera arrived in West Africa in 1970, the first such report in over 100 years, and quickly became an endemic disease in various African countries. The 1991 cholera epidemic in Latin America began with reported outbreaks in Peru and has subsequently spread to more than 10 countries in the region.

Until 1992, *V. cholerae* O1 was the only serogroup believed to cause epidemic cholera. However, large outbreaks of cholera were reported in Bangladesh in 1992 that were linked to a new serogroup of *V. cholerae*, designated O139, synonym Bengal. To date, this serogroup has been reported from 11 countries in South and East Asia.

For 1999, 61 countries notified the World Health Organization (WHO) of a total of 254,310 cases and 9175 deaths. It was reported that this represented a 13% decrease in reported cases compared to 1998. The overall case–fatality rate continued to be 3.6%. A total of 206,746 cases in Africa represented 81% of the worldwide total. For 1999, Asia reported 39,417 total cases, representing a 61% increase compared to 1998. The total number of cases for the Americas was reduced by 86% from 1998 to 1999. It is widely believed that the actual number of cases and outbreaks are higher due to underreporting and limitations in surveillance systems.

14.3.5. Detection of Organism

Methods for the isolation, detection, and enumeration of *V. cholerae* are found in the FDA *Bacteriological Analytical Manual* (Elliott et al., 1998). The methods for preparing and enriching environmental and food samples for *V. cholerae* are similar to those described for *V. vulnificus*, above, with some modification. Isolation begins by diluting the sample in APW. The APW is incubated at 35–37°C for 6–8 and 18–24 h. After incubation, a loopful of surface growth is transferred to a plate of TCBS agar and incubated for 18–24 h at 35–37°C. On TCBS agar, *V. cholerae* (El Tor and classical) appears as large, smooth, yellow (sucrose-positive) and slightly flattened colonies with opaque centers and translucent peripheries. *Vibrio cholerae* can be differentiated from non-*Vibrio* using TSI, KIA, and arginine glucose slants, oxidation fermentation tests, and the oxidase test. In addition, tolerance to salt is an simple method of differentiating *V. cholerae* from other *Vibrio* spp. Specifically, by using agar or broth containing 0 and 3% NaCl, *V. cholerae* and *Vibrio mimicus* will grow in the presence of both salt concentrations, whereas most other *Vibrio* spp. will only grow in 3% NaCl.

Diagnostic antisera and monoclonal antibodies can be used to identify the O1 group of *V. cholerae* as well as the subgroup Inaba (factors AC) and Ogawa (factors AB). Other antisera are available to identify serotype O139. For serology, colonies should originate from nonselective agar to avoid false-positive and false-negative reactions. Cultures confirmed as O1 can be further identified as classical and El Tor biotypes using polymyxin B sensitivity, hemolysin test, Voges–Proskauer test, and bacteriophage susceptibility.

These characteristics are considered to be presumptive of *V. cholerae*:

27. Gram-negative asporogenous rods
28. TSI or KIA: acid slant/acid butt; gas production, negative; H_2S, negative
29. Hugh–Leifson test: glucose fermentation and oxidation, positive
30. Cytochrome oxidase, positive
31. Arginine dihydrolase test, negative
32. Lysine decarboxylase test, positive
33. Growth at 42°C, positive
34. NaCl test: 0% NaCl, positive; 3% NaCl, positive; 6% NaCl, normally negative
35. Sucrose fermentation, positive
36. ONPG test, positive
37. Arabinose fermentation, negative
38. O/129 sensitivity, sensitive to 10 and 150 g O/129

14.3.6. Prevention and Control

Cholera epidemics typically occur in geographical regions where there is an increased risk of human feces contaminating drinking water and food. In these situations, it is very important to ensure that there is a safe drinking water supply, that food is handled safely, and that human feces is safely disposed.

REFERENCES

Barua, D., and W. B. Greenough, III. 1992. *Cholera*. New York: Plenum.

Colwell, R. R., and D. Jay Grimes. 2000. *Nonculturable Microorganisms in the Environment.* American Society for Microbiology. Washington, DC.

Elliott, E. L., C. A. Kaysner, L. Jackson, and M. L. Tamplin. 1998. *Vibrio cholerae, V. parahaemolyticus, V. vulnificus*, and other *Vibrio* spp. In: *U.S. Food and Drug Administration Bacteriological Analytical Manual*, 8th ed. Gaithersburg, MD: AOAC International, Chapter 9.

Kaper, J. B., A. Fasano, and M. Truckis. 1994. Toxins of *V. cholerae*. In: I. K. Wachsmuth, P. A. Blake, and O. Olsvik (Eds.), *Vibrio cholerae and Cholera: Molecular to Global Perspectives.* Washington, DC: American Society for Microbiology.

Wachsmuth, I. K., P. A. Blake, and O. Olsvik. 1994. *Vibrio cholerae* and cholera: molecular to global perspectives. American Society for Microbiology, Washington, DC.

BIBLIOGRAPHY

Amako, D., K. Okada, and S. Miake, 1984. Presence of a capsule in *Vibrio vulnificus. J. Gen. Microbiol.* **130**, 2741–2743.

Amaro, C., E. G. Biosca, B. Fouz, and E. Garay. 1992. Electrophoretic analysis of heterogeneous lipopolysaccharide from various strains of *Vibrio vulnificus* biotypes 1 and 2 by silver staining and immunoblotting. *Curr. Microbiol.* **25**, 99–104.

Anonymous. 2000. Weekly epidemiological record. *World Health Organization* **75**, 249–256.

Bag, P. K., N. Suvobroto, R. K. Bhadra, T. Ramamurthy, S. K. Bhattacharya, M. Nishibuchi, T. Hamabata, S. Yamasaki, Y. Takeda, and G. B. Nair. 1999. Clonal diversity among recently emerged strains of *Vibrio parahaemolyticus* O3:K6 associated with pandemic spread. *J. Clin. Microbiol.* **37**, 2354–2357.

Bassler, B. L., E. P. Greenberg, and A. M. Stevens. 1997. Cross-species induction of luminescence in the quorum-sensing bacterium *Vibrio harveyi. J. Bacteriol.* **179**, 4043–4045.

Basu, A., P. Garg, S. Chakraborty, T. Bhattacharya, A. Khan, S. K. Bhattacharya, S. Yamasaki, Y. Takeda, and G. B. Nair. 2000. *Vibrio cholerae* O139 in Calcutta, 1992–1998: Incidence, antibiograms, and genotypes. *Emerg. Infect. Dis.* **6**, 139–147.

Biosca, E. G., C. Amaro, J. L. Larsen, and K. Pedersen. 1997. Phenotypic and genotypic characterization of *Vibrio vulnificus*: Proposal for the substitution of the subspecific taxon biotype for serovar. *Appl. Environ. Microbiol.* **63**, 1460–1466.

Bisharat, N., V. Agmon, R. Finkelstein, R. Raz, G. Ben-Dror, L. Lerner, S. Soboh, R. Colodner, D. N. Cameron, D. L. Wykstra, D. L. Swerdlow, and J. J. Farmer. III. 1999. Clinical, epidemiological, and microbiological features of *Vibrio vulnificus* biogroup 3 causing outbreaks of wound infection and bacteraemia in Israel. Israel vibrio study group. *Lancet* **354**, 1421–1424.

Boyce, T. G., E. D. Mintz, K. D. Greene, J. G. Wells, J. C. Hockin, D. Morgan, and R. V. Tauxe. 1995. *Vibrio cholerae* O139 Bengal infections among tourists to Southeast Asia: An intercontinental foodborne outbreak. *J. Infect. Dis.* **172**, 1401–1404.

Brennt, C. E., A. C. Wright, S. K. Dutta, and J. G. Morris. 1991. Growth of *Vibrio vulnificus* in serum from alcoholics: Association with high transferrin iron saturation. *J. Infect. Dis.* **164**, 1030–1032.

Buchreiser, C., V. V. Gangar, R. L. Murphree, M. L. Tamplin, and C. W. Kaspar. 1995. Multiple *Vibrio vulnificus* strains in oyster as demonstrated by clamped homogenous electric field gel electrophoresis. *Appl. Environ. Microbiol.* **61**, 1163–1168.

Chang, B., H. Taniguchi, H. Miyamoto, and S. Yoshida. 1998. Filamentous bacteriophages of *Vibrio parahaemolyticus* as a possible clue to genetic transmission. *J. Bacteriol.* **180**, 5094–5101.

Chang, T. M., Y. C. Chuang, J. H. Su, and M. C. Chang. 1997. Cloning and sequence analysis of a novel hemolysin gene (vllY) from *Vibrio vulnificus. Appl. Environ. Microbiol.* **63**, 3851–3857.

Chen, F., G. M. Evins, W. L. Cook, R. Almeida, N. Hargrett-Bean, and K. Wachsmuth. 1991. Genetic diversity among toxigenic and nontoxigenic *Vibrio cholerae* O1 isolated from the Western hemisphere. *Epidemiol. Infect.* **107**, 225–233.

Douet, J. P., M. Castroviejo, A. Dodin, and C. Bebear. 1996. Study of the haemolytic process and receptors of thermostable direct haemolysin from *Vibrio parahaemolyticus. Res. Microbiol.* **147**, 687–696.

Espat, N. J., T. Auffenberg, A. Abouhamze, J. Baumhofer, L. L. Moldawer, and R. J. Howard. 1996. A role for tumor necrosis factor-alpha in the increased mortality associated with *Vibrio vulnificus* infection in the presence of hepatic dysfunction. *Ann. Surg.* **223**, 428–433.

Faruque, S. M., M. N. Saha, Asadulghani, P. K. Bag, R. K. Bhadra, S. K. Bhattacharya, R. B. Sack, Y. Takeda, and G. B. Nair. 2000. Genomic diversity among *Vibrio cholerae* O139 strains isolated in Bangladesh and India between 1992 and 1998. *FEMS Microbiol. Lett.* **184**, 279–284.

Fasano, A., B. Baudry, D. W. Pumplin, S. S. Wasserman, B. D. Tall, J. M. Ketley, and J. B. Kaper. 1991. *Vibrio cholerae* produces a second enterotoxin, which affects intestinal tight junctions. *Proc. Natl. Acad. Sci. USA.* **88**, 5242–5246.

Fiore, A., U. Hayat, S. S. Wasserman, A. Wright, C. A. Bush, and J. G. Morris, Jr. 1996. Antibodies that react with the capsular polysaccharide of *Vibrio vulnificus* are detectable in infected patients, and in persons without known exposure to the organism. *Diagn. Microbiol. Infect. Dis.* **24**, 165–167.

Fyfe, M., M. T. Kelly, S. T. Yeung, P. Daly, K. Schallie, S. Buchanan, P. Waller, J. Kobayashi, N. Therien, M. Guichard, S. Lankford, P. Stehr-Green, R. Harsch, E. DeBess, M. Cassidy, T. McGivern, S. Mauvais, D. Fleming, M. Lippmann, L. Pong, R. W. McKay, D. E. Cannon, S. B. Werner, S. Abbott, M. Hernandez, C. Wojee, J. Waddell, S. Waterman, J. Middaugh, D. Sasaki, P. Effler, C. Groves, N. Curtis, D. Dwyer, G. Dowdle, and C. Nichols. 1998. Outbreak of *Vibrio parahaemolyticus* infections associated with eating raw oysters—Pacific Northwest, 1997. *MMWR* **47**, 457–462.

Genthner, F. J., A. K. Volety, L. M. Oliver, and W. S. Fisher. 1999. Factors influencing in vitro killing of bacteria by hemocytes of the eastern oyster (*Crassostrea virginica*). *Appl. Environ. Microbiol.* **65**, 3015–3020.

Harris-Young, L., M. L. Tamplin, J. W. Mason, H. L. Aldrich, and J. K. Jackson. 1995. Viability of *Vibrio vulnificus* in association with hemocytes of the American oyster (*Crassostrea virginica*). *Appl. Environ. Microbiol.* **61**, 52–57.

Hayat, U. K., G. P. Reddy, C. A. Bush, J. A. Johnson, A. C. Wright, and J. G. Morris, Jr. 1993. Capsular types of *Vibrio vulnificus*: An analysis of strains from clinical and environmental sources. *J. Infect. Dis.* **168**, 758–762.

Hlady, W. G., and K. C. Klontz. 1996. The epidemiology of *Vibrio* infections in Florida, 1981–1993. *J. Infect. Dis.* **173**, 1176–1183.

Huq, A., P. A. West, E. B. Small, M. I. Huq, and R. R. Colwell. 1984. Influence of water temperature, salinity, and pH on survival and growth of toxigenic *Vibrio cholerae* serovar O1 associated with live copepods in laboratory microcosms. *Appl. Environ. Microbiol.* **48**, 420–424.

Huq, A., R. R. Colwell, M. A. Chowdhury, B. Xu, S. M. Moniruzzaman, M. S. Islam, M. Yunus, and M. J. Albert. 1995. Coexistence of *Vibrio cholerae* O1 and O139 Bengal in plankton in Bangladesh. *Lancet* **345**, 1249.

Iida, T., O. Suthienkul, K. S. Park, G. Q. Tang, R. K. Yamamoto, M. Ishibashi, K. Yamamoto, and T. Honda. 1997. Evidence for genetic linkage between the *ure* and *trh* genes in *Vibrio parahaemolyticus*. *J. Med. Microbiol.* **46**, 639–645.

Jackson, J. K., R. L. Murphree, and M. L. Tamplin. 1997. Evidence that mortality from *Vibrio vulnificus* infection results from single strains among heterogeneous populations in shellfish. *J. Clin. Microbiol.* **35**, 2098–2101.

Johnson, J. A., J. G. Morris, Jr, and J. B. Kaper. 1993. Gene encoding zonula occludens toxin (*zot*) does not occur independently from cholera enterotoxin genes (*ctx*) in *Vibrio cholerae*. *J. Clin. Microbiol.* **31**, 732–733.

Kaper, J. B., J. G. Morris, and M. M. Levine. 1995. *Cholera Clin. Microbiol. Rev.* **8**, 48–86.

Kaspar, C. W., and M. L. Tamplin. 1993. The effects of temperature and salinity on the survival of *Vibrio vulnificus* in seawater and shellfish. *Appl. Environ. Microbiol.* **59**, 2425–2429.

Kraffert, C. A., and D. J. Hogan. 1992. *Vibrio vulnificus* infection and iron overload. *J. Am. Acad. Dermatol.* **26**, 140.

Kreger, A., L. DeChalet, and P. Shirley. 1981. Interaction of *Vibrio vulnificus* with human polymorphonuclear leukocytes: Association of virulence with resistance to phagocytosis. *J. Infect. Dis.* **144**, 244–248.

Lee, K. K., H. T. Chiang, K. C. Yii, W. M. Su, and P. C. Liu. 1997. Effects of extracellular products of *Vibrio vulnificus* on *Acanthopagrus schlegeli* serum components in vitro and in vivo. *Microbioscience* **92**, 209–217.

Levine, M. M., J. B. Kaper, D. Herrington, G. Losonsky, J. G. Morris, M. L. Clements, R. E. Black, B. Tall, and R. Hall. 1988. Volunteer studies of deletion mutants of *Vibrio cholerae* O1 prepared by recombinant techniques. *Infect. Immun.* **56**, 161–167.

Lobitz, B., L. Beck, A. Huq, B. Wood, G. Fuchs, A. S. Faruque, and R. Colwell. 2000. From the cover: Climate and infectious disease: Use of remote sensing for detection of *Vibrio cholerae* by indirect measurement. *Proc. Natl. Acad. Sci. USA* **97**, 1438–1443.

Marshall, S., C. G. Clark, G. Wang, M. Mulvey, M. T. Kelly, and W. M. Johnson. 1999. Comparison of molecular methods for typing *Vibrio parahaemolyticus*. *J. Clin. Microbiol.* **37**, 2473–2478.

Matsumoto, C., J. Okuda, M. Ishibashi, M. Iwanaga, P. Garg, T. Rammamurthy, H. C. Wong, A. Depaola, Y. B. Kim, M. J. Albert, and M. Nishibuchi. 2000. Pandemic spread of an O3:K6 clone of *Vibrio parahaemolyticus* and emergence of related strains evidenced by arbitrarily primed PCR and toxRS sequence analyses. *J. Clin. Microbiol.* **38**, 578–585.

McDougald, D., S. A. Rice, and S. Kjelleberg. 2000. The marine pathogen *Vibrio vulnificus* encodes a putative homologue of the *Vibrio harveyi* regulatory gene, luxR: A genetic and phylogenetic comparison. *Gene* **248**, 213–221.

Mead, P. S., L. Slutsker, V. Dietz, L. F. McCraig, J. S. Bresee, C. Shapiro, P. M. Griffin, and R. V. Tauxe. 1999. Food-related illness and death in the United States. *Emerg. Infect. Dis.* **5**, 607–625.

Miyoshi, S., H. Nakazawa, K. Kawata, K. Tomochika, K. Tobe, and S. Shinoda. 1998. Characterization of the hemorrhagic reaction caused by *Vibrio vulnificus* metalloprotease, a member of the thermolysin family. *Infect. Immun.* **66**, 4851–4855.

Motes, M. L., A. DePaola, D. W. Cook, J. E. Veazey, J. C. Hunsucker, W. E. Garthright, R. J. Blodgett, and S. J. Chirtel. 1998. Influence of water temperature and salinity on *Vibrio vulnificus* in Northern Gulf and Atlantic Coast oysters (*Crassostrea virginica*). *Appl. Environ. Microbiol.* **64**, 1459–1465.

Okuda, J., and M. Nishibuchi. 1998. Manifestation of the Kanagawa phenomenon, the virulence-associated phenotype, of *Vibrio parahaemolyticus* depends on a particular single base change in the promoter of the thermostable direct haemolysin gene. *Mol. Microbiol.* **30**, 499–511.

Olsvik, O., J. Wahlberg, B. Petterson, M. Uhlen, T. Popovic, I. K. Wachsmuth, and P. I. Fields. 1993. Use of automated sequencing of PCR-generated amplicons to identify three types of cholera toxin subunit B in *Vibrio cholerae* O1 strains. *J. Clin. Microbiol.* **31**, 22–25.

O'Malley, S. M., S. L. Mouton, D. A. Occhino, M. T. Deanda, J. R. Rashidi, K. L. Fuson, C. E. Rashidi, M. Y. Mora, S. M. Payne, and D. P. Henderson. 1999. Comparison of the heme iron utilization systems of pathogenic Vibrios. *J. Bacteriol.* **181**, 3594–3598.

Osawa, R., and S. Yamai. 1996. Production of thermostable direct hemolysin by *Vibrio parahaemolyticus* enhanced by conjugated bile acids. *Appl. Environ. Microbiol.* **62**, 3023–3025.

Pace, J. L., T. J. Chai, H. A. Rossi, and X. Jiang. 1997. Effect of bile on *Vibrio parahaemolyticus. Appl. Environ. Microbiol.* **63**, 2372–2377.

Pelon, W., R. J. Siebeling, J. Simonson, and R. B. Luftig. 1995. Isolation of bacteriophage infectious for *Vibrio vulnificus. Curr. Microbiol.* **30**, 331–336.

Peterson, J. W., and L. G. Ochoa. 1989. Role of prostaglandins and cAMP in the secretory effects of cholera toxin. *Science* **245**, 857–859.

Raimondi, F., J. P. Kao, J. B. Kaper, S. Guandalini, and A. Fasano. 1995. Calcium-dependent intestinal chloride secretion by *Vibrio parahaemolyticus* thermostable direct hemolysin in a rabbit model. *Gastroenterology* **109**, 381–386.

Speelman, P., G. H. Rabbani, K. Bukhave, and J. Rask-Madsen. 1985. Increased jejunal prostaglandin E_2 concentrations in patients with acute cholera. *Gut* **26**, 188–193.

Tamplin, M. L., A. L. Gauzens, A. Huq, D. A. Sack, and R. R. Colwell. 1990. Attachment of *Vibrio cholerae* serotype O1 to zooplankton and phytoplankton of Bangladesh waters. *Appl. Environ. Microbiol.* **56**, 1977–1980.

Tamplin, M. L., J. K. Jackson, C. Buchreiser, R. L. Murphree, K. M. Portier, V. Gangar, L. G. Miller, and C. W. Kaspar. 1996a. Pulsed-field gel electrophoresis and ribotype profiles of clinical and environmental *Vibrio vulnificus* isolates. *Appl. Environ. Microbiol.* **62**, 3572–3580.

Tamplin, M. L., K. S. Robinson, V. M. Garrido, and V. V. Gangar. 1996b. A linear regression model to predict *Vibrio vulnificus* levels in US estuaries. Abstracts of the Annual Meeting of the American Society for Microbiology. New Orleans, LA, p. 455.

Tamplin, M. L., S. Spector, G. E. Rodrick, and H. Friedman. 1985. *Vibrio vulnificus* resists phagocytosis in the absence of serum opsonins. *Infect. Immun.* **49**, 715–718.

Trucksis, M., J. E. Galen, J. Michalski, A. Fasano, and J. B. Kaper. 1993. Accessory cholera enterotoxin (Ace), the third toxin of a *Vibrio cholerae* virulence gene cassette. *Proc. Natl. Acad. Sci. USA* **90**, 5267–5271.

Trucksis, M., J. Michalski, Y. K. Deng, and J. B. Kaper. ■. The *Vibrio cholerae* genome contains two unique circular chromosomes. *Proc. Natl. Acad. Sci. USA* **95**, 14464–14469.

Vanoy, R. W., M. L. Tamplin, and J. R. Schwarz. 1992. Ecology of *Vibrio vulnificus* in Galveston Bay oysters, suspended particulate matter, sediment and seawater: Detection by monoclonal antibody-immunoassay-MPN procedures. *J. Indust. Microbiol.* **9**, 219–233.

Wechsler, E., C. D'Aleo, J. Hopper, D. Myers-Wiley, E. O'Keeffe, J. Jacobs, F. Guido, A. Huang, S. N. Dodt, B. Rowan, M. Sherman, A. Greenberg, D. Schneider, B. Noone, L. Fanella, B. R. Williamson, E. Dinda, M. Mayer, M. Backer, A. Agasan, L. Kornstein, F. Stavinsky, B. Neal, D. Edwards, M. Haroon, D. Hurley, L. Colbert, J. Miller, B. Mojica, E. Carloni, B. Devine, M. Cambridge, T. Root, D. Schoonmaker, M. Shayegani, W. Hastback, B. Wallace, S. Kondracki, P. Smith, S. Matiuck, K. Pilot, M. Acharya, G. Wolf, W. Manley, C. Genese, J. Brooks, Z. Dembek, and J. Hadler. 1999. Outbreak of *Vibrio parahaemolyticus* infection associated with eating raw oysters and clams harvested from Long Island Sound–Connecticut, New Jersey, and New York, 1998. *J. Am. Med. Assoc.* **281**, 603–604.

Wright, A. C., J. L. Powell, M. K. Tanner, L. A. Ensor, A. B. Karpas, J. G. Morris, Jr, and M. B. Sztein. 1999. Differential expression of *Vibrio vulnificus* capsular polysaccharide. *Infect. Immun.* **67**, 2250–2257.

CHAPTER 15

Yersinia enterocolitica

SAUMYA BHADURI

15.1. INTRODUCTION

Yersinia enetrocolitica has the dubious distinction of being termed the pathogenic bacterium of the 1980s. The new human pathogen was discovered 60 years ago in the United States by Schleifstein and Coleman, who named it *Bacterium enterocoliticum*. Little was known about this bacterium until the early 1960s when clinical and veterinary microbiologists started reporting the isolation of "new" isolates of bacteria similar to *B. enterocoliticum*. Consequently, these isolates were classified and named *Y. enterocolitica* (i.e., pertaining to the intestine and colon). The reclassification of this organism makes possible a more meaningful evaluation of the distribution of this species and its pathogenicity. Subsequent studies showed a ubiquitous distribution and a wide range of *Y. enterocolitica* strains in food and clinical samples.

15.2. NATURE OF ILLNESS

Yersinia enterocolitica is recognized as a foodborne pathogen, and the disease caused by this bacterium is called yersiniosis. Symptoms of yersiniosis include severe abdominal pain that suggests an appendicitis-like attack, as well as fever, diarrhea, headache, and vomiting. The very young and very old are most susceptible to *Y. enterocolitica* infection. The incubation period for yersiniosis is 24–36 h or longer. The duration of illness is usually 1–3 days. A large number of food-associated outbreaks of yersiniosis have been reported. In developed countries, *Y. enterocolitica* can be isolated from 1–2% of all human cases of acute enteritis. According to the U.S. Centers for Disease Control and Prevention, although there have been an estimated 96,000 cases of yersiniosis in North America, there have been no reported deaths.

Guide to Foodborne Pathogens, Edited by Ronald G. Labbé and Santos García
ISBN 0-471-35034-6

15.3. CHARACTERISTICS OF AGENT

Yersinia enterocolitica belongs to the Enterobacteriaceae family. It is a gram-negative, oxidase-negative, catalase-positive, nitrate-reductase-positive, cold-tolerant facultative anaerobic rod 0.5–0.8 × 1–3 μm in size that exhibits significant pleomorphism. The organism is not motile when grown at 37°C but motile at 22–25°C, with relatively few, peritrichous flagella when it is grown at less than 30°C. In addition, the bacterium is urease positive, and it ferments mannitol and produces gas from glucose. *Yersinia enterocolitica* differs from most members of the family Enterobacteriaceae by virtue of its slower growth at 37°C.

Yersinia enterocolitica has been called a cold-tolerant pathogen. It can grow at temperatures as low as 0°C. However, the bacterium can grow at temperatures as high as 44°C, with the optimum temperature being 32–34°C. The bacterium is inactivated at 50°C. The pH range for growth is pH 4.5–8.5, with an optimum of pH 7–8. The organism can grow in the presence of 0.5–5% of NaCl. *Yersinia enterocolitica* can survive in frozen conditions for 12 weeks.

Human pathogenic strains of *Y. enterocolitica* are endowed with a number of properties that confer virulence on the organism. Several of these strains exhibit a marked temperature-dependent expression of genes that are correlated with the presence of a 70- to 75-kbp plasmid that is directly involved with the virulence of the organism. A number of plasmid-mediated phenotypic characteristics including colony morphology, low-calcium response (LcR), Congo red (CR) uptake, crystal violet (CV) binding, autoagglutination (AA), serum resistance, tissue culture detachment, and hydrophobicity (HP) have been applied to the determination of virulence in strains of *Y. enterocolitica*. At a low level of calcium ions, the virulence plasmid also encodes for "Yops" (i.e., a set of proteins secreted by plasmid-bearing virulent strains), which are important virulence factors. The delivery of Yops into the host cell subverts or modulates normal host cell signal transduction and cytoskeletal functions. Another plasmid-encoded, calcium-independent, outer membrane protein, YadA, on the bacterial surface mediates cellular attachment and entry. These physiological traits associated with the virulence plasmid are expressed only at 37°C. However, growth at 37°C also fosters the loss of the virulence plasmid and the concomitant disappearance of the associated virulence characteristics. Elements encoded by the chromosome are also necessary for virulence. Pathogenic *Y. enterocolitica* share two chromosomal loci, *inv* (i.e., the invasion loci that mediates the penetration of host cells) and *ail* (i.e., the attachment-invasion loci that mediates bacterial attachment to host cells), which are involved in the first step of pathogenesis.

15.4. EPIDEMIOLOGY

Transmission of *Y. enterocolitica* is by the fecal–oral route, facilitated by ingesting fecally contaminated food or water. Pigs are the principal reservoir for virulent strains, and the organisms are often isolated from the oral cavity and gastrointestinal tract of apparently healthy animals. *Yersinia enterocolitica* colonize the tongue and tonsil areas of the pigs. Common food vehicles in outbreaks of yersiniosis are meat (particularly pork), milk, dairy products, powdered milk, cheese, tofu, and raw vegetables. Since *Yersinia* can grow at low temperatures, even refrigerated foods are potential vehicles for the growth and dissemination of these organisms. The majority of food isolates differ in biochemical (based on biochemical reactions, termed biovars) and serological (based on lipopolysaccharide surface O antigens, termed O serovars) characteristics from "typical" clinical strains and are usually called "nonpathogenic" or "environmental" *Yersinia* strains.

Outbreaks of yersiniosis are uncommon considering the widespread occurrence of *Y. enterocolitica* in the environment, its ability to colonize and persist within animals, and its ability to grow at refrigerated temperatures. During the mid 1970s, two outbreaks of yersiniosis were reported. The first outbreak was caused by the consumption of raw milk by 138 Canadian schoolchildren, but the organism was not recovered from the suspected source. The second and most highly publicized outbreak, which occurred in 1976, involved 220 grade school children in a small community in New York. The source of infection was chocolate-flavored milk that was culture positive. Thirty-six children were hospitalized with apparent acute appendicitis. Before it could be established that the patients were suffering from *Yersinia*-mediated pseudoappendicitis, 16 of the children had already undergone emergency appendectomies. A 1981 outbreak affected 35% (159 persons) of 455 individuals at a diet camp in New York. Five of seven patients hospitalized underwent appendectomies. The source of infection was reconstituted powdered milk and/or chow mein contaminated by an infected food handler. In the same year, an outbreak was reported among 50 individuals in Washington State involving tofu and spring water. In a 1982 outbreak, 172 cases of infection occurred in Arkansas, Mississippi, and Tennessee due to consumption of pasteurized milk that may have been contaminated with pig manure during transport. Water was the putative source of infection for an individual in New York State in 1974 and for a small outbreak in Ontario, Canada, in 1986. That same year, a Pennsylvania outbreak caused by bean sprouts and well water affected 16 individuals. An outbreak in 1989 affecting 15 infants and children in Atlanta, Georgia, was transmitted from raw pork chitterlings by the food handlers. More recently, an outbreak in 1992 was reported in Los Angeles County, California, involving seven persons. The source of the outbreak remains unknown. No other yersiniosis outbreaks have been reported in the 1990s in the United States. The paucity of yersiniosis outbreaks in the 1990s as compared to the 1980s has no clear explanation.

15.5. DETECTION OF ORGANISM

15.5.1. Virulence Determinants

The plasmid-associated virulence determinants have been used to differentiate between virulent and avirulent strains of *Y. enterocolitica*. Thus, these virulence determinants provide a rapid, reliable, and simple method for isolation and detection of plasmid-bearing virulent *Y. enterocolitica* (YEP^+) strains from foods. The main disadvantage of the use of these plasmid-borne virulence determinants is the instability of the virulence plasmid. Incubation of strains at 37°C for isolation fosters the loss of the virulence plasmid resulting in plasmidless avirulent (YEP^-) strains; however, the plasmid-associated phenotypes are only expressed at 37°C. Because of the instability of the virulence plasmid at 37°C, it is difficult to isolate YEP^+ strains after initial detection. As a consequence, detection has been hampered in clinical, regulatory, and quality control laboratories that employ an incubation temperature of 37°C for isolation/detection of the organism.

15.5.2. Congo Red Binding

Congo red binding has been used to screen *Y. enterocolitica* strains for virulence. When YEP^+ and YEP^- strains were cultivated at 37°C for 24 h on a CR-

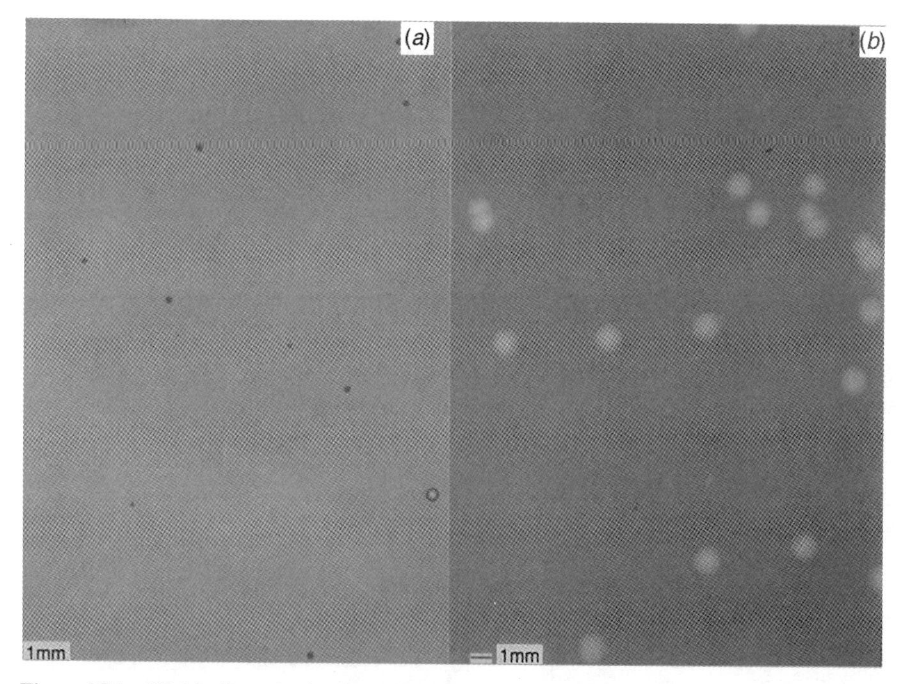

Figure 15.1. CR binding of colonies of *Y. enterocolitica* cells grown on CR-BHO for 24 h at 37°C: (*a*) YEP^+ cells showing pinpoint red colonies; (*b*) YEP^- cells showing large white or light orange colonies. The concentration of CR used in the binding assay was 75 μg/mL.

containing, low-calcium brain heart infusion agarose (CR-BHO) medium, two types of readily discernible colonies were observed. The YEP$^+$ cells absorbed CR and formed red pinpoint colonies (CR$^+$) (Fig. 15.1*a*). The YEP$^-$ cells failed to bind the dye and formed much larger white or light orange colonies (CR$^-$) (Fig. 15.1*b*). The size and colony morphologies of YEP$^+$ strains on CR-BHO also showed a LcR. The CR binding test was correlated with the presence of the virulence plasmid, with a number of virulence-associated properties and with mouse virulence, for a wide variety of pathogenic serotypes of *Y. enterocolitica* (Table 15.1). Thus, the binding of CR by YEP$^+$ strains consistently and efficiently allows differentiation of virulent and avirulent strains of *Y. enterocolitica*. In an investigation of a *Yersinia* outbreak in Los Angeles County, California, in 1992, the CR binding technique detected 1 YEP$^+$ isolate per 300 colonies recovered on CR-BHO from each of the seven patients. These data highlight the sensitivity of the CR binding technique for detection of YEP$^+$ cells in clinical samples that contained predominantly YEP$^-$ cells due to incubation of samples at 37°C during the initial isolation of the organism.

Since incubation at 37°C causes the loss of the virulence plasmid, an additional advantage of the CR binding technique is that it can be used to isolate *Y. enterocolitica* cells carrying the virulence plasmid. The ability of the CR binding technique for recovery of YEP$^+$ cells varied from 5 to 95%, indicating strain variation in the stability of the virulence plasmid. The YEP$^+$ strains showed all of the expected plasmid-associated properties, including virulence in a mouse (Table 15.1). By using the CR binding recovery technique mentioned above, Food and Drug Administration (FDA) investigators recovered and enhanced the level of plasmid carriage of YEP$^+$ strains from 0.3% to over 92% from clinical samples obtained during a 1992 outbreak of yersiniosis in Los Angeles County, California. Thus, the recovery technique is useful to isolate and enrich for viable YEP$^+$ cells, even if they are present at very low levels among a mixture of cells.

15.6. ISOLATION OF PATHOGENIC YEP$^+$ STRAINS FROM FOODS

The increasing incidence of *Y. enterocolitica* infections and the role of foods in some outbreaks of yersiniosis has led to the development of a wide variety of methods for the isolation of this bacterium from foods. The unstable nature of the virulence plasmid during passage at 37°C complicates the isolation of YEP$^+$ strains by causing the overgrowth of virulent cells with plasmidless derivatives and can eventually lead to a completely avirulent culture. Since the population of *Y. enterocolitica* in foods is usually low and since the natural microflora suppresses the growth of this organism, isolation methods usually involve enrichment followed by plating onto selective media.

Several approaches have been taken to enrich and isolate *Y. enterocolitica* from food. One method employs prolonged enrichment for 2–4 weeks at refrigeration temperatures to take advantage of the psychrotrophic nature of *Y.*

TABLE 15.1. Correlation among CR Binding Technique, Virulence, and Virulence-Associated Properties of Original and Recovered Plasmid-Bearing Strains of *Y. enterocolitica*

Strains[a]	Serotype	CM[b]	CV[c] Binding	LCR[d]	CR[e] Binding	AA[f]	HP[g]	Plasmid (70–75 kbp)	Diarrhea in Mice[h]
GER	O:3	+	+	+	+	+	+	+	+
GER-RE	O:3	+	+	+	+	+	+	+	+
GER-C	O:3	−	−	−	−	−	−	−	−
EWMS	O:13	+	+	+	+	+	+	+	+
EWMS-RE	O:13	+	+	+	+	+	+	+	+
EWMS-C	O:13	−	−	−	−	−	−	−	−
PT18-1	O:5:0:27	+	+	+	+	+	+	+	+
PT18-1-RE	O:5:0:27	+	+	+	+	+	+	+	+
PT18-1-C	O:5:0:27	−	−	−	−	−	−	−	−
O:TAC	O:TACOMA	+	+	+	+	+	+	+	+
O:TAC-RE	O:TACOMA	+	+	+	+	+	+	+	+
O:TAC-C	O:TACOMA	−	−	−	−	−	−	−	−
WA	O:8	+	+	+	+	+	+	+	+
WA-RE	O:8	+	+	+	+	+	+	+	+
WA-C	O:8	−	−	−	−	−	−	−	−

[a] Strains are from the FDA. Recovered strains are designated as RE. Plasmidless avirulent YEP^- strains are designated as C (cured).

[b] CM: Colony morphology. In a calcium-adequate brain heart infusion agar (BHA) medium, YEP^+ cells appeared as small colonies (diameter 1.13 mm) as compared to larger YEP^- colonies (diameter 2.4 mm).

[c] CV binding: Crystal violet binding. YEP^+ cells appeared as small dark violet colonies on BHA.

[d] LcR: Low-calcium response. Calcium-dependent growth at 37°C. YEP^+ cells appeared as pin point colonies of diameter 0.36 mm compared to the larger YEP^- colonies of diameter 1.37 mm on CR-BHO.

[e] CR binding: Congo red binding. YEP^+ cells appeared as red pinpoint colonies on CR-BHO.

[f] AA: Autoagglutination.

[g] HP: Hydrophobicity.

[h] Fecal material consistency was liquid; diarrhea was observed starting on days 3 and 4 postinfection.

enterocolitica and to suppress the growth of any background flora. Due to the extended time period needed for this method, efforts have been made to devise selective enrichment techniques employing shorter incubation times and higher temperature, thus making them more practical for routine use. However, high levels of indigenous microorganisms can overgrow and mask the presence of YEP^+ and nonpathogenic *Y. enterocolitica* strains. Enrichment media containing selective agents such as Irgasan, ticarcillin, and potassium chlorate are effective for enhancing recovery of a wide spectrum of *Y. enterocolitica* strains from meat samples. However, no single enrichment procedure is adequate for recovery of a broad spectrum of pathogenic *Y. enterocolitica* from foods. Since there is no specific plating medium for the isolation of YEP^+ strains, cefsulodin–irgasan–novobiocin (CIN) and MacConkey (MAC) agars are commonly used to isolate presumptive *Y. enterocolitica* from foods. The initial isolation of presumptive *Y. enterocolitica* from enriched samples on CIN and MAC agars adds an extra plating step, and the picking of presumptive *Y. enterocolitica* requires skilled recognition and handling of the colonies. The unstable nature of the virulence plasmid further complicates the detection of YEP^+ strains since isolation steps may lead to plasmid loss and the loss of associated phenotypic characteristics for colony differentiation. Moreover, the presumptive *Y. enterocolitica* colonies isolated should be confirmed as YEP^+ strains. Biochemical reactions, serotyping, biotyping, and virulence testing are essential for differentiation among YEP^+, YEP^-, environmental *Yersinia* strains, and other *Yersinia*-like presumptive organisms. Biochemical tests using commercially available systems such as analytical profile index (API) 20E test strips, give similar reactions among these organisms and are not conclusive. Serotyping of major O and H antigens differentiates between pathogenic and environmental *Y. enterocolitica* but fails to discriminate between YEP^+ and YEP^- strains. Likewise, biotyping does not confirm the presence of the virulence plasmid in YEP^+ strains. Several plasmid-associated phenotypic virulence determinants including colony morphology, AA, serum resistance, tissue culture detachment, HP, LcR, and CV binding have been used to portend the virulence of *Yersinia* isolates. These methods require specific reagents and conditions and do not give definite results. In addition, most of these procedures are costly, time consuming, complex, and impractical for routine diagnostic use in field laboratories. Although virulence can be demonstrated effectively using laboratory animals, this test is not suitable for routine diagnostic use. Molecular techniques such as deoxyribonucleic acid (DNA) colony hybridization, DNA restriction fragment length polymorphisims, and the polymerase chain reaction (PCR) have also been successfully applied to the detection of virulent strains. However, these techniques are complex and time consuming. These methods detect only the presence of a specific gene, not the actual presence of the live organism. Although virulence is plasmid mediated in all strains examined, the plasmids involved differ in molecular weight. Thus, in epidemiological studies, it is not sufficient to search for plasmids of a particular molecular weight as an indicator of *Y. enterocolitica* virulence. Unfortunately, methods described in the literature for the isolation of pathogenic *Y. enterocolitica* from foods do not treat confirmation of virulence in presumptive or known *Y. enterocolitica* isolates recovered from selective agars as an integral part of the detection method. The most rapid enrichment procedure available for the isolation of a wide spectrum of *Y. enterocolitica* serotypes does not include the identification of isolates as YEP^+ strains.

An improved homogenization-based procedure for selective enrichment, identification and maintenance of various pathogenic YEP^+ serotypes from pork samples was developed. This procedure is suitable for ground and liquid food samples wherein bacterial contamination is distributed throughout. In some cases, food slurries are used for enrichment, and the presence of food in the enrichment medium increases the background microflora and requires increased time for enrichment of YEP^+ strains. The CR binding technique is also not applicable for both detection and isolation, because the appearance of a red pin point colony is masked by the background microflora. Hence, it is necessary to first isolate presumptive *Y. enterocolitica* on selective agar plates and then subsequently identify any YEP^+ strains by the CR binding and LcR techniques. This approach takes 6 days to complete from sample enrichment through confirmation of YEP^+ strains, and as few as 9 colony-forming units (CFU/g) of YEP^+ strains of spiked ground pork can be recovered.

Recently, a method for simultaneous detection and isolation of pathogenic YEP^+ serotypes from enriched swab samples of various foods was reported. This procedure is applicable to foods that have a physical surface for swabbing, and such surfaces are often the primary site of contamination. Since the actual food sample was not used and since there was a low level of competing microflora, the time for enrichment of YEP^+ strains and their subsequent confirmation by the CR binding and LcR techniques was appreciably reduced (Fig. 15.2). This technique allowed for recovery of YEP^+ serotypes from various foods spiked with as low as 0.5 CFU/cm^2 within 4 days.

The above-mentioned homogenated slurry and swabbing techniques were effective for recovery of YEP^+ strains from naturally contaminated porcine tongues. The PCR assays validated these methods for detection of YEP^+ strains (Fig. 15.3). The YEP^+ serotypes isolated by these two procedures expressed plasmid-associated virulence characteristics and were positive in the mouse virulence test.

15.7. PREVENTION AND CONTROL

The patterns observed in foodborne *Y. enterocolitica* outbreaks indicate that postprocessing contamination is the main cause of illness. Specific prevention and control measures of foodborne yersiniosis include the following:

1. Special care should be taken during incision and removal of the intestines, tongue, pharynx, and tonsils of pigs to avoid cross-contamination.

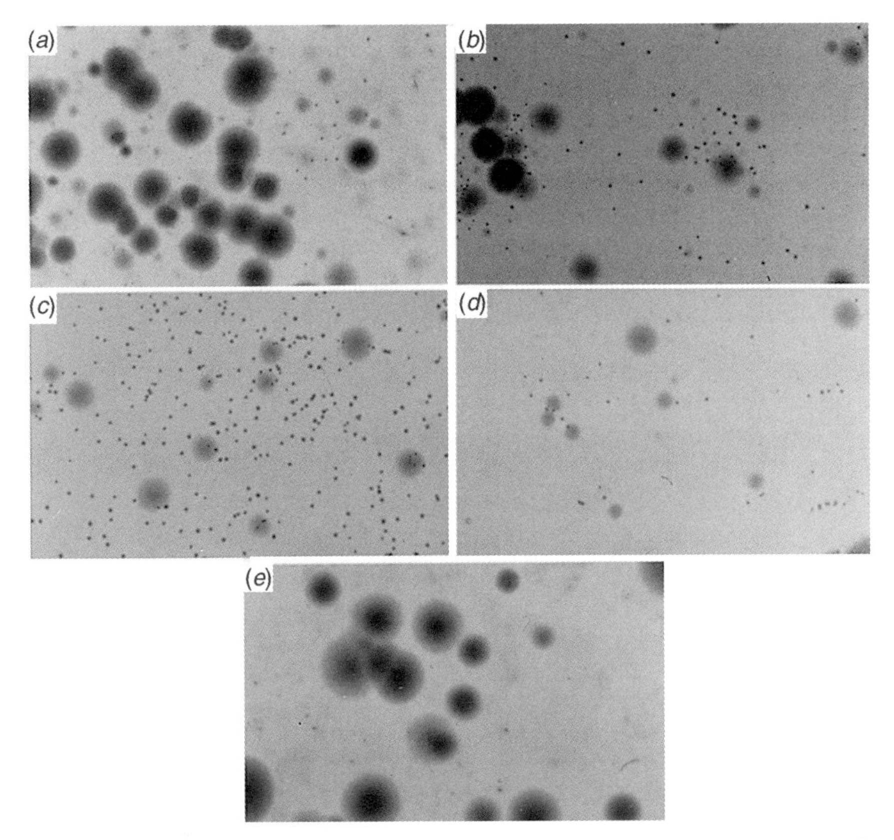

Figure 15.2. Recovery of YEP^+ strains by swabbing technique as red pinpoint colonies on CR-BHO from artificially contaminated pork chops (*a*), ground pork (*b*), cheese (*c*), and zucchini (*d*) and naturally contaminated porcine tongue (*e*).

2. Meat products, particularly pork, should be handled with care; food utensils, equipment, and countertops should be thoroughly cleaned to prevent cross-contamination.
3. Refrigerated foods should be thoroughly cooked or heated at temperatures sufficient to kill the pathogen. Likewise, cooked and refrigerated foods should be heated to a steaming temperature to kill *Yersinia* before consumption.
4. Precautions for prevention of fecal–oral spread of the pathogen should be practiced. Water supplies should be free from animal and human fecal waste.
5. Hands should be washed with soap and hot water after handling raw foods, including pork, as well as before serving food and eating food.

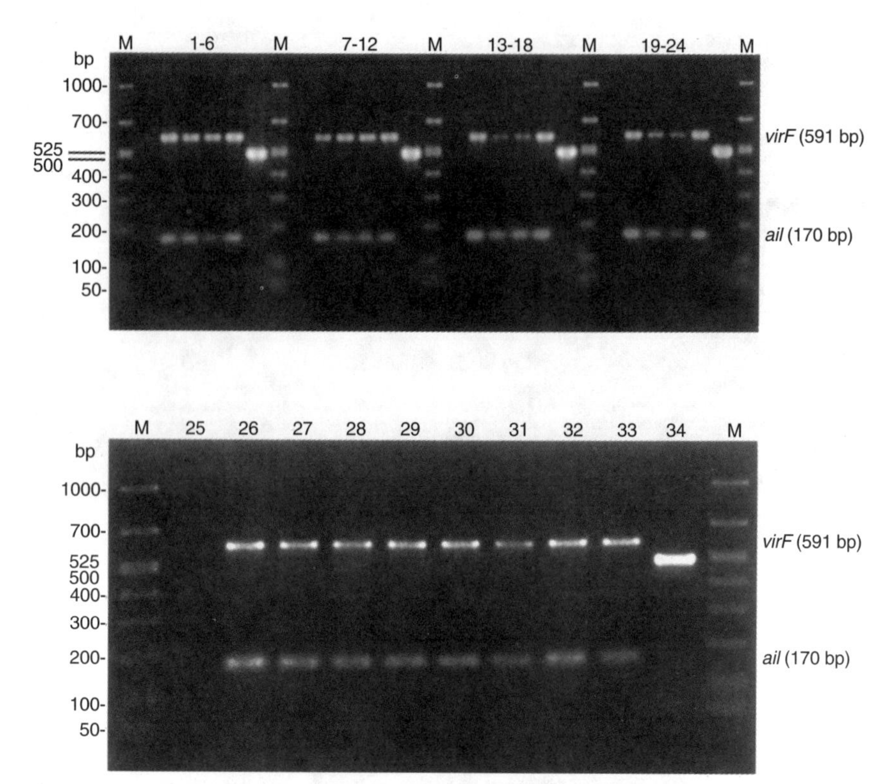

Figure 15.3. Confirmation of CR^+ clones isolated by swabbing technique from artificially contaminated various foods and from naturally contaminated porcine tongue as YEP^+ strains by multiplex PCR using chromosomal *ail* gene and *virF* gene from virulence plasmid. Lane M, 50- to 1000-bp ladder marker. Negative control with no template: lanes 1, 7, 13, 19, and 25. CR^+ colony showing the presence of 170- and 591-bp products with mixture of both *ail* and *virF* primers from chromosome and virulence plasmid, respectively: pork chops (lanes 2–4), ground pork (lanes 8–10), cheese (lanes 14–16), zucchini (lanes 20–22), and porcine tongues (lanes 26–32). Positive control with purified DNA from YEP^+ strain showing the presence of 170- and 591-bp products with mixture of both *ail* and *virF* primers from chromosome and virulence plasmid, respectively: lanes 5, 11, 17, 23, and 33. Positive control for PCR assay with λ as DNA template: lanes 6, 12, 18, 24, and 34.

In summary, these control measures require cooking and reheating to destroy YEP^+ strains since they are sensitive to heat. Cleanliness of the kitchen area, proper personal hygiene, and hand washing are very important in preventing infection.

BIBLIOGRAPHY

Bhaduri, S., and B. Cottrell. 1997. Direct detection and isolation of plasmid-bearing virulent serotypes of *Yersinia enterocolitica* from various foods. *Appl. Environ. Microbiol.* **63**, 4952–4955.

Bhaduri, S., and B. Cottrell. 1998. A simplified sample preparation from various foods for PCR detection of pathogenic *Yersinia enterocolitica*: A possible model for other food pathogens. *Molec. Cell. Probes* **12**, 79–83.

Bhaduri, S., B. Cottrell, and A. L. Pickard. 1997. Use of a single procedure for selective enrichment, isolation, and identification of plasmid-bearing virulent *Yersinia entero-*colitica of various serotypes from pork samples. *Appl. Environ. Microbiol.* **63**, 1657–1660.

Bhaduri, S., C. Turner-Jones, and R. V. Lachica. 1991. Convenient agarose medium for the simultaneous determination of low calcium response and Congo red binding by virulent strains of *Yersinia enterocolitica*. *J. Clin. Microbiol.* **29**, 2341–2344.

Bottone, E. J. 1997. *Yersinia enterocolitica*: The charisma continues. *Clin. Microbiol. Rev.* **10**, 257–276.

Cornelis, G. R. 1998. The *Yersinia* deadly kiss. *J. Bacteriol.* **180**, 5495–5504.

Cornelis, G. R., A. Boland, A. P. Boyd, C. Geuijen, M. Iriarte, C. Neyt, M-P. Sory, and I. Stainer. 1998. The virulence plasmid of *Yersinia*, an antihost genome. *Microbiol. Molec. Biol. Rev.* **62**, 1315–1352.

Feng, P., and S. D. Weagant. 1994. *Yersinia*. In: Y. H. Hui, J. R. Gordon, K. D. Murrell, and D. O. Cliver (Eds.), *Foodborne Diseases Handbook*, Vol. 1. New York: Marcel Dekker, pp. 427–460.

Kapperud, G., and S. B. Slome. 1998. *Yersinia enterocolitica* infections. In: A. S. Evans and P. Brachman (Eds.), *Bacterial Infections of Humans, Epidemiology and Control*. New York: Plenum Medical Book Company, pp. 859–873.

Mead, P. S., L. Slutsker, V. Dietz, L. F. McCaig, J. S. Bresee, C. Shapiro, P. M. Griffin, and R. V. Tauxe. 1999. Food-related illness and death in the United States. *Emerg. Infec. Dis.* **5**, 607–625.

Ravangnan, G., and C. Chiesa. 1995. Yersiniosis: Present and future. In: J. M. Cruse, and R. E. Lewis (Ser. Eds.), *Microbiology and Immunology*, Vol. 13. New York: Karger.

Robins-Browne, R. M. 1997. *Yersinia enterocolitica*. In: M. P. Doyle, L. R. Beuchat, and T. J. Montville (Eds.). Washington, DC: American Society for Micro Biology, pp. 192–215.

Weagant, S. D., P. Feng, and J. T. Stanfield. 1998. *Yersinia enterocolitica,* and *Yersinia pseudotuberculosis*. In: L. Tomlinson (Ed.), *Food and Drug Administration Bacteriological Analytical Manual,* 8th ed. (Revision A), Arlington, VA: AOAC International, pp. 8.01–8.13.

CHAPTER 16

Viruses

DEAN O. CLIVER

16.1. INTRODUCTION

Although viruses are often overlooked in discussions of foodborne disease agents, they cause a significant portion of the reported foodborne disease in the United States. During the period 1983–1987, gastroenteritis caused by the Norwalk virus ranked fifth among foodborne illnesses (in terms of total number of individuals affected), and hepatitis A ranked sixth. Hepatitis A moved to fourth in this ranking during 1988–1992, whereas gastroenteritis caused by Norwalk-*like* viruses dropped to ninth. No comparable data are yet available from other countries or for more recent years in the United States. A continuing problem in assessing the role of viruses as foodborne disease agents is the rarity of virus diagnosis and testing when outbreaks of foodborne disease are investigated. In that over half of recorded foodborne disease outbreaks, comprising nearly half of outbreak-associated foodborne illnesses, are of unknown etiology, one might well suppose that viruses, especially the Norwalk-like viruses, play an even larger role than these numbers would suggest.

16.2. NATURE OF ILLNESS

Viruses that are presently foodborne cause two types of illnesses: gastroenteritis and hepatitis. At one time, polioviruses were often transmitted via food and water and caused illnesses involving the central nervous system; the polioviruses have been eradicated in the Americas and will, one would hope, soon be gone from the earth. Other members of the human enterovirus group (coxsackieviruses and echoviruses) may also cause disease involving the central nervous system but have rarely caused outbreaks involving vehicular transmission. Viral illnesses transmitted via food and water are seldom lethal.

Guide to Foodborne Pathogens, Edited by Ronald G. Labbé and Santos García
ISBN 0-471-35034-6

Viral gastroenteritides have relatively rapid onsets (one to a few days, depending on the agent) compared to other virus diseases but somewhat slower onsets than many infectious bacterial gastroenteritides. They are characterized by diarrhea, accompanied in specific instances by vomiting, and a self-limiting course of illness. When vomiting occurs, the causal virus may be shed in vomitus as well as in the feces.

The hepatitides transmitted by the enteric (fecal–oral) route are A and E. Hepatitis A is common worldwide and has often been transmitted via food and water, whereas hepatitis E is rare in the more developed countries and seems often to be transmitted via water but seldom or never via food. Vaccines are available to evoke active immunity against hepatitis A; none has yet been developed for hepatitis E.

16.3. CHARACTERISTICS OF AGENTS

Two principal taxonomic groups of viruses (picornaviruses and caliciviruses) are regularly involved in causing foodborne disease, with others (rotaviruses and astroviruses) less often implicated. Taxonomy of viruses is based on size, whether there is a lipid envelope, the appearance of the protein coat, and the type of nucleic acid inside the protein coat. All of the virus groups listed lack the lipid envelope and contain ribonucleic acid (RNA). Those that have been tested withstand pH 3 for at least limited periods at room temperature.

Picornaviruses are ~28 nm in diameter and contain single-stranded RNA. The coat protein is relatively smooth, lacking distinctive surface features. Most have been isolated in or adapted to primate cell cultures. The hepatitis A virus (Fig. 16.1) and the polioviruses, coxsackieviruses, and echoviruses are members of this group.

Caliciviruses are 25–35 nm in diameter and contain single-stranded RNA. The coat protein has a distinctive pattern of indentations. The Norwalk-like viruses (Fig. 16.2) and other unnamed caliciviruses are members of this group, as was the hepatitis E virus until recently.

Rotaviruses are ~70 nm in diameter and contain double-stranded RNA. The double-layered coat protein shows some indication of icosahedral (20 equilateral triangles) symmetry.

Astroviruses are ~28 nm in diameter and contain single-stranded RNA. The surface of the coat protein has projections that appear to form five- or six-pointed star patterns.

16.4. EPIDEMIOLOGY

Essentially all of these agents are enteric viruses, transmitted by the fecal–oral route. Most are produced in the lining of the small intestine, but the hepatitis A virus (and probably the hepatitis E virus) is produced in the liver and drains into the intestine via the bile duct. In either case, the virus is shed in feces and infects when ingested, either directly with fecal material or with contaminated

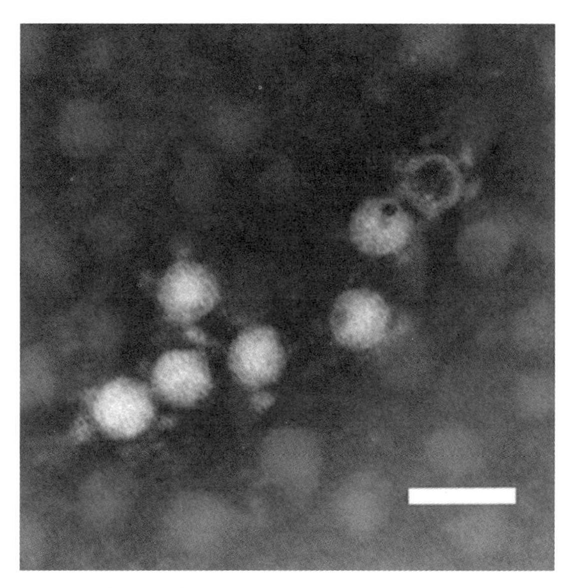

Figure 16.1. Particles of hepatitis A virus, a leading cause of serious foodborne and waterborne disease (bar = 50 nm). (Courtesy of F. P. Williams, U.S. Environmental Protection Agency.)

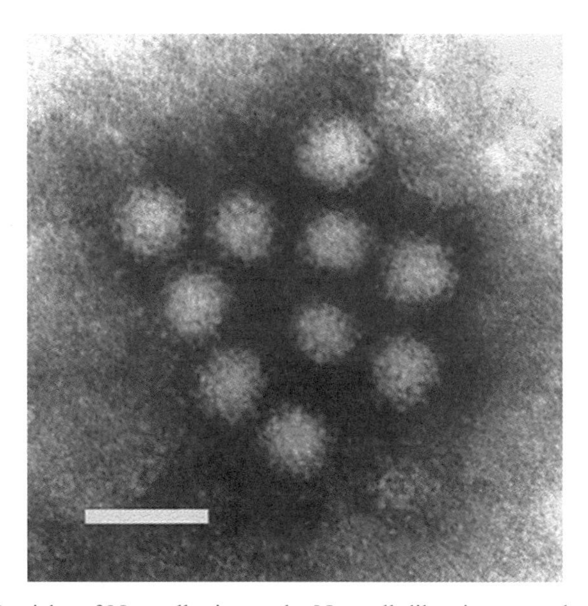

Figure 16.2. Particles of Norwalk virus—the Norwalk-like viruses are leading causes of transient foodborne gastroenteritis, often undiagnosed (bar = 50 nm). (Courtesy of F. P. William, U.S. Environmental Protection Agency.)

food or water. These viruses are generally specific for humans—no animal sources have been identified.

What is called direct contamination of food with enteric viruses typically occurs by way of fecally soiled hands. Water may be directly contaminated with feces by means as simple as flushing a toilet. Fecally contaminated water is sometimes the means of indirect contamination of food, either via irrigation or washing of the food or, in the case of shellfish, environmentally. Bivalve mollusks, such as clams, oysters, and mussels, feed by filtration of large volumes of their environmental water and are able to concentrate viruses in the process. Although the mollusks are not infected by the virus, they may retain it for extended periods, even when, say, indicator bacteria have been removed by a depuration process. Because shellfish are often eaten with little or no cooking, they are the most frequent vehicle of foodborne virus disease. Other food vehicles include foods contaminated in handling and eaten without cooking or contaminated in handling after cooking. Inevitably, the manner in which contamination occurred cannot always be determined, but the ultimate source is certainly human feces.

16.4.1. Norwalk-Like Viruses

Norwalk-like viruses, as stated earlier, have been assigned to the calicivirus group. Because of their distinctive surface features, they have been called small round structured viruses (SRSVs) by electron microscopists. Several serotypes, comprising two or more serological groups, are known. The original Norwalk virus derived its name from its causal role in a gastroenteritis outbreak in 1972 in Norwalk, Ohio. Infection follows ingestion of the virus. The illness is characterized by diarrhea, often accompanied by vomiting, with a typical incubation period of 24–48 h and a duration of 24–48 h. Virus is shed in vomitus, as applicable, and in feces for perhaps a week after onset. Direct, person-to-person transmission is common, but food (especially shellfish) and occasionally water are alternate routes. Although infection evokes an antibody response, those who have been ill are typically susceptible to the same virus serotype a year or more later. Diagnosis may be based on detection of virus in stool by enzyme-linked immunosorbent assay (ELISA), immune or conventional electron microscopy, or molecular methods or by demonstration of a diagnostic antibody response by immune electron microscopy or perhaps other methods. Not all diagnostic tests are available for all members of the group; the difficulties of diagnosis are likely responsible for the decline in rank of these agents among causes of foodborne disease, as mentioned earlier. Because even infection does not induce durable immunity, and because of the many serotypes involved, there is little incentive to attempt development of a vaccine.

16.4.2. Hepatitis A Virus

Hepatitis A virus is a member of the picornavirus group described above. The disease it causes is characterized by a wide range of severities, from mild or hardly noticeable to highly debilitating, including fever, malaise, nausea, abdominal discomfort, and anorexia, often followed by jaundice. Infection results from ingestion of the virus. The incubation period ranges from 15 to 50 days, with a median of 28 days. Virus is produced in the liver and drains into the intestine via the common bile duct, being shed in feces for 10–14 days before onset of illness and seldom more than a week after onset of jaundice. There is only one serotype worldwide. Infection evokes durable immunity, although relapses occur occasionally. Exposed persons are sometimes protected by injection of pooled immune serum globulin within 2 weeks; protection may last up to 3 months. An inactivated virus vaccine is now available for general prophylactic use. Transmission of the disease is often by direct, person-to-person contact, but transmission via foods (especially shellfish) and water are fairly common. Because hepatitis A is a reportable disease in the United States, it is possible to compare the total reported incidence with that associated with vehicular transmission. On this basis, only about 1% of hepatitis A is transmitted via food and water; but the incidence of sporadic cases (not perceived as associated with outbreaks) from vehicular exposures cannot be estimated on the basis of available data. Control of transmission could be accomplished by widespread vaccination, especially of food handlers, by prevention of fecal contamination of food and water, or by thorough cooking of food and disinfection of water. Unfortunately, shellfish are often eaten raw (and are very difficult to decontaminate by "depuration" processes), and some other foods are not cooked or may be contaminated in handling after cooking. What may have been the largest outbreak of foodborne disease on record occurred in Shanghai, China, in 1988, when local clams consumed during the winter holiday season led to nearly 300,000 cases of hepatitis A over a fairly short period. Sandwiches, salads, frozen strawberries, and many other foods have served as vehicles in other outbreaks of hepatitis A over time.

16.4.3. "Minor" Foodborne Viruses

The viruses to be discussed here are those that seldom are reported to be transmitted via food or water at this time. It should be recognized that problems that have been alleviated in developed countries may continue, undetected, in less affluent areas.

Poliomyelitis was the first virus disease reported to be transmitted via food in the United States Because the polioviruses have been eradicated from the Americas and from most developed countries in the Eastern Hemisphere, foodborne poliomyelitis is essentially no longer being reported. The other enteroviruses (principally coxsackieviruses and echoviruses) that infect humans have only rarely been reported to be transmitted via food or water, and not at all in very recent years.

Caliciviruses that cause gastroenteritis—other than the Norwalk-like, small round structured viruses—are only occasionally foodborne. Hepatitis E, which resembles the caliciviruses, occasionally occurs as waterborne outbreaks but for some reason seems not to be transmitted via foods. Other than a somewhat longer incubation period, the principal feature distinguishing hepatitis E from hepatitis A is the frequent death of pregnant women infected with the former.

Rotaviruses and astroviruses have been associated with occasional outbreaks of foodborne gastroenteritis. The rotaviruses are highly significant causes of infant mortality in less affluent parts of the world, but this is not necessarily associated with transmission via food.

16.5. DETECTION

Detection of viruses in foods may be done in several ways, depending on which virus is sought (if known) and which food is to be tested. In general, testing is undertaken either in the course of surveys or other situations in which human illness has not occurred or in the investigation of an outbreak. A food may be selected for testing because it is perceived to be at high risk of contamination; given that the foodborne viruses are all shed in human feces, one would not know in this situation which virus was most likely to be present. During an outbreak investigation, the virus sought may be known, but the choice of foods to be tested may be controlled by circumstances—which foods may have been eaten by those who became ill (and perhaps not by those who remained well) and which foods are still available for testing by the time the investigation is being conducted.

None of the methods available for detecting foodborne viruses is directly applicable to any food. Each sample must be processed in a manner that is determined by the food and by the putative viral contaminant before testing can be attempted. Typically, the sample is liquefied if not already a liquid, separated from food components (microbial contaminants, solids, and possibly solutes), and concentrated to reduce the volume to be tested.

16.5.1. Liquefaction

If the sample is solid, a liquid suspension should be made, incorporating no more food solids than necessary. For example, a solid food that might have been contaminated only superficially might simply be swabbed or rinsed to recover any external virus. If virus is within a solid food, liquefaction may be accomplished with sterile sand in a mortar and pestle or with a high-speed rotary homogenizer. It may be possible, however, to limit the extent to which food solids are homogenized in the suspension by using a Stomacher or other less rigorous apparatus. Given that the food solids must later be removed from the sample suspension, the latter option is preferable unless it leaves the virus associated with the food solids.

16.5.2. Clarification

Food solids are often removed from sample suspensions by centrifugation; however, filtration may also serve this purpose. Separation of solids from the liquid phase can be enhanced with polycation sewage coagulant or with an immiscible, high-density fluid such as trichlorotrifluoroethane. Whatever method is used, it is important that controls be included that show that the virus will not be eliminated with the food solids in clarification. Bacteria that might interfere with later steps are removed by filtration or combated with antibiotics or organic solvents such as chloroform.

16.5.3. Concentration

There is no absolute standard for the sensitivity of a test for foodborne virus. All the same, available test methods will accommodate only limited quantities of sample extract, so it is usually necessary to concentrate the extract before testing if any reasonable level of sensitivity is to be attained. This entails reducing the volume of fluid in which the extracted virus is suspended, ideally by removing both water and a proportionate amount of any solute present. Ultracentrifugation and ultrafiltration have often been used in this way—both are costly and impose limits on the number or volume of the sample extracts that can be processed. Phase separations and hydroextraction methods are useful in some instances. Adsorption of viruses to filter membranes, with later elution of the virus in a small volume of fluid, is used to concentrate samples of water whose volumes sometimes exceed 100 L. Once again, it is important to demonstrate by some means that the fractionation method will collect the virus in the portion that is retained rather than in what will be discarded. Immune capture (using antibody adsorbed to the interior of a tube or onto paramagnetic beads) can provide a further, specific means of concentrating virus from a small volume of suspension, assuming that the type of virus is known, so that the capture antibody can be selected accordingly.

16.5.4. Detection Methods

Viruses in foods are a concern because they may cause infections in those who eat the food. Therefore, detection methods based on viral infectivity are ideal. In current practice, infectivity testing usually entails inoculating cultures of established primate cell lines and observing either for viral cytopathology or for areas of cell killing under agar medium (plaques). Although infectivity methods are well established, they are essentially inapplicable to detection of the hepatitis and gastroenteritis viruses because the agents of interest do not infect available cell cultures demonstrably or at all. Therefore, alternate methods directed to the virus particle or its coat protein or RNA are used instead. In most instances, these alternate methods are so specific that they cannot detect any previously unrecognized virus.

Virus particles can be seen with transmission electron microscopy (TEM). Unfortunately, many nonviral particles may look a great deal like viruses. Specificity can be gained by means of antibody on the TEM grid, but at the cost of the potential to detect unknown viruses. Electron microscopy also suffers from the lack of sensitivity that results from ability to examine only miniscule samples. Presently, most viral detection methods are directed to the RNA, rather than the coat protein, and are based on amplification by the polymerase chain reaction (PCR) or similar procedures. Viral coat protein may serve as antigen in immune capture methods, but immunological detection methods, such as enzyme immunoassay, are generally not as sensitive as those directed to the RNA. A perceived liability of all of these alternate detection methods is their inability to discriminate against virus that is not infectious, say, for having been inactivated by heat or chemical disinfection. The extent to which inactivated virus gives false-positive PCR results seems not to have been well determined as yet. Beyond detection of viruses by RNA-specific methods, it is now possible to use nucleotide sequences to "fingerprint" viruses associated with a single outbreak. Polymerase chain reaction methods of diminished specificity have been devised to permit detection of more than one type of virus in a single test.

16.5.5. Indicators

Given the cost and difficulty of detecting foodborne viruses, there has been a continuing quest for "indicators," from whose presence in foods virus contamination could be inferred. Fecal or thermotolerant coliform bacteria have long served as indicators of contamination of water from which shellfish are harvested and of the effectiveness of shellfish depuration processes, but the presence of these bacteria has been shown to be poorly correlated with virus contamination of shellfish. Surveys of source waters for drinking water in the United States are presently directed to human enteroviruses that, although surely present as a result of human fecal contamination, are seldom transmitted to consumers of the water. Aside from bacterial indicators and human enteroviruses, most indicator studies have been directed to bacteriophages—either small, round phages that infect *Escherichia coli* (coliphages) or phages that infect *Bacteroides fragilis*. Each of these groups has its advantages and disadvantages but may prove useful in carefully selected situations. Their greatest advantages are low cost and read-out times of a few hours.

16.6. PHYSICAL METHODS FOR DESTRUCTION

Viruses multiply only in appropriate living host cells, so they cannot multiply in food but can only remain infectious or be inactivated (lose their infectivity). In food and water, they may be inactivated by heat, typically at times and temperatures that would kill vegetative bacterial cells. Viruses are preserved by chilling or freezing, so foods held in the refrigerator or freezer are likely to remain contaminated throughout their useful lives. In water or on food surfaces, viruses may be inactivated with oxidizing disinfectants (e.g., hypochlorite) or by ultraviolet light; these agents are of little use against virus inside of a solid food.

Because viruses contain only a small quantity of nucleic acid, by contrast to living cells, they present a very small target to ionizing rays and are inactivated to only a limited extent by most food irradiation processes. Some viruses are inactivated by drying, but others (most notably hepatitis A virus) are not and are able to persist for long periods in the dry state if temperatures are not too high.

16.7. PREVENTION AND CONTROL

It needs to be reiterated that viruses transmitted to humans via food and water come from other people's intestines. These include gastroenteritis and hepatitis viruses that are transmitted by the fecal–oral route. The obvious way to prevent such viruses from being transmitted via these vehicles is to prevent fecal contamination (in any quantity or dilution) of food and water by rigorous sanitation. Where appropriate, thorough cooking of a food will inactivate any virus that may be present; however, vehicles in outbreaks have often been foods eaten raw or, less frequently, foods that became contaminated after having been cooked. Vaccination against the hepatitis A virus, especially of food workers, could help prevent that agent from occurring in foods. Testing the food or water for virus contamination is costly and is seldom of use in preventing transmission, in that virus contamination is likely to be highly uneven. Indicators whose presence is supposed to imply the presence of virus are of limited use at this time. Finally, because viruses that cause some very frightful diseases have never been known to be transmitted via food or water (indeed, some cannot infect perorally), it is important to know which of them are "non-problems" and to deal with false alarms as such.

BIBLIOGRAPHY

Appleton, H. 1994. Norwalk virus and the small round viruses causing foodborne gastroenteritis. In: Y. H. Hui, J. R. Gorham, K. D. Murrell, and D. O. Cliver (Eds.), *Foodborne Disease Handbook*, Vol. 2: *Diseases Caused by Viruses, Parasites, and Fungi*. New York: Marcel Dekker, pp. 57–79.

Cliver, D. O. 1994. Other foodborne viral diseases. In: Y. H. Hui, J. R. Gorham, K. D. Murrell, and D. O. Cliver (Eds.), *Foodborne Disease Handbook*, Vol. 2: *Diseases Caused by Viruses, Parasites, and Fungi*. New York: Marcel Dekker, pp. 137–143.

Cliver, D. O. 1997a. Foodborne viruses. In: M. P. Doyle, L. R. Beuchat, and T. J. Montville (Eds.), *Food Microbiology: Fundamentals and Frontiers*. Washington, DC: American Society for Microbiology, pp. 437–446.

Cliver, D. O. 1997b. Virus transmission via foods, an IFT Scientific Status Summary. *Food Technol.* **51**, 71–78.

Cliver, D. O. 1997c. Foodborne viral infections. *World Health Stat. Q.* **50**(1/2), 90–101.

Cliver, D. O., R. D. Ellender, G. S. Fout, P. A. Shields, and M. D. Sobsey. 1992.

Foodborne viruses. In: C. Vanderzant and D. F. Splittstoesser (Eds.), *Compendium of Methods for the Microbiological Examination of Foods*, 3rd ed. Washington, DC: American Public Health Association, pp. 763–787.

Cromeans, T., O. V. Nainan, H. A. Fields, M. O. Favorov, and H. S. Margolis. 1994. Hepatitis A and E viruses. In: Y. H. Hui, J. R. Gorham, K. D. Murrell, and D. O. Cliver (Eds.), *Foodborne Disease Handbook*, Vol. 2: *Diseases Caused by Viruses, Parasites, and Fungi.* New York: Marcel Dekker, pp. 1–56.

Sattar, S. A., V. S. Springthorpe, and S. A. Ansari. 1994. Rotavirus. In: Y. H. Hui, J. R. Gorham, K. D. Murrell, and D. O. Cliver (Eds.), *Foodborne Disease Handbook*, Vol. 2: *Diseases Caused by Viruses, Parasites, and Fungi.* New York: Marcel Dekker, pp. 81–111.

CHAPTER 17

Seafood Toxins

JAMES M. HUNGERFORD

17.1. INTRODUCTION

Although seafood remains very popular in the United States and around the world, considerable effort is required to ensure a safe supply. In spite of these efforts, seafood toxins cause a significant number of illnesses each year. Rapid onset of symptoms is often the first indication of chemical intoxication versus microbial infection. Toxins take effect in minutes to hours while microbial infections typically require many hours. In addition to progressing rapidly, these illnesses can be very serious. Many of the naturally occurring toxins in fish and shellfish are potent neurotoxins, and some of them can be fatal to humans. When outbreaks do occur, doctors and other health professional must know how to recognize symptoms and quickly choose treatments. As a result of the threat posed by these toxins, extensive monitoring efforts are used when necessary and when possible, as in the harvest of molluscan shellfish.

As they filter-feed on toxic plankton, molluscan shellfish can be particularly toxic as they accumulate toxins over time. Most of these toxins are produced by phytoplankton, but there is some evidence to support bacterial origins also. The main shellfish toxins to be discussed here are those implicated in paralytic shellfish poisoning (PSP), amnesic shellfish poisoning (ASP), neurotoxic shellfish poisoning (NSP), and diarrhetic shellfish poisoning (DSP). Although they are primarily considered molluscan shellfish toxins, it is important to remember that fish feeding on plankton, as well as other invertebrates, can and sometimes do become toxic also. Fish intoxications include Scombroid poisoning, ciguatera, and fugu poisoning.

Control measures for fish and shellfish are fundamentally different. Their monitoring is also performed by different groups and with differing levels of oversight. Due to their mobile nature and also due to fundamental differences

Guide to Foodborne Pathogens, Edited by Ronald G. Labbé and Santos García
ISBN 0-471-35034-6

between fish toxins, strategies for fish toxins will be discussed separately from shellfish and, later in the text, specifically for Scombroid poisoning and ciguatera.

17.2. CONTROL MEASURES FOR SHELLFISH TOXINS

In the United States, monitoring shellfish for most toxins is performed at the harvest level by the individual state health and natural resources agencies. Federal oversight is by the National Shellfish Sanitation Program. Historical data have been accumulated over many years, and the seasonal nature of the plankton blooms is used to scale down sampling during cold months and increase it during warm weather. Yearly meetings are used to review data and make changes in requirements and are organized by the Interstate Shellfish Sanitation Commission. Japan, members of the European Union (EU), and other countries harvesting shellfish have their own procedures and requirements for monitoring shellfish, and the reader is encouraged to consult on-line resources such as EU documents and Internet sites.

Successful control measures require reliable analytical methods. Development of many of the most successful techniques in seafood toxin detection follow recurring patterns. As knowledge of a given toxin family increases (often due to improvements in chemical separation methods), it is sometimes found that the same separation method that improved our knowledge may not be practical for routine monitoring. Also, of those techniques that are eventually used in routine monitoring, most were first used as research tools. Many research methods, however, never become officially accepted for monitoring. This pattern repeats for many different toxins. For these reasons detection methodology will first be discussed from a broader viewpoint. Specifics of detection methods, as well as the symptoms and biological origins of each type of toxin, will be discussed in detail later in the text. Also, since analytical methods for seafood toxins have been reviewed extensively elsewhere, the focus here will be on those methods that are used currently for monitoring, are part of a general trend in seafood toxin detection, or are undergoing validation testing.

17.3. DETECTION STRATEGIES

With a few exceptions the most potent seafood toxins are monitored by animal bioassays. This approach has been used for decades, often because there was not yet sufficient information to develop alternative methods. The development of modern detection methods to replace bioassays is gaining momentum. This work is proceeding in three primary areas, including (1) chemical methods based on separations such as high-pressure liquid chromatography and electrophoresis; (2) biochemical methods such as immunoassays, enzymes, or competitive binding; and (3) cytotoxicity assays based on the mechanism of toxicity.

In the past decade, most efforts to develop alternative analytical methods for seafood toxins target biochemical assays or cytotoxicity assays. These techniques offer two primary advantages: (1) inherently high sensitivity and (2) reduced requirements for expensive or rare toxin standards. In some cases there is also (3) correlation with health risk and (4) the feasibility of developing a field test format for screening samples. One other important characteristic is common to both cytotoxicity and biochemical assays: They both give a lumped average response of the toxin forms they detect. This is both a curse and a blessing, since they require only one toxin standard and yet it may be complicated during development to predict or fine-tune the desired overall response. In the best case the relative mammalian toxicities at least parallel the relative response of the assay, so that the lumped average response is representative.

17.4. CHEMICAL SEPARATION METHODS

Separation methods are, in some cases, used in the monitoring and control of seafood toxins. This use is restricted to the less potent seafood toxins such as domoic acid and the diarrhetic shellfish toxins. On the other hand, there is no shortage of research methods based on chemical separations for studying the seafood toxins. Traditional chemical separations yield considerable detail by providing toxin profiles. This information comes at high cost, however, since expensive, and even very rare, toxins are often used in this research. As often occurs, having access to these toxins, even for research purposes, often involves lengthy and expensive efforts by some of the leading laboratories and firms. Limited availability of purified and commercially available toxin standards for many years stifled the development of chemical separation methods. Today some of the most lethal, and therefore the most worrisome, toxins are still unavailable, and those that can be purchased are generally expensive. Another limiting factor in the development of these methods is the need for very high sensitivity detection. This is due to the high toxicity of some of the toxins (such as the ciguatoxins) and the resulting need to detect minute quantities. Currently the most exciting research work on developing chemical separations of seafood toxins is centered on improved and less expensive liquid-phase mass detection. Liquid chromatography–mass spectrometry instruments, originally having price tags near $500,000, are now less expensive and offered by several vendors.

17.4.1. Biochemical Assays

Immunoassays and other competitive binding assays are for the most part very sensitive methods and are able to detect toxins below levels of concern. In the future, solid-phase "field test" immunoassays might be developed. At the present time they are all laboratory tests, as are another variety of binding assay, the competitive binding assay. Many of the seafood toxins exert neurotoxic effects by first binding to voltage-gated sodium channels on neurons. Competitive binding assays use sodium channel "synaptosome" preparations in which radiolabeled compounds compete with sample toxins for binding at sodium channel binding sites. This methodology can be used to detect saxitoxins, tetrodotoxins, brevetoxins, and ciguatoxins. Sample purification is required to prevent interference due to nonspecific binding. They also require handling and disposal of radioactive materials.

17.5. CYTOTOXICITY ASSAYS

In living neurons, specific binding to nerve cell receptors results in changes in ion permeability, and specific, directed cytotoxicity assays for seafood toxins are based on this. At the voltage-gated sodium channel, binding by brevetoxins or ciguatoxins (or toxins with similar toxic mechanism) leads to an increase in sodium channel permeability. These compounds are often called sodium channel enhancers. In the presence of veratridine, which also increases sodium ion permeability (by binding to a different site on the sodium channel than brevetoxin or ciguatoxin), the effect is synergistic. Ouabain is also added to poison the sodium/potassium adenosine triphosphatase (ATPase). This prevents the neurons from reducing the osmotic pressure resulting from ion flux and imbalance. In contrast, saxitoxins and tetrodotoxins (or toxins with similar toxic mechanism) cause a decrease in sodium ion permeability, and they are described often as sodium channel blockers. Cell death occurs in a dose-dependent manner, allowing quantitation with precision similar to PSP mouse bioassays. Using this approach, ciguatoxins can be detected at subpicogram levels in seafoods, brevetoxins can be detected in the low-nanogram range, and saxitoxins and tetrodotoxins can be detected down to picograms. No purification of crude extracts is required, and the assays can be completed in one work day.

17.6. PARALYTIC SHELLFISH POISONS

Paralytic shellfish poisoning is often first noticed as a tingling or burning sensation in the lips. Paralysis of the extremities and loss of motor coordination occur at high doses, and without treatment death by respiratory paralysis can occur. Since the toxic effect is temporary (due to reversible toxin binding), a respirator is often all that is needed to save the life of a victim.

The potent neurotoxins responsible for PSP are known as the saxitoxins. It was first believed that a single compound (the parent compound saxitoxin) caused PSP. At this time 21 different saxitoxins have been discovered (16 of the most common forms are shown in Fig. 17.1). Many make a significant contribution to the total toxicity, although their toxicities vary widely between certain groups. More complexity is introduced by interconversions among the various forms after shellfish ingest them. Many species of dinoflagellates are now associated with the saxitoxins. At northern latitudes these include *Alexandrium catenella*, *A. excavatum*, *A. fundyense*, and *A. tamarensis*. *Gymnodinium cate-*

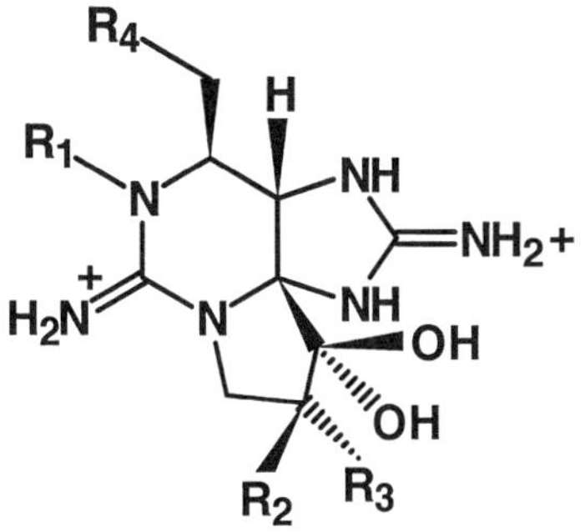

	R_1	R_2	R_3	R_4
Carbamoyl saxitoxins				
STX	**H**	**H**	**H**	**$OCONH_2$**
GTX II	**H**	**OSO_3^-**	**H**	"
GTX III	**H**	**H**	**OSO_3^-**	"
NEO	**OH**	**H**	**H**	"
GTX I	**OH**	**OSO_3^-**	**H**	"
GTX IV	**OH**	**H**	**OSO_3^-**	"
Decarbamoyl saxitoxins				
dc-STX	**H**	**H**	**H**	**OH**
dc-GTX II	**H**	**OSO_3^-**	**H**	"
dc-GTX III	**H**	**H**	**OSO_3^-**	"
dc-NEO	**OH**	**H**	**H**	"
Sulfamate saxitoxins				
B_1	**H**	**H**	**H**	**$OCONHOSO_3^-$**
C_1	**H**	**OSO_3^-**	**H**	"
C_2	**H**	**H**	**OSO_3^-**	"
B_2	**OH**	**H**	**H**	"
C_3	**OH**	**OSO_3^-**	**H**	"
C_4	**OH**	**H**	**OSO_3^-**	"

Figure 17.1. Structures of the known saxitoxins.

natum, *Pyrodinium bahamense*, and *Gonyaulax polyedra* are found at more southerly latitudes. Dinoflagellate exist in free-swimming and cyst forms, and although classic "red tide" blooms are caused by the free-swimming forms, it is the cysts (deposited in bottom sediments) that can remain toxic for several months. Toxin profiles produced by a particular dinoflagellate strain in a fixed location are remarkably unchanging but vary widely between the various dinoflagellate species.

The mouse bioassay remains the predominant method of detection in the management of PSP in shellfish. For alternative bioassays using intact organisms, house flies (*Musca domestica*), chicks, and brine shrimp have been suggested. Detecting the many different saxitoxins responsible for PSP is a classic example of the differences between research methods and methods applicable to routine monitoring. This is perhaps the best-understood suite of toxins, since most of the toxin structures were elucidated years ago, and some are now commercially available. On the other hand, the states with most active monitoring programs for preventing PSP analyze thousands of samples over the course of a year. At the present time the cost of buying these multiple toxin standards to run so many samples prohibits routine monitoring by the chemical separation methods that require them.

17.7. TETRODOTOXINS

Pufferfish (members of order Tetradontiformes) are popular food items in Japan, in spite of the fact that some of these fish contain potentially fatal levels of tetrodotoxin (TTX), a potent sodium channel blocker. Tetrodotoxin is the most toxic form of a series of compounds containing the guanidinium group (Fig. 17.2). With much higher toxicity than the other forms, TTX itself accounts for most of the total toxicity observed. Other members of the TTX family of compounds have been detected (as minor components) in species of *Fugu* pufferfish (*F. niphobles*), *F. pardalis*, and *F. poecilonotus*. This group includes 4-epi-TTX, 6-epi-TTX, 4,9-anhydro-TTX, 11-deoxyTTX, and 11-norTTX-6(R)-ol. Tetrodonic acid has also been reported. The flesh of pufferfish caught in temperate waters is believed to have low or nonexistent toxicity, but improper handling can easily result in release of the toxin from highly toxic tissues.

Symptoms of pufferfish, or tetrodotoxin, poisoning include depressed core body temperature, sweating, weak and rapid pulse, tremors, cyanosis, muscle twitching and tremors, and poor muscle coordination. At lethal doses death occurs by respiratory paralysis. In Japan alone, pufferfish poisoning is implicated in more than 10 deaths per year. Certain exotic gastropods have also caused TTX poisonings and even death. Tetrodoxin occurs in the ivory shell *Bablyonia japonica*, the frog shell *Tutufa lissostoma*, the trumpet shell *Charonia sauliae*, and the lined moon shell *Natica lineata*. The TTXs are also found in the eggs of the blue-ringed octopus (*Hapalochlaena maculosa*), newts, and certain tropical and subtropical crabs.

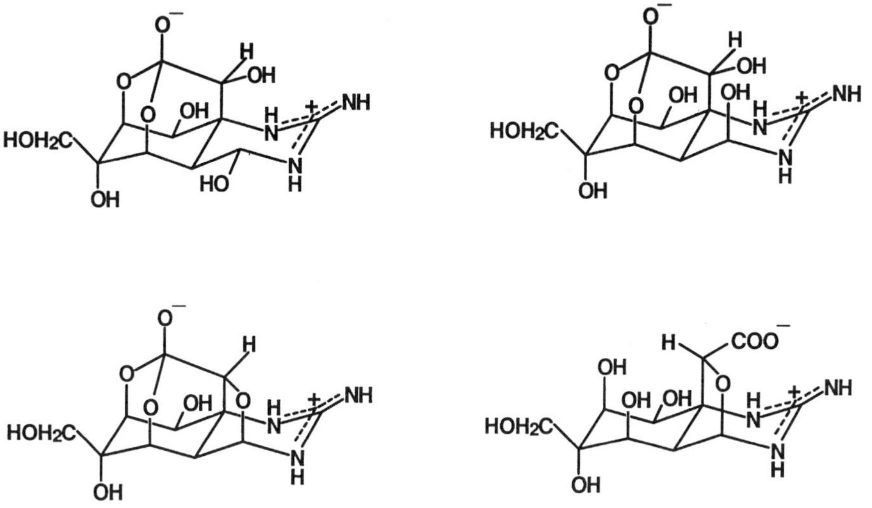

Figure 17.2. Structures of tetrodotoxin (TTX, upper left) and related compounds.

Many species of bacteria are capable of producing TTX and/or anhydro-TTX. It is now well established that these toxins accumulate in marine sediments. The TTX-producing bacteria have also been found in deep sea sediment. It has been suggested that TTXs found in marine sediments are synthesized only by bacteria and are concentrated and deposited by benthic organisms in the food web. It appears that toxin-producing bacteria might be responsible for much of the TTX toxicity of several marine finfish and shellfish. For example, TTX-producing bacteria have been isolated from the intestines of the trumpet shell *C. sauliae*, which has been implicated in poisoning incidents, and from the intestines of the xanthid crab *Atergatis floridus*. One study suggests that TTX found in pufferfish is a result of their exposure to TTX-producing marine bacteria. It would seem that the production of tetrodotoxins by a wide variety of bacteria, many of which are prevalent in the marine environment, could lead to TTX contamination of finfish other than pufferfish.

17.7.1. Detection Methods

The mouse bioassay for TTX is similar to that for the PSP toxins. Both toxin groups cause blockage of the sodium channel with death by respiratory paralysis. Since bioassays provide only a composite response representing the total potency of samples, initial efforts at detecting the toxins by chemical means were predicated on an assumption that only TTX was present since the other forms had not yet been discovered. As with the saxitoxins, this was soon proven incorrect as high-performance liquid chromatography (HPLC) and bioassay studies soon revealed the presence of multiple toxin forms.

17.8. NEUROTOXIC SHELLFISH POISONING

Striking in their appearance, the classic "red tides" most often associated with toxic plankton blooms are caused by the marine dinoflagellate *Ptychodiscus brevis* in the Gulf of Mexico and particularly Florida. Shellfish impacted by an intense *P. brevis* bloom can be toxic to humans consuming them. The illness is termed neurotoxic shellfish poisoning (NSP). Some of the symptoms of NSP are similar to ciguatera, although there is no documented case of a human fatality due to NSP (compounds causing NSP are much less toxic to mammals than ciguatoxin). Reversal of hot and cold sensation, a noteworthy paresthesia caused by ciguatera poisoning, can also occur with NSP. Fortunately NSP is also of much shorter duration than ciguatera at 1–72 h (17 h duration is often observed). Another characteristic is the occurrence of massive fish kills during a bloom of *P. brevis.* This occurs when the cell densities reach about 5×10^5 cells/L. The fish die so rapidly that there is no chance for fish exposed to *P. brevis* to become toxic to humans.

The toxicity of *P. brevis* is due to a group of polyether lactones known as the brevetoxins (Fig. 17.3). Six of the brevetoxins (type 1) are based on a single structural polyether backbone and an additional three (type 2) are based on a different polyether backbone.

Although the Gulf Coast is all impacted, within the United States, most blooms of *P. brevis* occur in Florida, and this state has historically had the greatest impact from brevetoxins. A 1987 red tide reached North Carolina waters due to transport by currents. Control measures for the brevetoxins are somewhat different than the other toxins in that plankton densities are also used. When *P. brevis* cells reach 5000 L^{-1}, The Florida Department of Natural Resources (FDNR) closes waters to shellfish harvesting. Mouse bioassay is used to reopen waters to harvesting following the bloom. Alternative assays are being explored. New Zealand has also been impacted by brevetoxins.

Several different methods for the detection or quantitation of brevetoxins are available or are under development, including (a) animal bioassays (mouse, fish), (b) molecular pharmacological assays using voltage-dependent sodium channel preparations, (c) immunoassays such as radioimmunoassays (RIA) and enzyme-linked immunosorbent assay (ELISA), and (d) HPLC. Many of these methods have potential for application to shellfish monitoring in the management of NSP. At this time, only the mouse bioassay has been applied to the detection of the brevetoxins in shellfish. Brevetoxins are detected in shellfish meats by a standardized mouse bioassay. Toxicity is determined by using the relationship of dose to death time of mice injected (intraperitoneally) with toxic residues extracted from shellfish with diethyl ether.

17.9. DIARRHETIC SHELLFISH POISONING

Diarrhetic shellfish poisoning was first reported in Japan in 1978. In Europe the toxins causing DSP are also found and have caused human illness. Predom-

		R_1	R_2	R_3	
Brevetoxin	2	H	CH_2	CHO	
	3	H	CH_2	CH_2OH	
	5	$COCH_3$	CH_2	CHO	
	6	H	CH_2	CHO	(27, 28-β-epoxide)
	8	H	O	CH_2Cl	

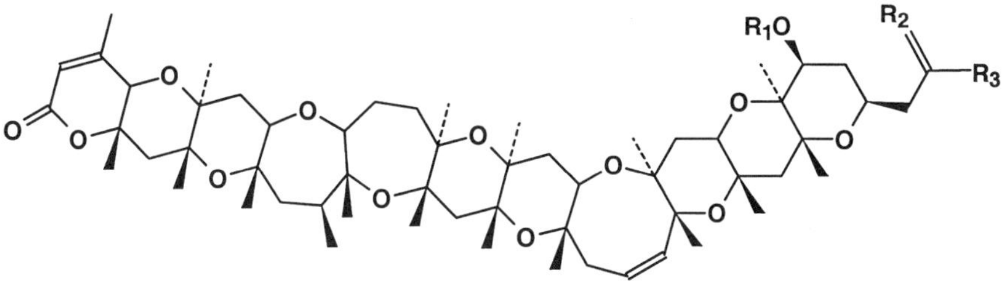

(*a*)

		R_1	R_2	R_3
Brevetoxin	1	H	CH_2	CHO
	7	H	CH_2	CH_2OH

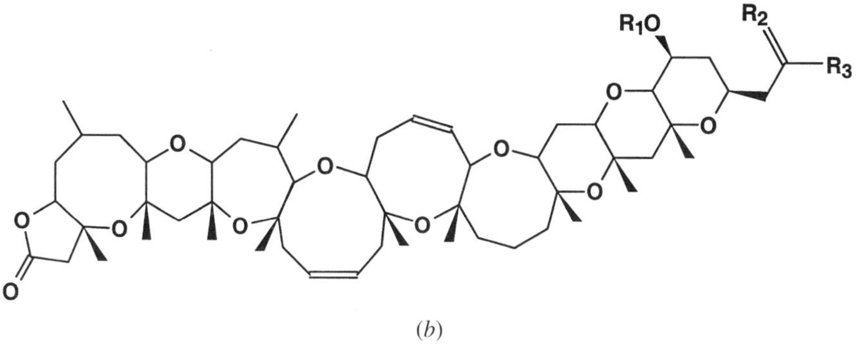

(*b*)

Figure 17.3. Major brevetoxins: (*a*) type 1 skeleton; (*b*) type 2 skeleton.

inantly self-limiting, DSP nonetheless causes severe gastroenteritis. It is now recognized as a worldwide threat to human health. Many thousands of consumers of mulluscan shellfish are affected, and considerable economic hardships result in the shellfish industry. Human symptoms include diarrhea (92%), nausea (80%), vomiting (79%), and abdominal pain (53%). Because these symptoms are not unique to DSP, it is thought that this illness is very likely one of the most underreported maladies. The absence of unique symptoms,

	R_1	R_2
Okadaic acid	H	H
Dinophysistoxin 1	H	CH_3
Dinophysistoxin 3	acyl	CH_3

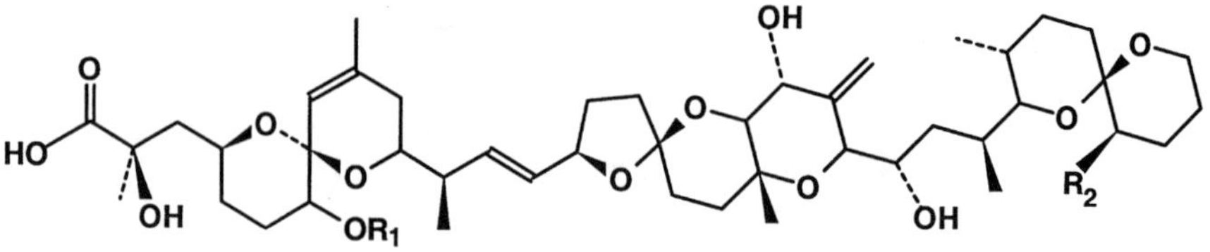

Figure 17.4. Okadaic acid group of diarrhetic shellfish toxins.

especially those indicating neurotoxicity (e.g., paralysis), often implicates other marine intoxications rather than DSP. Diarrhetic shellfish poisoning is distinguished from bacterial infection by its rapid onset time and heat stability. The first symptoms occur a few hours after ingestion. The onset time is as short as 30 min in severe cases. After 3 days victims recover with or without medical treatment. Although DSP-implicated shellfish harvested in Japan have been reported to contain many different toxins (Fig. 17.4; only okadaic acid derivatives are shown), in most areas of the world the predominant toxins are okadaic acid (OA) and its derivatives dinophysistoxin-1 (DTX-1) and dinophysistoxin-3 (DTX-3). These are also the only forms of the group that cause severe diarrhea. Other known forms include the macrolides pectenotoxin-1 (PTX-1), pectenotoxin-2 (PTX-2), pectenotoxin-3 (PTX-3), and pectenotoxin-6 (PTX-6) and a sulfated toxin-yessotoxin (YTX). The latter has some structural similarities with the brevetoxins (the presence of contiguous transfused ether rings). Okadaic acid and derivatives may cause diarrhea by stimulating the phosphorylation of a protein that controls sodium secretion by intestinal cells. This is similar to that caused by cholera toxin. One form, DTX-1, causes, in addition to diarrhea, severe injuries to intestinal mucosa. The form PTX-1 is nondiarrheagenic but is hepatotoxic in a way similar to that caused by fungal toxins such as cyclochlorotine (from *Penicillium islandicum*) and phalloiodine (from *Amanita phalloides*).

Although YTX is also nondiarrheagenic, it is cardiotoxic. Okadaic acid and DTX-1 were recently found to be tumor promoters. Okadaic acid and derivatives are potent inhibitors of protein phosphatases-1 and -2A. Tumor promotion may stem from increased phosphorylation proteins that are substrates for protein kinase C and dephosphorylation of these enzymes. Mussels collected in Sweden, the Netherlands, France, and Spain contain OA as the major toxin, and in Norway mussels collected from one area were found to contain OA as the main toxic constituent, while DTX-1 was predominant in another area. Scallops from Japan continue to show the most complicated toxin profile; for example, pectenotoxins have been detected and confirmed only in Japanese shellfish.

The toxins are produced by dinoflagellates belonging to the genera *Dinophysis* and *Prorocentrum*. Difficulties in culturing *Dinophysis* sp. under laboratory conditions and their low population densities in the sea make the assignment of a species to a particular DSP outbreak challenging and somewhat controversial. Okadaic acid or DTX-1, and in some cases both toxins, have been found by HPLC in *Dinophysis acuminata, D. acuta, D. fortii, D. mitra, D. norvegica, D. rotundata*, and *D. tripos* and in *Prorocentrum lima*; in addition, PTX-2 was found in *D. forti*). In addition, DSP toxins have been associated with *D. sacculus*. Cell densities of *D. forti* as low as 200 cells/L have been associated with shellfish toxic to humans.

Most of the methods developed for the detection or quantification of DSP toxins detect the presence or the physiological action of OA toxins. Reliance on detection methods for DSP toxins specific for only OA and its derivatives, although better than no methods at all, may not be adequate if resident phytoplankton in shellfish-growing areas are capable of producing the pectenotoxins and/or yessotoxins without producing the OA toxins. Hopefully, future monitoring methods can be devised that will measure all of the components of DSP toxins.

The first analytical method developed for DSP was a mouse bioassay. An official mouse bioassay scheme published by the Ministry of Health and Welfare, Japan (1981), is cited by many investigators, and this method or a variation of it was first used for monitoring shellfish toxicity in many of the areas impacted by DSP. Presently HPLC methods are used most frequently. These methods use preseparation labeling of the toxins with a fluorescent tag to allow detection.

17.10. AMNESIC SHELLFISH POISONING

Domoic acid is a neurotoxic amino acid with the structure shown in Figure 17.5. It binds to glutamate receptors in the brain, thereby causing continuous stimulation of nerve cells and eventually lesions are formed. Victims experience headache, loss of balance, disorientation, and the usual gastrointestinal symptoms typical of most food-poisoning episodes. The most characteristic symptom, however, is persistent and apparently permanent loss of memory—thus the term amnesic shellfish poisoning. In December 1987, 156 individuals became ill after ingesting blue mussels (*Mytilus edulis*) cultivated and harvested at Prince Edward Island, Canada. Four elderly people died. Domoic acid (up to 900 ppm) was found in the implicated mussels. It is believed that the domoic acid was produced by the diatom *Nitzschia pungens forma multiseries* and then

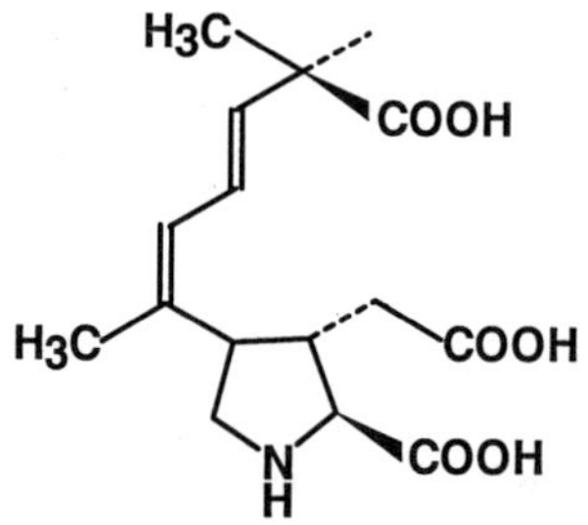

Figure 17.5. Domoic acid.

retained by the mussels. As a result of this incident, shellfish are now monitored for domoic acid in Canada and to a limited extent in the United States. The relatively low toxicity of domoic acid [median lethal dose (LD_{50}), intraperitoneal (i.p.) of 3.6 mg domoic acid/kg mouse] and lethality at 4 mg/kg in the monkey *M. fascicularis* have led to the establishment of an action level of 20 ppm.

During initial efforts to monitor mussels for ASP toxicity, aqueous extracts were injected i.p. into mice. Mice so injected show a scratching behavior that aids in distinguishing domoic acid poisoning from PSP. Immediately following the scratching behavior, mice injected with domoic acid show uncontrolled twisting and rolling motions and loss of the righting reflex. Tremors and cyanosis occur next, followed by death. Time elapsed from injection to the onset of scratching is inversely related to dose, as is death time. The bioassay is performed using essentially the same procedure as in the official bioassay for PSP, except that observation times are extended. Since the mouse bioassay is not sufficiently sensitive to monitor domoic acid at the 20-ppm action level, instrumental methods were developed to provide the required sensitivity.

The presence of two conjugated double bonds in the molecule results in a strong absorbance at 242 nm, which allows for detection by monitoring absorbance at this wavelength. Acidobasic groups at the carboxylate and imino positions of domoic acid allow separation based on charge. Capillary electrophoresis (CE) has been used to the detect domoic acid in contaminated mussels. Domoic acid was readily separated from components of the mussel sample matrix in 10 min. A detection limit (signal-to-noise ratio 5:1) of 2 ppm was achieved. With excellent mass detection limits, the CE method requires samples of only 3–15 nL and will find applications where sample size is severely limited. The HPLC methods described below have superior concentration detection limits.

Domoic acid is readily determined in mussels using conventional reversed-phase HPLC. The official Association of Official Analytical Chemists (AOAC) International procedure for extracting PSP toxins from shellfish has been applied to a modification of the method of Quilliam for domoic acid based on reversed-phase separation with UV detection. Recently this method was collaborated. As for the mouse bioassay careful timing of the acid extraction step is crucial since domoate is acid labile. (Neutral aqueous extraction stabilized with methanol has been suggested as an alternative to the acid extraction.)

It had been proposed to chemically convert domoic acid to phenyliso-thiocyanate derivatives and trialkyl esters as a means of confirming domoic acid in mussels. The derivatives are quantitated by reversed-phase HPLC with absorbance detection at 242 nm. Trialkyl derivatives were also determined by gas chromatography–mass spectrometry (GC–MS). Mussel tissue is extracted with HCl, as in the official HPLC method, except that additional sample conditioning is required. For this purpose a (strong) cation exchange cartridge and a C-18 cartridge are used to retain and wash the extract. The detection limit is expected to be similar to the 2-ppm limit of the direct-detection method.

A precolumn derivatization HPLC method for domoic acid has been described. Reaction of domoic acid with 9-fluorenylmethylchloroformate produces a fluorescent derivative, allowing sensitive fluorescence detection. The derivative is separated by conventional reversed-phase chromatography on a microbore (2.1 mm internal diameter) column and detected at excitation and emission wavelengths of 264 and 313 nm, respectively. The detection limit in seawater in cultures of *N. pungens forma multiseries* and in naturally occurring phytoplankton was 15 pg/mL (50 p*M*).

17.11. CIGUATERA

Ciguatera is a term designating an illness caused by eating a variety of reef fishes and carnivorous fish that feed on them. It is most prevalent in the Caribbean and South Pacific. Symptoms include gastrointestinal disturbances and (more diagnostic) neurological problems (e.g., paresthesia and dysesthesia-temperature reversal), and cardiovascular disorders. The toxic effect can last from several days to several months with resurgence of some symptoms after several years. These effect are also cumulative. It is estimated that ciguatera causes annually over 20,000 illnesses in the United States and more than one-third of all finfish-borne illness outbreaks. Levels as low as 1 ppb can cause intoxication in adults. The vast majority of these illnesses are caused by recreational harvesting.

Until the structures of several ciguatoxins were deduced several years ago, the term ciguatera has been defined more by the epidemiological aspects and symptomatology of the illness than by the chemistry or structure of the toxins. Although to some extent this holds for most of the other marine toxins, ciguatera has been one of the most difficult of the marine toxins to study. Difficulties include the unpredictable and variable nature of fish toxicity, the scarcity of toxic fish to study, the logistics problems encountered in working in some of the endemic areas, the diversity of fish species implicated, the tedious isolation procedures required due to the extremely low toxin concentration in the fishes (ppb), and the complex nature and multiplicity of the toxin(s). Finally, research

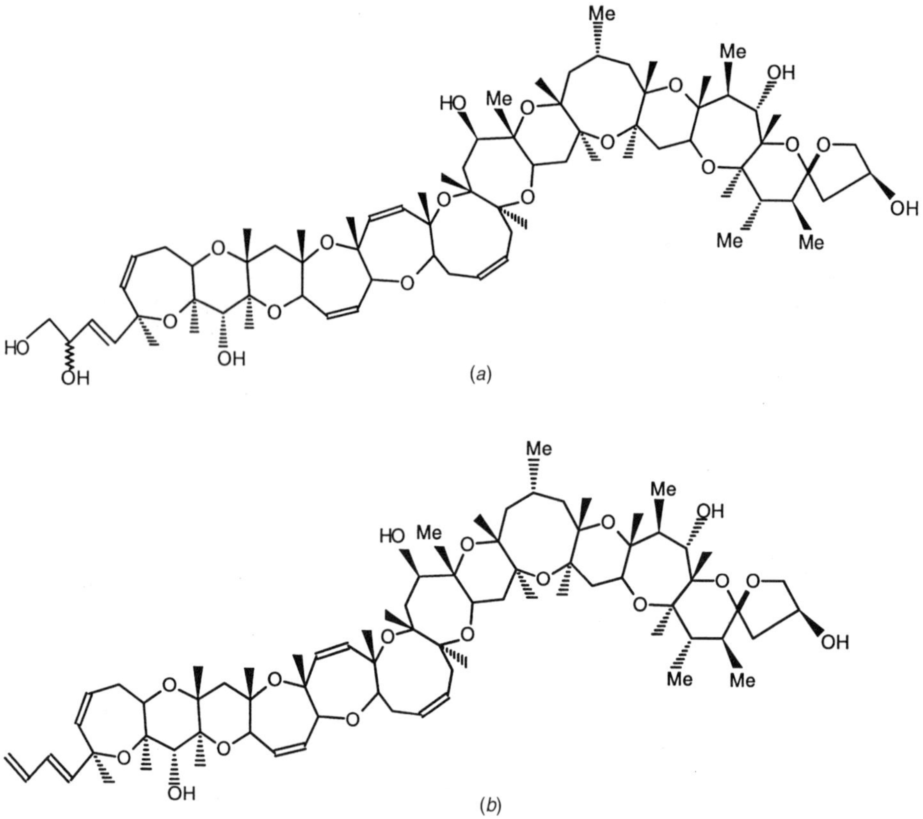

Figure 17.6. (*a*) Ciguatoxin (CTX-1B) and (*b*) precursor (CTX-4B).

on detection methods for ciguatoxins has been impossible for many due to the absence of any toxin standards. Among the many other polyether marine toxins, the characteristic feature of the most potent ciguatoxin form (CTX-1B; see Fig. 17.6) is the presence of 13 contiguously transfused ether rings resembling the brevetoxins and YTX. Several structural analogs of CTX have been structurally defined (Fig. 17.6 shows just two of these structures). The toxin(s) are first produced by an epiphytic dinoflagellate and then transmitted to various fishes through the food chain. *Gambierdiscus toxicus* is the toxin-producing dinoflagellate. Difficulties in laboratory studies result from the fact that this alga produces little or variable quantities of toxin in culture and also shows variation in toxin production in the natural environment. So far *G. toxicus* has not produced CTX-1B directly. It is believed that instead it produces at least four less polar toxins and that some (CTX-4B, shown in Fig. 17.6) are converted to CTX-1B by partial metabolism in fish. *Gambierdiscus toxicus* also produces maitotoxin (MTX), a second important toxin first detected in surgeonfish. A

Caribbean form of ciguatoxin has been discovered and a detailed structural study has nearly been completed.

Ciguatoxin has historically been detected by mouse bioassay. However, this assay is time consuming and difficult to use quantitatively. Several alternative methods have been proposed for the detection of ciguatoxins, including HPLC, immunoassays, binding assays at sodium channels, and sodium channel–directed cell bioassays. For relatively rapid detection without the need for multiple toxin standards, the latter two methods are very promising and undergoing validation studies. For confirmation of toxins and in detailed studies, powerful HPLC–MS methods are becoming more popular and more affordable. Their major drawback remains the need for multiple toxin standards that are hard or impossible to obtain for many laboratories.

17.12. SCOMBROID POISONING

The symptoms of Scombroid poisoning, besides including the usual gastrointestinal problems, resemble an allergic response. The main difference is that Scombroid poisoning will have a virtually 100% attack rate while allergies are much lower. Another similarity is that the victims also respond to antihistamines.

Scombroid intoxication results from ingestion of certain species of fish that have not been adequately chilled. Only those species that have high levels of free (nonpeptide) L-histidine are implicated. Some bacteria are known to have enzymes that can cause decarboxylation of free L-histidine to histamine. This, the frequent finding of elevated histamine levels in implicated fish, and finally the above response to antihistamines have led to scombroid poisoning often being called histamine poisoning. Although histamine is somehow involved in scombroid intoxications, there may be other agents, either inhibitors that interfere with human enzymatic detoxification of histamine or even potent histamine-like compounds that bind to the same receptors as histamine.

The most commonly used and officially approved method for detecting histamine in seafoods is a wet chemical procedure that is performed manually. Alkaline condensation of histamine with *o*-phthalaldehyde (OPA) and subsequentl dehydration with phosphoric acid produces a fluorescent product. In the batch-mode procedure used in the official method, ion exchange cleanup is necessary prior to performing the chemical reaction. In this cleanup procedure, L-histidine, the major interference found in the fish sample matrix, is removed as the anion. Using flow injection analysis (FIA), the chemical kinetics of the OPA condensation reaction are manipulated to enhance selectivity (histamine versus L-histidine) 25-fold. Reaction timing is controlled using fixed reaction geometry and flow rates. The FIA procedure, which can process 60 injections/h is also automated on commercially available instrumentation. This method will soon be collaboratively studied for approval. Other methods for histamine determination in seafoods include commercial immunoassays, liquid chromatography, and electrophoresis.

17.13. CONCLUSION

Seafood toxins are a diverse group. The most potent (and occasionally fatal) seafood toxins are neurotoxins. Many of these neurotoxins exert their toxicity via the voltage-gated sodium channels of nerve cells. Fish toxins are produced by dinoflagellates and bacteria. Shellfish toxins are produced by a variety of dinoflagellates and diatoms. Shellfish toxins are managed by harvest-level monitoring. Ciguatera is one of the most challenging fishborne illnesses to manage due to the mobility of the fish, sporadic distribution of the toxin, and challenges in developing detection methodology. For the purpose of health protection the potency of a given toxin determines the degree of sensitivity and selectivity required to detect it. Domoic acid, for example, can be monitored adequately by HPLC with UV detection due to relatively low toxicity. In contrast, ciguatoxins and saxitoxins require much higher sensitivity. Instrumental methods for monitoring these potent neurotoxins are less practical due to both the scarcity and/or cost of standards and the high sensitivity required. The most promising technologies for monitoring high-potency seafood toxins appear to be competitive binding assays such as immunoassays and receptor assays and cell assays based on modes of toxicity. The development of reliable and rapid screening methods for both laboratory and field use will greatly improve seafood safety and allow more effective use of seafood resources.

BIBLIOGRAPHY

Baden, D. G. 1988. Public health problems of red tides. In: A. Tu (Ed.), *Marine Toxins and Venoms*. New York: Marcel Dekker, p. 259.

Doucette, G. J., M. M. Logan, J. S. Ramsdell, and F. M. Van Dolah. 1997. Development and preliminary validation of a microtiter plate-based receptor binding assay for paralytic shellfish poisoning toxins. *Toxicon* **35**, 625.

Dyckman, L. 2000. *Food Safety: FDA's Use of Faster Tests to Assess the Safety of Imported Foods, Report to Congressional Requestors*. Washington, DC: United States General Accounting Office.

Hungerford, J. M., and M. M. Wekell. 1992. Analytical methods for marine toxins. In: A. Tu (Ed.), *Seafood Poisoning*. New York: Marcel Dekker, p. 415.

Hungerford, J. M., and M. M. Wekell. 1993. Control measures in U.S.A. In: *Toxic Algae in* P. Krogh and B. Hald (Eds.), *Food and Drinking Water*. New York: Marcel Dekker, p. 117.

Manger, R. L., L. Leja, S. Lee, J. M. Hungerford, and M. M. Wekell. 1993. Tetrazolium-based cell bioassay for neurotoxins active on voltage-sensitive sodium channels: Semiautomated assay for saxitoxins, brevetoxins, and ciguatoxins. *Anal. Biochem.* **214**, 190.

Poli, M. A., R. J. Lewis, R. W. Dickey, S. M. Musser, C. A. Buckner, and L. G. Carpenter. 1997. Identification of Carribbean ciguatoxins as the cause of an outbreak of fish poisoning among U.S. soldiers in Haiti. *Toxicon* **35**, 733.

Vernoux, J. P., and R. J. Lewis. 1997. Isolation and characterization of Caribbean ciguatoxins from the horse-eye jack (*Caranx latus*). *Toxicon* **35**, 889.

Wekell, M. M., and J. M. Hungerford. 1994. Microbiological quality of seafoods: Marine toxins. In: F. Shahidi and J. R. Botta (Eds.), *Seafoods: Chemistry, Processing Technology and Quality*. London: Blackie Academic and Professional, p. 220.

CHAPTER 18

Parasites

GEORGE J. JACKSON

18.1. INTRODUCTION

Parasites, in the past, had a special niche in biology. Too big and anatomically complex to be considered microbes and too small and host dependent to be considered predators, the potential of these animals to elicit immunity in a host, be cultured in the laboratory, or display behavioral traits was viewed skeptically by many scientists during the first decades of the twentieth century. That some parasites show susceptibilities to the same metabolic poisons as their hosts and are sexually differentiated, yet also have the capacity for asexual reproduction in their frequently complex life cycles, were among the additional reasons they came to be studied separately from the disciplines of microbiology and zoology.

Human infections with parasites were commonly regarded as originating in places that fell far short of the hygienic standards achieved by developed societies located in temperate climates. Consequently, parasitology was neglected at centers of biomedical research, except during times of colonial expansion or warfare in tropical regions. Due to these circumstances, medical parasitology was often paired with tropical medicine, despite many parasites' cosmopolitan distribution that includes their occurrence in host populations inhabiting arctic regions.

More recently, in the urban centers of developed nations there has been the suspicion—from clinical cases with no history of travel as well as from environmental surveys and sewage sampling—that certain "hot climate" or "rural" parasitoses are not absent, especially in slum settings. Nevertheless, attempts to document their incidence have received little support.

The current tendency to handle parasites as if they differ insignificantly from bacteria can be as impractical as considering them to be substantially different.

Guide to Foodborne Pathogens, Edited by Ronald G. Labbé and Santos García
ISBN 0-471-35034-6

Many parasites cannot simply be frozen and thawed for maintenance, and convenient laboratory hosts are hard to find. Culturing has been achieved only for certain species, and often for just part of the life cycle and at the expense of selectively eliminating major portions of the population spectrum in the culture inoculum. Sanitary precautions against contamination by feces or mucus are not effective against those parasites that infect humans because of host-finding behavior or through other hosts in their life cycle—hosts that are part of the human food chain.

While the study of parasites has contributed significantly to overall progress in biology by presenting the first apparent examples of an impressive array of basic phenomena (such as the reduction division in cell replication, antigenic modulation and mimicry, and, perhaps, the physiological state of "uncultur-ability"), attempts to merge parasitology with other biological specialties have been successful only in part. At the beginning of a new millennium, the parasite may indeed exemplify "just another model" at the cellular and molecular levels but is still distinct in respect to aspects of laboratory practice and epidemiology. Some of these aspects impinge particularly on the concerns of the food scientist.

18.2. PARASITES AND ZOOLOGICAL CLASSIFICATION

Most parasitic animals are invertebrates that belong to the protozoa (i.e., single-celled organisms) or metazoa. Among the metazoa, the helminths (i.e., worms) receive the most attention whereas the parasitic arthropods (e.g., insects) are often relegated to the discipline of entomology because the most common of these species exist externally on their hosts where the distinction between predator and parasite becomes blurred. Nevertheless, parasitic arthropods also exist internally, and those of interest to the food scientist are, principally, the fly maggots that inhabit the flesh wounds of animals used as meat. Other arthropods play an important role in infectious disease because prominent members of the group function as vectors or as links in a host chain for different types of pathogenic microorganisms including parasitic animals.

Vertebrates that may be considered parasites include fish that are exoparasitic on the gills of other fish. Mammals (besides being "parasitic" in the mother during embryonic development) have been rumored to be parasites. This is, most likely, a fantasy despite seafarer tales that tell of rats abandoning sinking whaling vessels, clambering onto live whales, and then burrowing beneath the skin to excavate passages and chambers in the blubber.

18.3. SPECTRUM OF FOODBORNE PARASITES

Depending on how one classifies, there are between 107 and 127 species of animals parasitic for humans that may be foodborne. Only those whose infective stages are encysted in or on food animals and plants are exclusively foodborne; examples include the meatborne nematode *Trichinella spiralis*, the seafood-

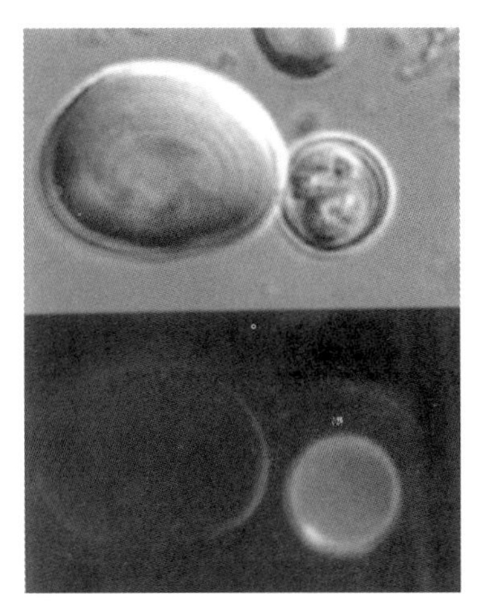

Figure 18.1. *Cyclospora cayetanensis* (Cc) has been difficult to detect in or on foods epidemiologically associated with outbreaks of cyclosporiasis. Initially, using DNA-based procedures and then confirming by microscopy, this emerging parasite was identified by the FDA and the CDC in frozen leftovers of a chicken–pasta–basil salad that caused illness in Missouri in the summer of 1999. The oocyst of Cc typically is of 8–10 μm in diameter and when developed to infectivity (sporulated) has two internal cysts (sporocysts) that are seen under white light (above); under UV light (below), the oocyst wall fluoresces. The larger oval body beside the oocyst is a starch grain. Lopez, A. S., D. R. Dodson, M. J. Arrowood, P. A. Orlando, Jr., et al. 2001. Outbreak of cyclosporiasis associated with basil in Missouri in 1999. (*Clin. Infect. Diseases.* **32**, 1010–1017.)

borne anisakid nematodes, the beef tapeworm *Taenia saginata*, and watercress-borne liver flukes of the genus *Fasciola*. Others—such as protozoa and helminths, whose cysts or eggs reach the outside environment by way of the definitive host's feces—contaminate a variety of fomites (diverse inanimate items or objects that transfer pathogens mechanically to a host by direct or indirect contact—a glove may serve as a fomes, as may fertilizer, aerosol, ice, and packaging). In the past, food scientists in any one part of the world were likely to encounter only a few of the many types of parasites. Now, given the increasing internationalization of the food market and the ever greater frequency of travel, one ought to familiarize oneself with the entire range of foodborne parasitic animals.

An example of a recently emerged foodborne pathogen that has caused many scattered cases and common-source outbreaks at sites distant from its origins is the coccidian protozoa *Cyclospora cayetanensis*. Not definitively described until the early 1990s, it has presumably, on the basis of epidemiological evidence, been transported into the United States and Canada on fresh produce such as raspberries, lettuce, and perhaps basil grown in regions to the south. The environmental survival stage of *C. cayetanensis*—the oocyst—is not directly infectious when it is shed in an infected individual's feces but must remain outside the host in order to mature (sporulate) (Fig. 18.1). Humans are the only known host for *C. cayetanensis*. Another foodborne coccidian protozoa, *Cryp-*

tosporidium parvum, has oocysts that are directly infectious (already sporulated) on being shed in the feces and can, therefore, pass directly from host to host. It infects a broad range of mammalian hosts, including herbivores and humans, and has a geographic range wider than that of *C. cayetanensis*. Waterborne infections with *C. cayetanensis*, and especially *C. parvum*, may be more important in human epidemiology than foodborne infections. *Toxoplasma gondii*, a coccidia that can be meatborne, is of increasing concern, too, as a waterborne pathogen.

18.4. DETECTION METHODS

The testing of foods for pathogens (whether they are viruses, bacteria, fungi, or parasites) is in many instances more difficult than the testing of patients. In or on a food or a fomes, the pathogen often exists in very low numbers and in an injured state (the damage may be due to heat and other food-processing techniques). If not injured, it at least subsists under conditions that are not optimal for its replication. Also, the presence in foods of a variety of other microorganisms that are or may become culture competitors often adds to the difficulty of detecting the pathogen outside the host.

Methods to overcome these obstacles for foodborne bacterial pathogens are not unknown. Moreover, refrigeration or freezing will maintain many bacteria readily; resuscitation media to repair injury and selective media to exclude competitors have helped to make culturing routine for many bacterial species. Food parasitology, however, as a newer discipline has yet to overcome the many difficulties in detecting, maintaining, and multiplying the parasites that may be resident in a food sample.

To analyze a food for all the live organisms it contains is a very different task from testing it for a specific pathogen. A total analysis yields information that is valuable for ecological understanding. However, the procedures involved are time consuming in the extreme. To focus an analysis only on those organisms that are of health significance when ingested by mammals, the principal tool is digestion by the "artificial stomach juice" method. The reasoning behind applying this technique is that an organism must, at minimum, survive passage through the stomach in order to establish itself and be infective.

The digestion method for food parasites, as standardized by the U.S. Food and Drug Administration (FDA), is performed in a vessel immersed in a $36 \pm 0.5°C$ water bath; the artificial stomach juice consists of a solution of 15 g pepsin powder and 750 mL physiological saline (0.85% NaCl) adjusted to pH 2 with a 1:1 solution of water and concentrated (37%) hydrochloric acid. Digestion proceeds with continuous stirring (250 rpm) until visually complete but should not exceed 24 h. Prior to digestion, proper preparation of the test portion (100 g) by cutting, homogenization, or predigestion with other enzymes (such as papain) may vary with the type of food being examined and the sorts of parasites expected. Digestion is followed by sieving and sedimentation of the digested mixture; the parasites tend to settle to the bottom of a cone, are withdrawn from there, and are examined macroscopically and microscopically for preliminary identification. Detailed identification procedures usually follow.

Minor variations of the digestion procedure do not seem to affect significantly the number of organisms recovered, according to a European collaborative study in which *T. spiralis* was the test parasite. An FDA comparison of digestion and elution for enumerating nematodes in fish found that although the two methods did not differ significantly in the total numbers of worms obtained from split portions of 470 flatfish, 1062 of the 1110 nematodes recovered by digestion were potential pathogens for human consumers of raw or semiraw fish, while with elution there was the significantly lower number of 608 pathogens from a total of 922 nematodes. In other words, digestion did serve as a screen for selecting the more invasive larval worms and reducing but not totally removing the nonpathogens.

To date, the digestion method has proved useful in screening for pathogenic helminths in mammalian meat, seafood, and escargot. However, not all worms that are nonpathogenic for mammals are eliminated or reduced in number. Obviously, there is a need to test the technique with more foods and other invertebrates, particularly the protozoa. Whether the digestion method preserves intact the metabolites and symbionts that are associated with parasites must also be investigated.

18.5. ARE ALL PARASITES PATHOGENS?

There is no simple answer to this question. The state of the host's immunity is as important a factor as a parasite's virulence characteristics. Since the discovery of the acquired immunodeficience syndrome (AIDS) epidemic in the early 1980s, we have come to realize that immunodeficient individuals are susceptible to a number of organisms (such as *Pneumocystis carinii* and various microsporidia) previously classified as nonpathogenic or rarely pathogenic in humans.

Those who distinguish between pathogenic and nonpathogenic parasites often point to the presumably clear-cut examples of *Entamoeba histolytica* and *Entamoeba coli*. However, the pathogenicity of the former is not constant. Rather, it varies among strains. Moreover, while there are no known instances of illness due to the latter (even in immunodeficients), one should remember that the role of *E. coli* as a host for internal symbionts that are potential pathogens has yet to be investigated.

18.6. PARASITE PRODUCTS AND SYMBIONTS

Protozoa, both parasitic and freeliving, do serve as hosts for an array of other life forms. It is believed that *Giardia lamblia* trophozoites and cysts, when first isolated, normally contain viral symbionts internally and fungal and bacterial symbionts internally and on the surface. These are often lost with recovery and culturing because the use of antibiotics and other cleansing agents to eliminate competitors tends also to eliminate the associates.

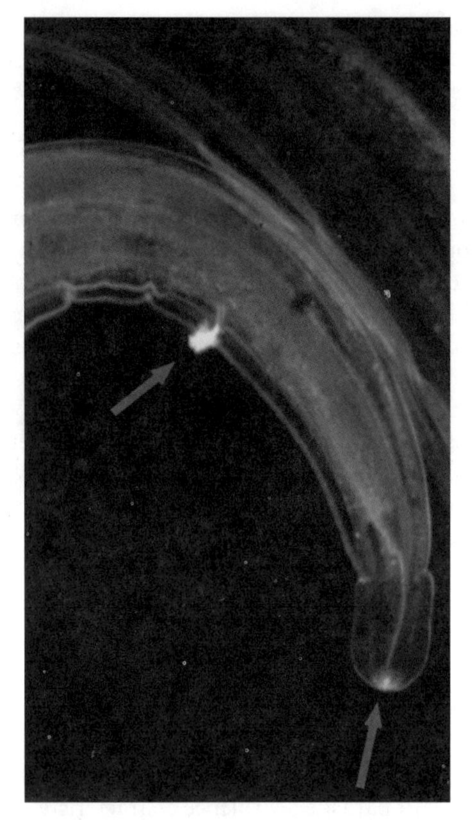

Figure 18.2. Parasite secretory and excretory products have diverse functions for the parasite and may also affect other organisms, including the parasite's host. Precipitated by antibody and highlighted by a fluorescent tag tied to the antibody, the parasite product is shown at two locations on this nematode—the oral aperature and the excretory pore (arrows).

"Wormy cod" (i.e., cod fillet with larval anisakid nematodes) is dewormed in some countries before being offered to the consumer. Even if the removal procedure were totally effective (which it is seldom or never), the worms' products would remain in the fish flesh. The same question arises with parasites that are not removed but killed before the food that contains them is eaten. Do *T. spiralis* or *Taenia* spp. produce substances detectable in cooked meat and, if so, are these products still capable of biological activity?

Parasites excrete and secrete diverse substances (Fig. 18.2). In some instances their function for the parasite is apparent, as when they act as external enzymes in invasion or digestion or when they are adhesive and anchor the parasite to a host site. For the host, parasite products have antigenic and allergic effects that are well known, but they also have consequences for host physiology and may be at least potentially toxic. For other microorganisms, parasite products (and those of freeliving invertebrates) may trigger morphological changes. Cooking (i.e., heat) inactivates some but not all of the potentially toxic effects of parasites' physiological products.

The microbial symbionts of parasitic animals are just beginning to be studied. It has been postulated that symbionts were crucial in the development of eukaryotic organisms from prokaryotes and that *G. lamblia* (or, more likely, one of its ancestors) represents an important link in this major step in evolution. Such protozoa as *Giardia* and *Entamoeba* spp. may contain not only unique viruses but also those that cause illness in humans. Even human immunodeficiency virus (HIV), the AIDS virus, has been shown to occur in the trophozoite stage (i.e., the active form of protozoa) of individuals with both an enteric protozoan infection and HIV positivity. However, to date, HIV has not been demonstrated in the cysts of enteric protozoa, and it is this stage that principally contaminates food and drink.

18.7. EPIDEMIOLOGY

The status of the population in respect to intestinal parasites was inferred for one North American country—the United States—by its Centers for Disease Control and Prevention (CDC). The data originated in the diagnostic laboratories of 49 of that nation's states. Summarized was information from the stool examinations of 216,275 ill individuals. Of these, 43,539 (20.1%) were positive for parasites. This is a significant increase over results in 1977, when 15.6% of illness stools were positive.

The predominant parasite species found in the 1987 stools were protozoa, with the most frequent being *G. lamblia* (15,497, or 7.2%), *E. coli* (9146, or 4.2%), and *Endolimax nana* (9086, or 4.2%). Among the helminths, the hookworm nematodes predominated (3299, or 1.5%), although these roundworms are not usually ingested, being able to penetrate skin directly from contaminated soil. Next in frequency were two nematodes that are ingested (including with food)—the whipworm *Trichuris trichiura* (2682, or 1.2%) and the large common roundworm *Ascaris lumbricoides* (1735, or 0.8%), followed by Far Eastern fishborne trematodes (or flukes) of the genera *Clonorchis* and *Opisthorchis* (jointly 1226, or 0.6%) and the common small cestode (or tapeworm) *Hymenolepis nana* (900, or 0.4%).

Infections with such foodborne parasites as the anisakid nematodes (from fish) (Fig. 18.3) or *T. spiralis* (from meat) would not usually be detected by routine stool examination. The former is frequently discovered when a worm migrates back up into the throat or mouth, whereas trichina larvae penetrate into the musculature.

Factors to which the CDC ascribed the increased frequencies of parasitic infections in the United States include immigrants (especially from the Far East for the flukes), improved testing techniques, opportunism due to the immunosuppressive AIDS epidemic, and the increase in numbers of child daycare

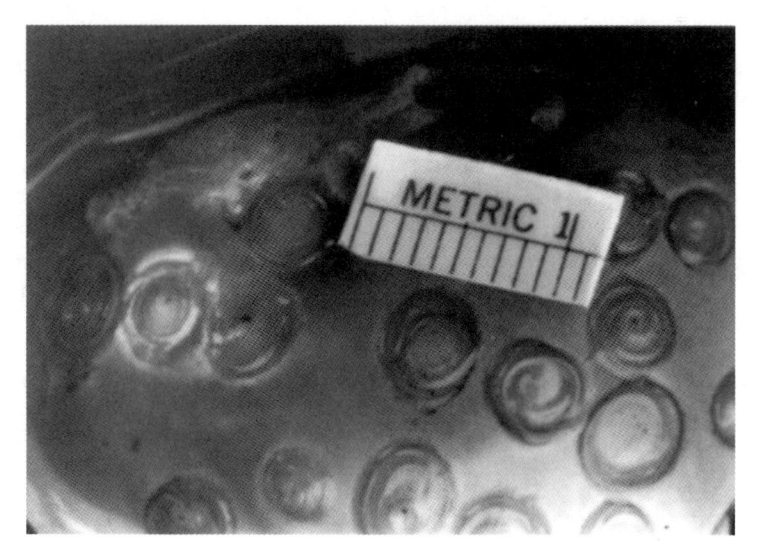

Figure 18.3. Anisakid nematodes encysted in a fish.

centers. Other factors, not mentioned by the CDC, include the increased popularity of semicooked and raw foods, increased importation of fresh and frozen foods, and more foreign travel. Whatever the causes, it is clear that foodborne parasites are not endangered species.

18.8. SYMPTOMATOLOGIES

Because of the diversity of foodborne parasitic animals, symptoms of infection and illness are varied.

Diarrhea is the most frequent symptom in protozoan infections of the gastrointestinal tract and is triggered by some amoebas (such as *E. histolytica*), flagellates (such as *G. lamblia*), ciliates (such as *Balantidium coli*), as well as coccidia (such as *Cryptosporidium parvum* and *Cyclospora cayetanensis*). The duration of these diarrheas is often longer (weeks) than of diarrheas associated with foodborne bacterial pathogens (days). Toward the end of a gastrointestinal protozoan infection, the opposite of diarrhea—constipation—can occur, and during the course of illness there may also be cramps, nausea, and other signs of digestive tract distress. Sharp pain may accompany instances in which *E. histolytica* causes ulceration of the walls of the digestive tract, and the formation of amoebomas (an even rarer event) may cause blockage of the digestive tract.

The symptoms caused by foodborne helminths vary with their final location in the human host as well as with their normal and aberrant paths of migration through the body. *Trichinella spiralis* larvae in the musculature may cause muscle pain. *Ascaris lumbricoides* migrates through the lungs, where it can cause asthmatic symptoms; ultimately the target site for this large roundworm is the intestinal tract where a bolus of multiple worms can occlude the lumen, particularly in children. The fishborne anisakid nematodes may produce ulcerative pain when they attach to the gastric or intestinal mucosa. Eosinophilia is often a sign of helminth infections.

At times, parasites produce no symptoms or only annoyingly vague ones, particularly when the infecting dose is low and the parasites do not multiply further within the host.

18.9. METHODS INACTIVATION

Most foodborne parasites are susceptible to higher temperature processing such as pasteurization and cooking. However, the required time–temperature combination must be obtained not just at the surface of a food but within, where the parasites may be situated. The effects of low temperatures on parasites are variable; freezing may preserve or kill parasites. The cysts or oocysts of food-and-drinkborne protozoa are fairly resistant to the routine chlorination of drinking water. Helminth eggs and some protozoan cysts and oocysts are rather resistant to secondary sewage treatment. Being tried against parasites is the use of ozone, chloramine, and such radiations as ultraviolet light, simply or in combinations. Large parasites are readily filtered from water; small parasites such as *C. parvum* and the microsporidia require very fine filters.

BIBLIOGRAPHY

Aggarwal, A., R. C. Gallo, T. E. Nash, L. S. Diamond, and G. Franchini. 1991. In vitro association of *Giardia lamblia* and *Entamoeba histolytica* with HIV-1. *AIDS Res. Human Enterovirus* **7**(2), 188.

Barrett, A. J. 1988. Cysteine endopeptidases and their inhibitors in tissue invasion. In: J. D. Lousdale-Eccles (Ed.), *Protein Traffic in Parasites and Mammaliam Cells*. Nairobi, Kenya: ILRAD.

Bier, J. W., and R. B. Raybourne. 1988. *Anisakis simplex*: Formation of immunogenic attachment caps in pigs. *Proc. Helminthol. Soc. Washington* **55**, 91–94.

Bier, J. W., G. J. Jackson, A. M. Adams, and R. A. Rude. 1998. Parasitic animals in foods. In: *FDA Bacteriological Analytical Manual*, 8th ed., rev. A. Gaithersburg, MD: AOAC International, Chapter 2.

Boczon, K., and J. W. Bier. 1986. *Anisakis simplex*: Uncoupling of oxidative phosphorylation in the muscle mitochondria of infected fish. *Exper. Parasitol.* **62**, 270–279.

Brown, M., S. Reed, J. A. Levy, M. Busch, and J. H. McKerrow. 1990. Detection of HIV-1 in *Entamoeba histolytica* without evidence of transmission to human cells. *AIDS* **5**, 93–96.

Bryan, R. T., A. Cali, R. L. Owen, and H. C. Spencer. 1991. Microsporidia: Opportunistic pathogens in patients with AIDS. *Prog. Clin. Parasitol.* **2**, 1–26.

Bundy, D. A. P. 1997. This wormy world—then and now. *Parasitol. Today* **13**(11), 407–408.

Jackson, G. J. 1983. Examining food and drink for parasitic, saprophytic and free-living protozoa and helminths. In: M. Rechcigl, Jr. (Ed.), *Handbook of Foodborne Diseases of Biological Origin*. Boca Raton, FL: CRC Press.

Jackson, G. J., D. E. Hanes, and J. W. Bier. 1998. Problems and progress in the detection, inactivation and cultivation of *Cyclospora. Dairy Food Environ. Sanit.* **18**(7), 480, 473.

Jackson, G. J., R. Herman, and I. Singer (Eds.). 1969 (Vol. 1), 1970 (Vol. 2). *Immunity to Parasitic Animals.* New York: Appleton Century Crofts.

Kappus, K. K., D. D. Juranek, and J. M. Roberts. 1992. Results of testing for intestinal parasites by state diagnostic laboratories, United States, 1987. *MMWR* **40**(SS-4), 25–45.

Pramer, D., and N. R. Stoll. 1959. Nemin: A morphogenic substance causing trap formation by predaceous fungi. *Science* **129**, 966–967.

Raybourne, R. B., T. L. Deardorff, and J. W. Bier. 1986. *Anisakis simplex*: Larval excretory–secretory protein production and cytostatic action in mammalian cell cultures. *Exper. Parasitol.* **62**, 9297.

Smyth, J. D. (Ed). 1990. In vitro *Cultivation of Parasitic Helminths.* Boca Raton, FL: CRC Press.

Stoll, N. R. 1947. This wormy world. *J. Parasitol.* **33**, 1–18.

Taylor, A. E. R., and J. R. Baker. 1987. In vitro *Methods for Parasite Cultivation.* London: Academic.

Teras, J. H. 1986. Protozoal viruses and the interaction of protozoa with mammalian viruses. *Insect Sci. Applicat.* **7**(3), 355–361.

Trager, W. 1970. *Symbiosis.* New York: VanNostrand Reinhold.

CHAPTER 19

Miscellaneous Agents: *Brucella*, *Aeromonas*, *Plesiomonas*, and *β*-Hemolytic Streptococci

EDUARDO FERNÁNDEZ-ESCARTÍN and SANTOS GARCÍA

19.1. INTRODUCTION

The microorganisms studied in the previous chapters are recognized pathogens in which foods play an important role in their transmission. Other bacteria known to be pathogenic to humans since the nineteenth century or those that have acquired recent interest due to the potential transferability through foods must also be discussed. Two of these, *Brucella* spp. and *Streptococcus pyogenes* have caused outbreaks with the particular characteristic that gastroenteritis is not present. *Aeromonas hydrophila* and *Plesiomonas shigelloides*, although associated with diarrhea, are relatively infrequent. Their involvement as etiologic agents in disease is more likely in the absence of the most common pathogens in suspect foods or clinical samples. This chapter will deal with the four mentioned microorganisms.

19.2. *Brucella*

The first species of the genus was described in 1887 by Bruce and named *Brucella melitensis*. Later, *Brucella abortus* was isolated from a cow that had miscarried, and *Brucella suis* was isolated from the miscarried fetus of pig. Another species, *Brucella canis*, is found in dogs and probably in cats. Brucellosis is a typical zoonotic illness. A number of cases of human infection are clearly related with the consumption of contaminated foods. The species names were designated in accordance to the host in which they were first isolated and characterized; however, recently associations with specific hosts are less appar-

Guide to Foodborne Pathogens, Edited by Ronald G. Labbé and Santos García
ISBN 0-471-35034-6

ent. The three important species that cause human brucellosis primarily affect livestock. *Brucella canis* is only moderately pathogenic to humans.

19.2.1. Nature of Illness

Brucellosis, also known as undulant fever or Malta fever, is an insidious illness with a varied symptomatology in humans. The incubation period ranges from 3 to 21 days and occasionally up to 7 months. Illness appears gradually or abruptly. Acute cases show fever, sweating, chills, weakness, chest pain, migraine, arthralgia, anorexia, and loss of weight. In 10–15% of patients infected with *B. melitensis* complications such as osteomyelitis are observed; the musculoskeletal, cardiovascular, genitourinary, and neurological systems are sometimes affected. Hepatosplenomegaly is also common. Chronic forms can result in permanent disability.

The illness is cured with antibiotics or resolves by itself in a few days, weeks, or years. Even with treatment the individual usually suffers the disease over several months, with frequent relapses. Lethality is uncommon. Nearly 90% of fully recovered people show resistance to reinfection. The severity of the disease is species dependent. That caused by *B. melitensis* is the most severe followed by *B. suis* and then by *B. abortus*. In the case of dairy products, fewer cells of *B. melitensis* are required to produce illness compared to *B. abortus*.

Whatever the route of entry, the invasive ability of the microorganism enables it to reach the lymph nodes. There, some microorganisms are destroyed, but others multiply inside macrophages. Through the thoracic duct they reach the cells of the reticuloendothelial system and colonize the liver, spleen, bone marrow, lymph nodes, and kidney.

Brucella is an intracellular parasite. Serum favors bacterial phagocytosis. Vacuoles with the phagocyted cell fuse with the lysosome, and the microorganism remains partially protected inside of the phagolysosome. The mononuclear cells of immune animals have the ability to destroy the ingested organisms. *Brucella* can persist for years or even for life in the host cells, causing relapsing episodes of illness.

Infection can result after consumption of contaminated milk or cheese, undercooked meat, or vegetables that have been in contact with feces or urine from infected animals. Cases associated with meat consumption are rare. Human brucellosis is related to the prevalence and distribution of the illness in animals.

Inhalation of contaminated aerosols and dust are other routes of ingestion. The skin is an effective barrier against the bacteria, although lesions of workers that handle infected animals can be a route to infection. The conjunctiva is susceptible to infection following contact with contaminated material or hands. Most ill individuals have previously been in contact with infected animals, or contaminated meat, as in the case of slaughterhouse workers. Some cases of infection also result from mishandling of laboratory samples. Congenital transmission as well as transmission following transfusions and transplants has also been reported. Person-to-person transmission is rare.

Brucellosis can be transferred venereally in animals. In susceptible ones it causes mastitis, miscarriage, and sterility. Low production of milk is evident among affected cows. *Brucella* can be excreted in milk for several months in successive periods of lactation, even if the animal shows no signs of infection.

19.2.2. Characteristics of Agent

The genus *Brucella* consists of small, gram-negative (0.5- to 0.7-μm) nonmotile, ovoid rods or coccobacilli. They are strict aerobes, although *B. abortus* grows better, especially in primary culture, under microaerophilic conditions. They are catalase and oxidase positive. Seven biovars of *B. abortus*, three of *B. melitensis*, and five of *B. suis* are recognized based on the ability to produce H_2S and urease, CO_2 requirement, sensitivity to thionine and basic fuchsin, phage susceptibility, and agglutination with specific antisera.

Brucella requires enriched media to grow; biotin, thiamine, nicotinamide, and traces of magnesium are also necessary. The bacterium grows better in liquid media. Growth of *Brucella* occurs between 20 and 42°C, with 37°C being the optimum. The most favorable pH is between 6.6 and 7.4, although it grows in the 5.8–8.7 range. Low temperature during storage allows survival of the organism. Survival at 5°C for 8 days at pH 4 and 54 h at pH 3 has been reported. It grows more slowly than common foodborne bacteria. Thus, for clinical samples, it is recommended that cultures not be discarded before 28 days. It does not grow (on liver agar) if the concentration of NaCl is greater than 3% (*B. suis*) or 4% (*B. melitensis*). Each is also inactivated after 12 or 6 days in culture broth containing 12 or 25% NaCl, respectively.

Brucella spp. is non–spore forming and does not exhibit special resistance to stressful environmental conditions; however, it survives considerable periods on damp floors, in manure, and on animal cadavers. In animal products (including milk) *Brucella* can remain viable for months in the environment. *Brucella melitensis* survives for 72 days in marsh lands, 17 days in milk, and 25 days in sea water. Freezing temperatures (−40°C) allows greater survival than, for example, at 25°C. Table 19.1 gives examples of the survival rate of *B. abortus* and *B. melitensis* in several materials and at different temperatures.

19.2.3. Epidemiology

Brucella is an animal parasite. The organism is released through the discharges of the uterus after abortion or normal delivery in infected animals. Twelve to 44% of infected cows and nearly 60% of ill goats excrete the bacterium in the milk. After delivery, counts of 1000 organisms/mL of milk are found for a variable period. In the later stages of lactation, excretion of bacteria may stop but will restart after a new delivery.

Brucellosis is a zoonosis of world distribution. *Brucella* has host specificity among animals: *B. abortus* in cattle, *B. melitensis* in goats and sheep, and *B. suis* in pigs. *Brucella melitensis* is prevalent in the Mediterranean area and in Mexico. The disease is mainly associated with consumption of dairy products

TABLE 19.1. Survival of *B. abortus* and *B. melitensis* in Environment and Food at Selected Temperatures

Material	Temperature (°C)	Survival Time
	B. abortus	
Water	4	114 days
Dry soil	Ambient	<4 days
Humid soil	Ambient	66 days
Mature	Ambient (summer)	1 day
	Ambient (winter)	53 days
	B. melitensis	
Goat cheese	37	2–3 days
	4	11 weeks
Milk	37	5–24 h
	4	18 months
Cream	4	46 weeks
Butter	8	142 days
Ice cream	0	30 days

Source: López-Merino, 1998.

and constitutes an important public health problem in those countries in which a hyperendemic condition is present. *Brucella suis* and *B. abortus* predominate in Europe, Canada, and the United States.

In 1969, 201 cases of human brucellosis were reported in the United States, 68% among meat packers. In this country, the incidence diminished from 6000 cases in 1947 to 200 cases in 1978 and to 105 in 1992. However, it is estimated that only 4–10% of cases are reported. In contrast, in 1998, 2641 human cases were reported in Mexico.

In developed countries, most of the cases are occupational illnesses. Only around 10% are related to the consumption of dairy products. In 1983, an outbreak of 29 cases occurred in Houston, Texas. The implicated food was fresh, unpasteurized goat cheese, made in Linares, NL, Mexico, and carried to the United States by Mexican immigrants. In other developed countries, brucellosis associated with the consumption of milk is also a minor problem. In England, 12 cases were registered between 1960 and 1971 and 20 between 1975 and 1980. In the latter time frame, consumption of raw cow's milk was associated with the disease. In 1997, dairy products accounted for 84% of the sources of infection of human brucellosis in Mexico.

19.2.4. Detection of Organism

Outbreaks of foodborne brucellosis are confirmed by (a) isolation of the organism from blood or bone marrow of two or more affected people, (b) an increase (fourfold or more) of the antibody titer in samples taken during the acute stage of illness and after 3–6 weeks, (c) a titer of at least 1:160 among people with symptoms of brucellosis who have consumed suspect food, or (d) isolation of *Brucella* from the food involved.

Brucella is usually present in very low levels in foods. It is possible to grow it from the sediment of homogenized food (e.g., cheese dispersed in 2% sodium citrate at 40°C). The organism can be recovered using a very rich medium, such as brain heart infusion, Bacto-tryptose plus sheep blood, or horse serum (inactivated at 56°C for 15 min). High selectivity is achieved by the use of antibiotics on these media. For this purpose a mixture of bacitracin (25 U/mL), polymixin (6 U/mL), and cycloheximide (100 μg/mL) is very effective. Petri plates are incubated at 37°C for 4 days under 10–15% CO_2 to promote growth of *Brucella* and for additional selectivity. For analysis of highly contaminated foods, the Farell medium is recommended. It is a nutritive base enriched with bovine or horse serum, and its selectivity is achieved by means of five antibiotics (bacitracine 25 U/mL, polymixin 5 U/mL, cycloheximide 100 μg/mL, vancomycin 20 μg/mL, nalidixic acid 5 μg/mL, and nystatin 100 U/mL).

Brucella colonies appear after 2–3 days of incubation and measure 2–3 mm or more by the fourth or fifth day. They are flat and translucent and have a smooth consistency. They are not hemolytic or pigmented. Microscopic examination of the organism, inoculation in Kligler medium (no sugar fermentation), urease production in Christensen medium, negative citrate utilization in Simmons medium, and positive-catalase and H_2S tests are useful characteristics for its identification. Isolates can be examined by slide agglutination using polyvalent agglutinating serum that is commercially available. Biovar identification (important for epidemiological studies) requires a specialized laboratory. Subcultures in the laboratory can produce rough colonies whose pathogenicity to guinea pigs, hamsters, and rabbits is diminished. Rough colonies are inadequate for agglutination tests.

Definitive diagnosis of brucellosis is only achieved by the isolation and identification of the etiologic agent in clinical specimens. Since its presence in blood is intermittent, it may be necessary to make blood cultures on several occasions. Success can be increased if blood sampling is done during the febrile stage of the illness. In humans, diagnostic serological tests can be carried out when diagnosis is uncertain. Agglutinating antibodies appear soon after the infection. Such serological tests include rose Bengal (quick agglutination using serum and a suspension of *B. abortus* stained with rose Bengal), agglutination (reaction of immunoglobulins with a bacterial suspension), and enzyme-linked immunosorbent assay (ELISA).

19.2.5. Prevention and Control

The application of common germicides at the recommended concentrations for plant sanitation reliably inactivates the microorganism. Chlorine- or iodine-based compounds are recommended for disinfection of areas exposed to infected animals.

TABLE 19.2. Time and Temperature Required to Inactivate *B. melitensis* and *B. abortus* in Milk

Temperature (°C)	*B. melitensis* (s)	*B. abortus* (s)
60	210	175
62	100	55
65	32	20
70	22	15
72	20	12
75	12	9
80	4	3

Source: Johnson et al., 1990.

Although it survives in foods, growth of *Brucella* in foods has not been reported. Control of the microorganism is achieved by avoiding food contamination and assuring its destruction, mainly using heat. *Brucella* is killed by pasteurization at 62.7°C for 30 min or at 71.6°C for 15 sec. Heating at 60°C for 5 min inactivates 10^7 *B. abortus* per milliliter of raw milk. *Brucella melitensis* is more resistant than *B. abortus* to thermal treatment in milk (Table 19.2). Pasteurized milk and dairy products, properly stored and handled, are safe for consumption.

Appropriate acidification of cheeses during maturation contributes to the elimination of *Brucella*. However, the traditional hand-made techniques in which many cheeses are manufactured in developing countries such as Mexico do not assure satisfactory acidification. The aging procedure of 90 days that some cheeses receive may not be enough to inactivate the organism. In areas of endemic brucellosis in particular, it is not advisable to use raw milk to make cheeses, even fermented ones. To prevent the disease, it is imperative to use only pasteurized milk for drinking and for making dairy products. If the milk cannot be pasteurized, it should be boiled.

Control of the disease resides essentially in eliminating the source of primary infection, such as ill animals. It must be determined that a new animal on a farm comes from a brucellosis-free herd; additionally isolation of new animals for 30 days and serological tests have to be carried out.

Although brucellosis due to *B. abortus* has been relatively well controlled in many industrialized countries, an increasing number of cases due to *B. suis* has been observed among hunters and people who handle boar carcasses. In these wild animals, it is difficult to control the microorganism.

Effective vaccines using attenuated bacteria are available. In the United States vaccination of livestock, killing of sick animals, and milk pasteurization are widespread practices. This had led to a very low annual incidence in humans (0.07 per 100,000). The illness has been eradicated in Norway, Sweden, Finland, Denmark, Czechoslovakia, and Holland. Elimination of human brucellosis is dependent on the eradication of the microorganism in livestock. Programs to control the disease are based on the diagnosis of illness followed by the elimination of ill or infected animals and massive vaccination. Animals showing a positive diagnostic test should not be used for commercial purposes or transported; these should be killed and the carcasses destroyed.

19.3. *Aeromonas hydrophila*

19.3.1. Introduction

Epidemiological, bacteriological, clinical, and immunological evidence indicate that some *Aeromonas* strains are enteropathogens. These show virulence properties such as enterotoxigenicity, cytotoxin and hemolysin production, and invasiveness. The genus was proposed in 1936 to incorporate bacteria similar to Enterobacteriaceae but motile by means of a single polar flagellum. Some species are pathogenic to fish; others are potential human pathogens. They have been associated with traveler's diarrhea and infantile diarrhea. Their involvement as a cause of foodborne or waterborne outbreaks has not been conclusive. However, *A. hydrophila* is considered a pathogen involved in gastrointestinal diseases (due to the infections caused) in immunocompromised individuals. It has the ability to produce toxins and other virulence factors similar to those exhibited by typical enteropathogens.

19.3.2. Nature of Illness

The involvement of *Aeromonas* as a foodborne pathogen is based on indirect evidence, such as (a) the frequency of isolation from diarrheic feces (in absence of other enteropathogens), (b) pathogenicity tests in animal models, (c) immunological response after the intestinal infection, and (d) presence in foods potentially implicated in gastroenteritis, in which no other pathogen is identified. On the other hand, experiments with human volunteers have not confirmed the pathogenicity of strains isolated from human cases. No diarrhea was observed in 55 out of 57 individuals who ingested up to 10^{10} organisms of five strains of putatively pathogenic *A. hydrophila*.

To establish the role of this bacterium in the etiology of diarrhea, it should be taken into account that the isolated strains can be the cause of infection or can be commensals. Certainly, both types of strains could coexist in the same individual. A high level of organisms in diarrheic feces cannot constitute the etiologic cause, since during these episodes, changes in intestinal flora may occur.

Enteritis caused by this bacterium usually occurs as a self-limiting watery diarrhea with or without light fever. Children younger than 2 years old can suffer vomiting, and in adults abdominal pain can be present. Duration of the illness can be more than 10 days. The infection can also present as dysentery-like. Several cases of life-threatening infection have been reported. *Aeromonas hydrophila* has also been involved in cases of infected wounds exposed to polluted water and in septicemia, meningitis, ocular infections, and endocarditis, especially in immunocompromised people. In the latter infections, the bacteria ingested with the food could have disseminated from the intestine.

Several *Aeromonas* species produce at least two enterotoxins, one heat labile (56°C, 10 min) and the other heat stable. Cell-free culture supernatants can cause fluid accumulation in rabbit ileal loops and diarrhea in mice. It is possible that factors in addition to the enterotoxins are required to cause diarrhea in humans.

19.3.3. Characteristics of Agent

The genus *Aeromonas* is currently classified in the family Vibrionaceae, although a new family Aeromonadaceae has been proposed for this genus based on molecular genetic comparisons. Features differentiating it from halophilic vibrios are its inability to grow in 6% NaCl and, from the *Vibrio cholerae* group, by its resistance to 2,4-diamino-6,7-diisopropylpteridine (O/129), a vibriostatic agent, although this characteristic does not apply to all strains. There is a tendency to form two groups, one nonmotile, represented by *Aeromonas salmonicida* (psychrophilic and pathogenic to fish), and the other which includes *Aeromonas hydrophila* (mesophilic, motile, and pathogenic to humans), *Aeromonas sobria*, and *Aeromonas caviae*, which are biochemically and genetically related. Another species, *Aeromonas veronii*, has been reported to cause diarrhea. It can elaborate toxins and virulence factors at refrigeration temperatures. According to deoxyribonucleic acid (DNA)–DNA hybridization profiles, up to 14 *Aeromonas* species have been described. However, the taxonomy of the genus is in flux; thus in this chapter, the species will be described according to the terms used in the current literature.

Aeromonas species are facultatively anaerobic, gram-negative rods, measuring 1.0–4.4 µm by 0.4–1 µm. Not all mesophilic aeromonads are motile. There is evidence of strains with inactive or absent flagella. Under CO_2 atmosphere, the organism forms filaments and does not show proteolytic or hemolytic activities; if incubated aerobically, both enzymatic activities appear in 24 h. In many strains of *A. hydrophila* and *A. sobria*, production of gas from carbohydrates occurs at 22°C but not at 30°C, the latter being the optimal for growth. It has been shown that more than 85% of human fecal strains correspond to phenotypes *A. hydrophila*, *A. caviae*, and *A. sobria*. Several characteristics allow differentiation among aeromonads *Vibrio* and *Plesiomonas* (Tables 19.3 and 19.4).

The general temperature range for growth is from 0 to 45°C with the optimal from 28 to 35°C. In several foods contaminated with the organism, it is possible to observe an increase from 1 to 3 $\log_{10}$ after 1 week at 5°C, while at 28 or 35°C the increase can be 7 $\log_{10}$ in 24 h. At 12°C the generation time is 4 h. At 28°C most of the strains grow in media with 4% NaCl; however, at 4°C there is very

TABLE 19.3. Differentiation of *Aeromonas* in Vibrionaceae Family[a]

Test	*Aeromonas*	*Vibrio*	*Plesiomonas*
Catalase	+	+	+
Oxidase	+	+	+
Gas from glucose	+	−	−
Inositol	−	−	+
Mannitol	+	+	−
Sucrose	+	V	−
Ornithine decarboxylase	−[b]	DB	+
NaCl requirement	−	+	−
Amylase	+	−	−
Gelatinase	+	+	−
Lipase	+	−	−
Growth in TCBS	−	+	−
O/129 inhibition, 10 μg	R	DB	DB
O/129 inhibition 150 μg	R	S	S

[a] DB = different biotypes; R = resistant; S = sensitive.
[b] *A. veroni* is positive.

little growth in the presence of 3% NaCl. The organism does not tolerate low pH. Usually, the microorganism will not be a problem in foods with pH < 6 or with 3% or more NaCl. The optimal pH for growth is 6.5–7.2. At pH 5.5 growth is slow and at 4.5 there is loss of viability. This could explain the decrease in cell numbers in oysters stored at 5°C, due to the fermentation of glucose and a resulting lower pH. Growth under CO_2 at 30°C is delayed, and at 5°C, recovery of viable cells is diminished.

Aeromonas hydrophila survives freezing and the organism has been recovered from oysters stored at −72°C for 18 months, without losing its ability to produce hemolysin, enterotoxin, and cytotoxin (the oysters involved had been involved in a gastroenteritis outbreak). Reducing the oxygen concentration in

TABLE 19.4. Differentiation of Three Species of *Aeromonas* and *P. shigelloides*[a]

Test	*A. hydrophila*	*A. caviae*	*A. sobria*	*P. shigelloides*
Esculin hydrolysis	+	+	−	−
KCN	+	+	−	?
Acid from salicin	+	+	−	V
Gas from glucose	+	−	+	−
Gas from arabinose	+	+	−	−
Voges Proskauer	−	−	V	−
H_2S from cysteine	+	−	+	?
Elastase production	+	−	−	?

[a] V = varies; ? = unknown.

Source: Popoff, 1984; Kirov, 1997.

refrigerated meats favors *Aeromonas* growth, probably as a result of the inhibition of aerobic microflora that competes with the pathogen. The shelf life of cauliflower, asparagus, and broccoli can be increased under controlled environments; however, *A. hydrophila* can survive and grow under these conditions.

19.3.4. Epidemiology

Aeromonas hydrophila is a ubiquitous bacterium. It is a common resident in aquatic environments, such as springs, chlorinated and nonchlorinated domestic water, and sewage-polluted waters. It and especially *A. caviae* also have been reported in sea water. In Australia, an increase in the number of these organisms in water during the summer coincides with a high incidence of gastroenteritis attributed to *Aeromonas*.

Aeromonas is not considered a normal inhabitant of the human intestine, since it is only isolated from 0–8% of feces. It is isolated from 11–12% of fecal sample from healthy farm animals such as pigs, horses, cows, and lambs. The organism is frequently isolated from fish, frogs, and axolotls. It is associated with infections that can cause miscarriage in cows and diarrhea in pigs. Motile aeromonads cause disease in reptiles, fish, bivalve mollusks, and snails. More than 70% of bivalve mollusks can be contaminated with the bacterium. Amphibians, reptiles, and fish can be reservoirs of *Aeromonas*. The organism is frequently found in raw foods of animal or vegetable origin, although the incidence varies in different countries.

Aeromonas sobria seems to be the common species in Japan and Australia, while *A. hydrophila* predominates in the United States. *Aeromonas* was isolated from 7% of 779 samples of processed foods obtained from supermarkets and from 28% of drinking water samples in Copenhagen, Denmark (*A. hydrophila* was the predominate species).

Naturally contaminated foods include crustaceans (71%), bivalve mollusks (53%), raw milk (50%), and fish (37%). In Brazilian dairy foods, *A. caviae* and *A. hydrophila* were isolated (59 and 13%, respectively) from pasteurized milk samples and (7 and 27%, respectively) from white cheese. Most *A. hydrophila* strains produced a cytotoxin and hemolysin. *Aeromonas hydrophila* has also been found in chicken carcasses and feces. Motile *Aeromonas* were detected in 33% of samples of lamb carcasses.

Aeromonas hydrophila grows well in several foods and competes with the associated flora. It is associated with the spoilage flora of meats, raw milk, nonpasteurized whole egg, and seafood. Indol has been detected in spoiled seafood. Strains of *Aeromonas* isolated from these products produced high levels of this compound.

Raw milk sustains growth of the organism. Levels of 10^8 CFU/mL of milk can be achieved without any signs of deterioration. Refrigerated storage of parsley, spinach, celery, alfalfa, cabbages, broccoli, endive, and lettuce can result in an increase in the initial level of *Aeromonas*.

Consumption of untreated water is a significant risk factor for infection with this bacterium in the United States. On the other hand, in Japan, an increased incidence of *Aeromonas* spp. in human feces correlated with high levels of aeromonads in beef, pig, and chicken meat, but not in tap water or rivers.

It was reported in 1982 that in Australia enterotoxigenic *A. hydrophila* was found in 10.2% of feces from 1156 children (mostly younger than 2 years) with diarrhea and in only 0.6% of 1156 healthy controls. In more than a third of ill individuals diarrhea lasted more than 2 weeks and in almost 25% was dysentery-like.

Several outbreaks of gastroenteritis due to consumption of oysters contaminated with *A. hydrophila* have been reported in the United States. In Sweden outbreaks have resulted after consumption of contaminated dishes composed of smoked sausage, shrimps, boiled ham, and liver paté. Other outbreaks have occurred after consumption of oysters and prawns in England, terrestrial snails in Nigeria, prefrozen fish in Russia, and soup in Hungary.

The epidemiological studies available suggest that incidence of gastroenteritis is higher during the warmer months, when higher levels of the organism are detected in aquatic environments.

19.3.5. Detection of Organism

Generally, the culture media used for analysis of clinical samples are not suitable for environmental samples or foods. *Aeromonas* grows well in common media for Enterobacteriaceae, such as MacConkey, *Salmonella–Shigella*, deoxycholate citrate lactose sucrose (DCLS), Hektoen, or cefsulodin–irgasan–novobiocin (CIN) agars for *Yersinia enterocolitica* and violet red bile agar for coliforms. A medium based on starch and ampicillin (SA agar) is satisfactory for enumeration of *Aeromonas* in vegetables and foods of animal origin and in animal feces. In general, SA agar works better than the media of Ryan and other commercial media. Addition of ampicilin and mannitol to MacConkey medium makes it as effective as the SA agar. For enrichment, tryptic soy broth plus 30 μg/mL ampicilin (TSBA) can be used.

The most probable number (MPN) technique can be used to enumerate the microorganism; dilution of samples is done in any of two enrichment media: alkaline peptone water (APW) or TSBA, depending on the number of associated microflora. For species differentiation, several minikits are commercially available.

19.3.6. Prevention and Control

The natural psycrophilic flora of ground meat is antagonistic to *Aeromonas* when stored at 5°C. The microorganism is destroyed by pasteurization. Under heat treatment (48 or 50°C) of a bacterial suspension, a biphasic inactivation pattern is present with *D* values much shorter in the first stage (15 min) than in the second stage, suggesting that a more thermoresistant subpopulation is present.

Sensitivity of *Aeromonas* to ionizing radiation is similar to that of common intestinal pathogens. The *D* values (between −15 and 22°C) are 0.131–0.274 kGy when present in phosphate buffer and ground fish meat.

Aeromonas is sensitive to common germicides; 10 mg/L iodophors can inactivate it in 10 min at 25°C, and 80 mg/L quaternary ammonium salts in 1 min or 10 mg/kg in 10 min. When in contact with 5 mg/L of free chlorine, it is inactivated in 1 min, or in 5 min at a free chlorine concentration of 2.5 mg/L. The bacterium is isolated from chlorinated waters, suggesting resistance to chlorine; however, it is inactivated at concentrations ordinarily used for municipal water treatment and for food industry sanitation. *Aeromonas hydrophila* can adhere to copper, stainless steel, and plastic pipes in less than 4 h at 4 or 20°C. In the case of recently formed biofilms, removal is achieved by treatment at 50°C for 1 min or by the use of 25 mg active chlorine/kg. Eight-day-old biofilms require more drastic treatments for removal such as 60°C or 75 mg chlorine/kg for 1 min.

Prevention of infection by *A. hydrophila* should lean primarily toward its inactivation in water and foods. This implies appropriate disinfection and complete cooking of products. Control of the initial contamination is not practical due to its ubiquitousness. However, recontamination should be avoided in processed foods that will not receive subsequent heat treatment before consumption. As in the case of other foodborne disease agents, reheating of leftover food before consumption is an additional control point. This is important due to the possible recontamination and potential for growth of *A. hydrophila* at low temperature. The organism is not a problem in foods that contain more than 3–3.5% NaCl in the aqueous phase or have a pH lower than 6.0 and stored at refrigerated temperatures.

19.4. *Plesiomonas shigelloides*

19.4.1. Introduction

Plesiomonas shigelloides is a potential pathogen for humans, and its enteropathogenicity is still a matter of controversy. However, it has been found as the only pathogen in feces from persons suffering gastroenteritis (some people showing severe symptoms), occasionally present as almost a pure culture from diarrheic feces. It has also been isolated as a pure culture from extraintestinal lesions and has been detected in water or foods suspected of being vehicles in foodborne outbreaks.

19.4.2. Nature of Illness

The illness can present as simple watery diarrhea with up to 30 bowel movements per day, as dysentery-like (feces with blood, mucus, and leukocytes), or cholera-like. Ill persons can also present with symptoms such as nausea, vomiting, abdominal pain, fever, chills, and migraine headaches. The incubation period lasts between 24 and 50 h, and the duration of symptoms ranges from 1 to 9 days.

19.4.3. Characteristics of Agent

Plesiomonas shigelloides has been previously grouped in the genera *Escherichia, Pseudomonas, Aeromonas, Fergusiona,* and *Vibrio*. It is a gram-negative rod, facultatively anaerobic, non–spore forming, measuring 2–3 μm by 0.1–1.0 μm, and belonging to the family *Vibrionaceae*. A positive-oxidase test is fundamental in differentiating this bacterium from the Enterobacteriaceae. Also, while the motile members of the latter have peritrichous flagella, *P. shigelloides* is lophotrichous with two to five flagella. Some cells are monotrichous and even atrichous. It is related to *A. hydrophila* from which it differs in G + C content 58–62% in *A. hydrophila* and 51% in *P. shigelloides*. Serogroup O:17 cross reacts with antiserum against *Shigella*.

There is a long list of biochemical tests involved in the identification and phenotypic differentiation of *P. shigelloides* from common bacteria of the intestine and the environment. It is separated from *Vibrio* and *Aeromonas* by its inability to produce gelatinase (Table 19.3). Also, in contrast with these two genera, it does not ferment mannitol but ferments inositol without gas production. Other important negative tests include urease, amylase, caseinase, lysine decarboxylase, and growth on citrate (as the only carbon source), in cyanide broth, in 6.5% NaCl as well as its inability to form gas from glucose or metabolize malonate. Positive tests include ornithine decarboxylase and nitrate reduction. Lactose fermentation is very slow (up to 14 days).

The minimum temperature for growth is 8°C, and the maximum is 44°C, while the optimum is 30–37°C. The bacterium grows at pH 5.0–8.0 and tolerates a maximum concentration of 5% NaCl.

Several virulence factors of *P. shigelloides* have been reported. These include extracellular enzymes, enterotoxins (thermostable and thermolabile), β-hemolysin, endotoxin, invasins (invasion of HeLa and Hep-2 cells), and adhesins. However, the role of these factors in foodborne illness is inconclusive.

19.4.4. Epidemiology

Plesiomonas shigelloides is a free-living aquatic organism found in fresh, estuarine, and sea water in many parts of the world. The prevalence of the bacteria is higher during the warmer seasons. It is also isolated from soil, food, domestic animals, and asymptomatic persons. Foods most commonly associated with infections due to *P. shigelloides* are of aquatic origin and include crab, shrimp, oyster, and fish. It is not a normal inhabitant of the human intestine since the rate of human carriers varies from 0.0078% to nearly 6%. The frequency of positive carriers is higher in mature individuals than those younger than 2 years. It has been isolated from dogs, cats, cows, lambs, goats, pigs, clams, crabs, oysters, turkeys, turtles, snakes, and monkeys. Survival in sea water is limited; thus, it is difficult to isolate the organism beyond 100 m from the contaminated coastal areas. The bacterium was isolated in Tokyo in only 3 out of 38,454 healthy individuals and in 3.8% of dogs, 10.3% of cats, 10.2% of river fish, 12.8% of river waters, and 10.5% of river sediments.

Plesiomonas is associated with cases of gastroenteritis following consumption of contaminated marine animals and polluted water. It is considered an opportunist, invasive pathogen that can be isolated in pure cultures from clinical samples of cholecystitis, septicemia, meningitis, endophtalmitis, and other infections, mainly from immunocompromised individuals. Elderly and debilitated persons are usually (but not exclusively) more likely to be affected.

Two outbreaks of acute diarrhea due to *P. shigelloides* occurred in Japan in 1973 and 1974. Of the 2141 persons at risk in the first outbreak, 978 became ill, most of them with diarrhea, abdominal pain, and fever. The microorganism was isolated in 21 out of 124 samples of feces. Serovar O17:H2 was identified in the patients and also in tap water. In the second outbreak, 24 out of 35 individuals were sick with similar symptoms. The implicated serotype was O24:H5, which was also isolated from samples of water, fish, and bivalve mollusks. *Plesiomonas shigelloides* was associated with several episodes of diarrhea among travelers in Bangladesh and in Thailand, suggesting that the bacterium can be considered as one agent of traveler's diarrhea.

Due to the high rate of isolation of *P. shigelloides* in individuals with diarrhea in a region of Zaire with intense fishing activity, the frequency of the microorganism in the intestinal contents of river fish captured in that locality was investigated. The bacterium was isolated from 59% of the samples. Since other common intestinal pathogens were not isolated, the conclusion was that fish constituted the main source of human infection.

An infection due to *P. shigelloides* occurred in a boy from contact with water from a domestic aquarium. *Plesiomonas shigelloides* was isolated from the individual's feces and from the aquarium water, although it was not confirmed that it was the same strain. Later, the presence of the microorganism was confirmed in the water of 18 aquaria in the vicinity of the involved home. Twenty-two percent of the samples were positive.

Risk factors associated with infections due to *P. shigelloides* include compromised immune systems or, in healthy individuals, exposure to polluted recreational aquatic facilities and consumption of raw bivalve mollusks.

19.4.5. Detection of Organism

Isolation of *P. shigelloides* is possible using certain culture media commonly used for Enterobacteriaceae, such as MacConkey, *Salmonella–Shigella*, Hektoen, and XLD agars. Two media have been developed for *P. shigelloides: Plesiomonas* (PL) agar and inositol brilliant green-bile salts (IBB) agar. Several authors have reported high recovery of the organism from fish feces when MacConkey and XLD media are used. The method recommended by Koburger consists of inoculating IBB agar (whitish to pink colonies) and PL agar (pink colonies with pink halo). For enrichment, tetrathionate broth without iodine incubated at 40°C is suitable. Identification of isolates is made by means of the gram stain, mobility test, triple-sugar iron agar reactions (glucose positive without gas, lactose negative, and H_2S negative) as well as the following tests: gelatinase, oxidase, lysine and ornithine decarboxylase, and inositol fermentation. Interference caused by growth of *Pseudomonas* can be minimized by anaerobic incubation, which does not affect *P. shigelloides*.

19.4.6. Prevention and Control

The microorganism will not grow at 5°C or at 50°C. Pasteurization (60°C for 30 min) has been shown to destroy the bacterium. As indicated for other foodborne pathogens, consumption of raw seafood should be avoided. This is especially true for immunocompromised or debilitated persons.

19.5. β-Hemolytic *Streptococcus*

19.5.1. Introduction

The group of β-hemolytic streptococci include several species of the genus *Streptococcus* that are pathogenic for humans and animals. *Streptococcus pyogenes* is the foodborne pathogen of most interest in this genus. Literature on its behavior in foods is very limited. Initially, interest in this bacterium arose as a result of outbreaks and cases of infections associated with consumption of raw milk. When sanitary practices for production were introduced, especially milk pasteurization, the incidence of streptococcal infections associated with consumption of foods diminished notably. Besides *S. pyogenes*, other species of the same genus have been involved in this problem. At present, outbreaks of infections in industrialized countries due to consumption of food contaminated with this organism occur only sporadically.

19.5.2. Nature of Illness

Streptococcus pyogenes is a well-known pathogen that causes acute and chronic infections of the pharynx (septic sore throat), nephritis, scarlet fever, rheumatic fever, septicemia, and skin infections. The digestive tract (except the pharynx) is not involved in the disease. However, there are two forms of illness associated with foods. As an occupational illness, infections of the skin of slaughterhouse workers and meat packers have been reported. Although decreasing in incidence in industrialized countries, pharyngitis caused by *S. pyogenes* still occurs after consumption of contaminated foods.

19.5.3. Characteristics of Agent

The genus *Streptococcus* consists of spherical, gram-positive bacteria typically grouped in chains. It includes nearly 40 species, some of which are pathogenic for humans and animals, other are commensals, and others are part of the lactic acid bacteria group. Pathogenic species usually do not colonize the digestive system; rather, they localize in the respiratory apparatus and can cause infections in other organs and tissues. The genus behaves as homofermentative, catalase negative (distinguishing it from *Staphylococcus*), and hemolytic, the latter being important for genus identification. Pathogenic streptococci that affect humans and animals belong to the β-hemolytic group. They are distributed in serologic groups according to their carbohydrate antigens in the cell wall; groups A, B, C, and G are the most common in humans. The cell surface of *S. pyogenes* contains the specific protein M. In the absence of antibodies, this protein confers resistance to phagocytosis and is useful for strain differentiation. It is interesting to note that this bacterium has the ability to survive for months in water and can use CO_2 as a carbon source.

19.5.4. Epidemiology

Scarlet fever and septic sore throat as a result of milk consumption have been reported since the late nineteenth and early twentieth centuries. The following more recent examples give an idea of the importance of illnesses caused by streptococci that are transmitted by food.

Among the attendants of a meeting in Oregon in the United States in 1981, nearly 300 people contracted pharyngitis. Five out of 10 kitchen employees that prepared the foods had pharyngitis and/or lesions in their skin. Group A β-hemolytic streptococci were isolated from several of them. In this outbreak, it was not possible to identify a specific food as the vehicle.

In 1984, an outbreak of pharyngitis in Puerto Rico occurred after consumption of carrucho (salad with marine snail). β-Hemolytic *Streptococcus* was isolated from the pharynx of 11 out of 47 affected people as well as from leftover food. The incubation period was 12–84 h.

In 1990 an outbreak of pharyngitis occurred in a military base affecting 61 persons. The implicated food was an egg salad. All presented with pharyngitis with a purulent exudate, difficulty in swallowing, and cervical lymphadenopathy. Also, 45 persons suffered headache, 26 nausea, 21 myalgia, 13 coughing, 12 vomiting, and 2 diarrhea. In 19 out of 20 cases *S. pyogenes* type T12 was isolated. The bacterium was also isolated from two chefs, one of them suffering from pharyngitis the day the food was prepared.

Implication of *S. pyogenes* as a cause of a foodborne illness outbreak is confirmed by the isolation of the same strain from two or more affected people (suffering fever, pharyngitis, or scarlatina) that consumed the food identified as the vehicle and/or the isolation of the same such strain from the food.

Streptococci from other groups are also associated to outbreaks of foodborne diseases transmitted by foods. For example, in Europe and the United States, *Streptococcus zooepidermicus* from group C has been involved in outbreaks originating from the consumption of unpasteurized milk and cheese. Pneumonia, endocarditis, meningitis, abdominal pain, septicemia, and death have occurred in several ill persons.

19.5.5. Detection of Organism

There are no official methods for the isolation of β-hemolytic streptococci from foods. However, microbiological media used for clinical samples that contain abundant associated flora can be used. It is important to use blood agar prepared in a rich medium such as Columbia or brain heart infusion. It is convenient to inoculate blood agar and Todd-Hewitt broth (with colistin and oxolinic acid) straight from homogenized food, followed by overnight incubation and restreaking of suspect colonies onto blood agar plates. For isolation, nalidixic acid–neomycin blood agar, gentamicin–blood agar, and sulfametoxazole–trimethoprim blood agar are satisfactory. In cases of high levels of associated flora in foods, these media should be used together, with the objective of suppressing other microbes present. Colonies of *S. pyogenes* are small and intensely β-hemolytic. The serologic group can be determined by capillary test, fluorescent antibodies, or coagglutination with commercial reagents. The catalase test and absence of growth at 45°C and in 40% bile are important in identifying hemolytic isolates. *Streptococcus pyogenes* differs from other β-hemolytic streptococci in its susceptibility to bacitracin and ability to hydrolyze PYR (L-pyrrolidonyl-β-naphthylamide).

19.5.6. Prevention and Control

Prevention of foodborne diseases caused by β-hemolytic streptococci is based on avoiding food contamination with pharyngeal discharges or contaminated hands. Persons with respiratory infections should be excluded from places where the food is prepared. Streptococci are thermoduric bacteria, but they can be eliminated from food by heating at 65–70°C. Pasteurization inactivates β-hemolytic streptococci in milk.

REFERENCES

Johnson, E. A., J. H. Nelson, and M. Johnson. 1990. Microbiological safety of cheese made from heat-treated milk, Part II. Microbiology. *J. Food Prot.* **53**, 519–540.

Kirov, S. M. 1997. *Aeromonas* and *Plesiomonas* species. In: M. P. Doyle, L. R. Beuchat, and T. J. Montville (Eds.), *Food Microbiology. Fundamentals and Frontiers*. Washington, DC: American Society for Microbiology, pp. 265–287.

Koburger, J. A., and C. I. Wei. 1992. *Plesiomonas shigelloides*. In: C. Vanderzant, and D. F. Splittstoesser (Eds.), *Compendium of Methods for the Microbiological Examination of Foods*. Washington, DC: American Public Health Association, pp. 517–522.

López-Merino, A. 1998. La brucelosis como una zoonosis de interés en México. En: Memorias. III Foro Nacional de Brucelosis. Secretaría de Sagricultura, Ganadería y Desarrollo Rural. Acapulco, Gro. México, pp. 53–62.

Popoff, M. 1984. Genus III, *Aeromonas*. In: N. Krieg and J. Holt (Eds.), *Bergey's Manual of Systematic Bacteriology*. Vol. 1. Baltimore: Williams and Wilkins.

BIBLIOGRAPHY

Abeyta, C., C. A. Kaysner, M. M. Wekell, et al. 1986. Recovery of *Aeromonas hydrophila* from oysters implicated in an outbreak of foodborne illness. *J. Food Prot.* **49**, 643–646.

Abeyta, C., and M. M. Wekell. 1988. Potential sources of *Aeromonas hydrophila*. *J. Food Safety* **9**, 11–22.

Alton, G. G., L. M. Jones, R. D. Angus, and J. M. Verger. 1988. *Techniques for the Brucellosis Laboratory*. Paris: Institut National de la Recherche Agronomique.

Anonymous. 1996. *Campaña Nacional contra la brucelosis en los animales*. Norma Oficial Mexicana. Mexico City NOM-041-ZOO-1995.

Arai, T., and N. Ikejima. 1980. A survey of *Plesiomonas shigelloides* from aquatic environment, domestic animals, pets and humans. *J. Hyg.* **84**, 203–211.

Brenden, R. A., M. A. Miller, and J. M. Janda. 1988. Clinical disease spectrum and pathogenic factors associated with *Plesiomonas shigelloides* in humans. *Rev. Inf. Dis.* **10**, 313–316.

Bryan, F. L., J. J. Guzewich, and E. C. Todd. 1997. Surveillance of foodborne disease. II. Summary and presentation of descriptive data and epidemiological patterns; their value and limitations. *J. Food Prot.* **60**, 567–578.

Buchanan, R. L., and S. A. Palumbo. 1985. *Aeromonas hydrophila* and *Aeromonas sobria* as potential food poisoning species: A review. *J. Food Safety* **7**, 15–29.

Callister, S. M., and W. A. Agger. 1987. Enumeration and characterization of *Aeromonas hydrophila* and *Aeromonas caviae* isolated from grocery stored produce. *Appl. Environ. Microbiol.* **53**, 249–253.

Centers for Disease Control. 1982. A foodborne outbreak of Streptococcal pharyngitis—Portland, Oregon. *MMWR* **31**, 3–5.

Centers for Disease Control. 1983. Brucellosis—Texas. *MMWR* **32**, 548–553.

Centers for Disease Control. 1986. Group-A, B hemolytic *Streptococcus* skin infections in a meat-packaging plant—Oregon. *MMWR* **35**, 629–630.

Centers for Disease Control. 1997. Case definitions for infectious conditions under public health surveillance. *MMWR* **46**(No. RR-10), 1–55.

Champusar, H., A. Andremon, D. Mathieu, et al. 1982. Cholera like illness due to *Aeromonas sobria*. *J. Inf. Dis.* **145**, 248–254.

Comité de Expertos de la OMS. 1976. *Aspectos Microbiológicos de la Higiene de los Alimentos*. Serie de informes Técnicos 598. Ginebra: Organización Mundial de la Salud.

Facklam, R. R., and R. B. Carey. 1985. Streptococci and aerococci. In: E. H. Lennette, A. Balows, W. J. Hausler, and H. J. Shadomy (Eds.), *Manual of Clinical Microbiology*, 4th ed. Washington, DC: American Society for Microbiology, pp. 154–175.

George, W. L., M. M. Nakata, J. Thompson, and M. L. White. 1985. *Aeromonas*-related diarrhea in adults. *Arch. Intern. Med.* **145**, 2207–2211.

Gracey, M., V. Burke, and J. Robinson. 1982. *Aeromonas*-associated gastroenteritis. *Lancet* **II**, 1304–1306.

Hackney, C. R., and A. Dicharry. 1988. Seafood-borne bacterial pathogens of marine origin. *Food Technol.* **42**(3), 104–109.

Hausler, W. J., N. P. Moyer, and L. A. Holcomb. 1985. *Brucella*. In: E. H. Lennette, A. Balows, W. J. Hausler, and H. J. Shadomy (Eds.), *Manual of Clinical Microbiology*, 4th ed. Washington, DC: American Society for Microbiology, pp. 382–386.

Halling, S. M., and E. J. Young. 1994. *Brucella*. In: Y. H. Hui, J. R. Gohrham, and K. D. Murrell (Eds.), Foodborne Disease Handbook. New York: Marcel Dekker, pp. 63–69.

International Commission on Microbiological Specifications for Foods. 1996. *Microorganisms in Foods*, Vol. 5: *Microbiological Specifications of Food Pathogens*. London: Blackie Academic.

Kaufmann, A. F. 1986. Brucellosis. In: J. M. Laste (Ed.), *Public Health and Preventive Medicine*, 12th ed. Appleton-Century-Crofts, pp. 400–403.

Khardori, N., and V. Fainstein. 1988. *Aeromonas* and *Plesiomonas* as etiological agents. *Ann. Rev. Microbiol.* **42**, 395–419.

Kirov, S. M. 1993. The public health significance of *Aeromonas* spp. in foods. *Int. J. Food Microbiol.* **20**, 179–198.

Merino, A. 1991. *Brucelosis*. Publicación Técnica No. 6. Instituto Nacional de Diagnóstico y Referencia Epidemiológicos. Secretaría de Salud, México.

McMahon, M. A., I. S. Blair, and D. A. McDowell. 1999. Filamentation in *Aeromonas hydrophila*. *Food Microbiol.* **15**, 441–448.

Miller, M. L., and J. A. Koburger. 1985. *Plesiomonas shigelloides*. An opportunistic food and waterborne pathogen. *J. Food Prot.* **48**, 449–457.

Nishikawa, Y., and T. Kishi. 1988. Isolation and characterization of motile *Aeromonas* from human, food and environmental specimens. *Epidemiol. Infect.* **101**, 213–223.

Palumbo, S. A., and R. L. Buchanan. 1988. Factors affecting growth or survival of *Aeromonas hydrophila* in foods. *J. Food Safety* **9**, 37–51.

CHAPTER 20

Disinfecting and Sterilizing Agents Used in Food Industry

MAFU AKIER ASSANTA and DENIS ROY

20.1. INTRODUCTION

In the food industry, the use of modern food production practices, especially the universal practice of sanitation, could minimize the potential threat of pathogenic organisms and the numbers of outbreaks and recalls involving food and food products. Adequate application of sanitation practices during transportation of animals and slaughtering, in processing meat, poultry, dairy, or others foods, and during storage or retailing will lead to a significant reduction in numbers of potential pathogenic microorganisms by both physical removal and inactivation. Therefore, disinfection is always considered an important and integral part of the measures necessary to prevent any biological contamination in order to increase, day by day, the safety of the food products and their shelf life. The high degree of success of this process is confirmed by the infrequent contamination of food products.

In this chapter, which cannot be more than a brief review, general principles of disinfection and descriptions of sanitizing agents that are widely recognized to be effective and safe for use in the food-processing industry are presented. Related subjects such as methods of testing or bases for choosing particular chemical agents and procedures for their use are also discussed.

20.2. CONTROL OF FOODBORNE PATHOGENS

During the processing, microbial cells may dislodge from food contact surfaces and contribute to product contamination, which may shorten product shelf-life

Guide to Foodborne Pathogens, Edited by Ronald G. Labbé and Santos García
ISBN 0-471-35034-6

TABLE 20.1. Efficacy of Three Universal Food Disinfectants against Several Microorganisms

Organism	Temperature (°C)	Exposure time	Disinfectant Concentration[a] (ppm)		
			Chlorine	Iodine[b]	QAC[c]
Aspergillus spp.	25	5 min	0.5	0.1	2.5
Aeromonas hydrophila	20	5 min	12.5	25	475
Bacillus cereus	20	5 min	100	ND	ND
Campylobacter spp.	25	5 min	100	50	250
Clostridium botulinum	25	30 s	75	100	425
Clostridium perfringens	25	1 min	0.30	ND	ND
Coxsackies virus	25	2 min	ND	0.31	ND
Escherichia coli	25	5 min	100	50	200
Entamoeba histolytica	25	150 min	ND	0.12	ND
Hepatitis virus	20	30 min	ND	3.5	ND
Listeria monocytogenes	20	10 min	130	30	425
Mycobacterium tuberculosis	50	30 s	ND	50	ND
Pseudomonas aeruginosa	20	20 min	65	50	425
Pseudomonas fluorescens	21	15 s	ND	5.0	ND
Salmonella spp.	25	5 min	100	50	200
Shigella dysenteriae	25	3 min	ND	0.055	ND
Staphylococcus aureus	20	5 min	100	50	200
Vibrio spp.	20	15 s	13	25	175
Yersinia enterocolitica	20	5 min	125	60	225

[a] Attached on metal strip. ND = not determined.

[b] Iodophors

[c] Quaternary ammonium compounds

and increase the potential of transmitting foodborne diseases. In order to minimize any physical, chemical, and biological contamination of the products, cleaning, sterilization, and disinfection are considered an important and integral part of the procedures necessary to assure the high hygienic quality of food products. Their adequate application (Table 20.1) on food-processing surfaces will lead to a significant reduction in numbers of pathogenic microorganisms. Also, the exact distinction among the three entities and the basic knowledge of how to achieve and monitor each state are extremely important if effective application of well-known principles is to be realized.

20.2.1. Cleaning

To minimize microbial contamination and colonization of industrial food-processing equipment, cleaning is the most important stage before applying a disinfectant. The cleaning is intended to eliminate, on food contact equipments, residues and dirt that may harbor food poisoning and spoilage organisms and act as a source of food contamination. Thus, the removal of debris from con-

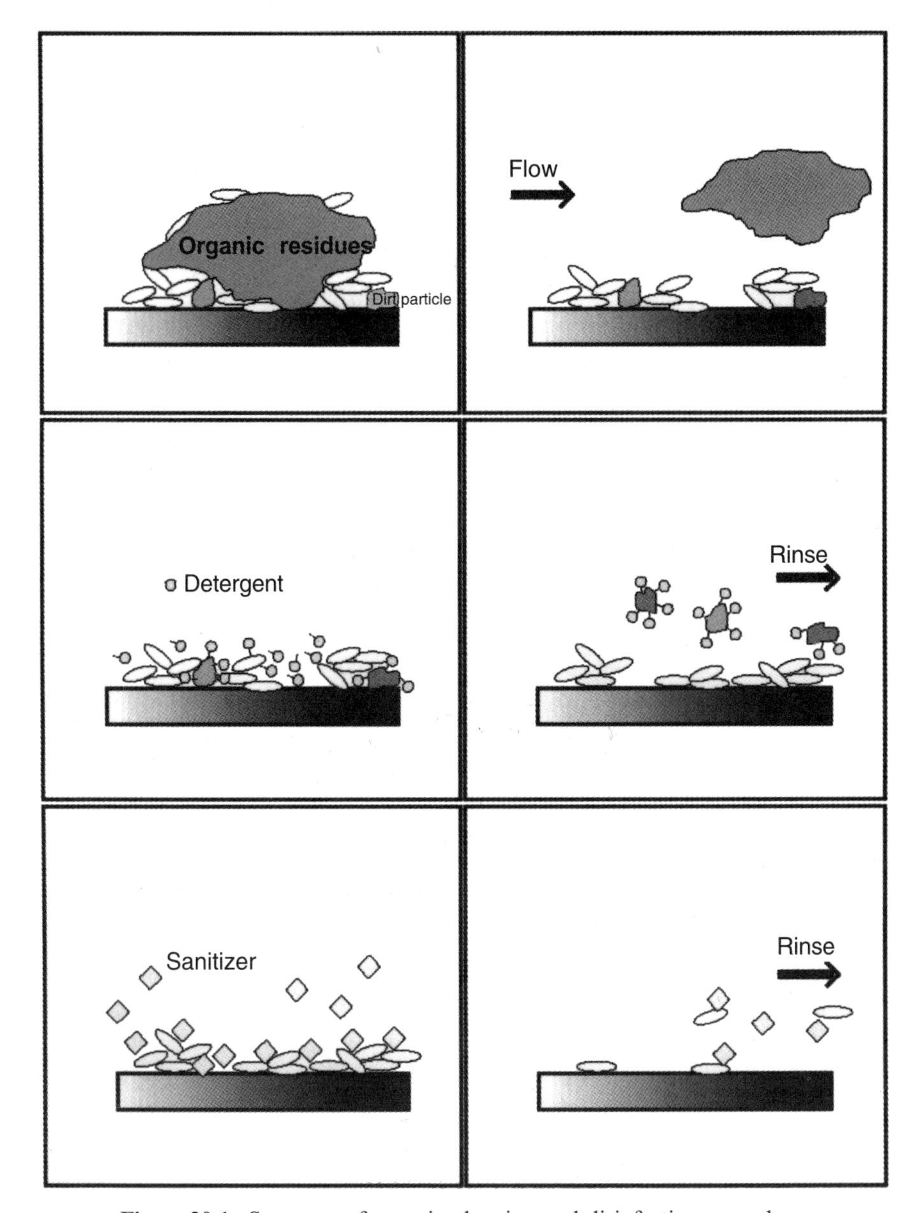

Figure 20.1. Sequence of steps in cleaning and disinfecting procedures.

tact surfaces by mechanical action (brushing, scraping of deposits, and using medium- and high-pressure jets) followed by the application of potable water (Fig. 20.1) is recognized as the best means in eliminating attached cells, especially on smooth surfaces. The temperature of the water used will depend upon the type of residues to be removed.

Cleaning is achieved by the separate or combined use of physical (scrubbing or turbulent flow) or chemical (alkalis or acid detergents) methods. For the chemical procedure, the detergent must have good wetting capacity and the ability to remove soil from surfaces. Thus, chlorinated alkaline detergents followed by nonchlorinated alkaline detergents are the more effective for food contact surface cleaning. Acid products are no more effective than hot water. Also, they must have good rinsing properties and should be noncorrosive and compatible with other materials, including disinfectant used in the hazard analysis critical control point (HACCP) program. Lastly, it is important to assure that equipment is left to dry as soon as possible after cleaning in order to avoid microbial growth that may occur on wet surfaces. It is vital to know that microorganisms are far more sensitive to disinfectant once they have detached from the surface to which they were adhered. Also, the extent to which the attached organisms are removed depends on the microbial strain implicated. Various substances are known to improve the performance in cleaning. Cations, calcium in particular, are thought to play a part in bonding polymer molecules in the biofilm, ensuring the cohesion of the polymer layer; conversely, absence of these ions or their chelation by ethylenediaminetetraacetic acid (EDTA) or ethyleneglycoltetraacetic acid (EGTA) can lead to biofilm becoming detached. Also, ultrasound, which reaches the parts where a conventional mechanical process cannot reach, has been used with some success.

20.2.2. Sterilization

Sterilization is defined as the use of a physical or chemical procedure to destroy all microbial life, including highly resistant endospores, especially on solid contact surfaces. Moist heat generated by a steam autoclave and ethylene oxide gas constitute the major sterilizing methods used, but some chemical compounds, normally considered as disinfectants, when used appropriately, can also be used for sterilization. To sterilize some materials, gamma and other irradiation methods could be employed.

20.2.3. Disinfection

Generally considered as a less lethal process than sterilization, disinfection consists of destroying virtually all recognized pathogenic microorganisms but not necessarily all microbial forms such as endospores on contact surfaces. By this approach, disinfection does not assure an overkill but reduces the level of microbial contamination to an acceptable degree. Thus, the effectiveness of many sanitizing agents depends on a numbers of factors (length of exposure, temperatures, nature of the organism, product concentration, amount of organic matters, solid surface type) that may have a significant effect on the end result.

20.3. RESISTANCE OF PATHOGENS TO SANITIZERS

Sanitizing agents are widely used in the food industry to destroy indigenous agents, but the mechanisms of resistance of microorganisms to disinfectants are

TABLE 20.2. Cleaners and Disinfectants Used on Food Contact Surfaces

Surface Types	As cleaning solution	As disinfectant solution
Glass and ceramic ware	Alkaline, nonionic detergents	Hypochlorite, iodophors, organic chlorine compounds, quaternary ammonium compounds, ampholyte solutions
Stainless steel	Alkaline, nonionic detergents, detergent disinfectants, acid detergent mixture (removal milk stone)	For glass and ceramic ware
Plastic, rubber, paint, wood	Alkaline, nonionic detergents, detergent disinfectants	For glass and ceramic ware
Cotton or synthetic fibers	Alkaline, nonionic detergents, hot water (77–80°C)	Hypochlorite, organic chlorine compounds, boiling or autoclaving

less well understood than for antibiotics. In addition, although many studies have attempted to evaluate the efficacy of various cleansers and sanitizing agents on microorganisms encountered in the food-processing industry, information on the effects of chemical agents on food pathogen cells attached to surfaces remains sparse. Thus, more knowledge on the inactivation of food pathogens on surfaces in the processing plant is needed.

Sanitizing agents such as sodium hypochlorite, iodophor, and quaternary ammonium compounds are widely used in the food industry, and evaluation of their bactericidal activity against such organisms as *Listeria monocytogenes* or *Salmonella* demonstrates that quaternary ammonium compounds (QACs) were more effective against *Salmonella* than *Listeria*. However, the efficacy of several sanitizing agents against *L. monocytogenes* on contaminated surfaces indicated that all the sanitizers were more efficient against cells attached to nonporous surfaces, such as stainless steel and glass, than to porous surfaces such as polypropylene and rubber. The concentration required to sanitize nonporous surfaces corresponded well to those recommended by the manufacturers. However, the concentration of disinfectants was at least 5–10 times higher for sanitizing rubber surfaces than those required for stainless steel or glass surfaces. At the same time, it is not possible to determine the exact concentration of each sanitizer needed for disinfecting plastic surfaces, despite the very high concentrations of disinfectant. It is evident that the sanitizer resistance of cells is related to the product concentration, the temperature, the exposure time, and the nature of the surface on which the microorganism is attached (Table 20.2). Also, cells dried onto surfaces were more resistant to disinfectants than those in suspension, and the efficacy of several disinfectants is better when increasing the concentrations. Thus, the longer exposure times are necessary to eliminate microorganisms harbored on porous surfaces than on nonporous surfaces.

Hence, the evaluation of sanitizing agents by standard laboratory methods frequently fails to take into consideration the influence of actual environmental conditions on efficacy. It is important to indicate that protective or surface-adhering substances such as exopolymer could also affect the efficacy of disinfectants. The biofilm matrix (cells and exopolysaccharides) may act as a barrier that may protect cells from the influence of a range of stress factors, including antimicrobial agents. In food-processing environments, surface-associated pathogens and food spoilage bacteria are likely to be protected by other microorganisms and their exopolysaccharides, by the substantial accumulation of organic materials of food-processing origins, and by surface irregularities of processing equipment. One hypothesis to explain resistance to sanitizers is that the microbial cells and their exopolysaccharide products significantly reduce the penetration of antimicrobial agents. In the case of a reactive chemical, like chlorine or monochloramine, the failure to penetrate the biofilm is presumably due to a reaction–diffusion interaction. Some authors reported that capsulated microorganisms were shown to have a 150-fold greater resistance to chlorine when they grew on glass surface rather than in suspension. Therefore, elimination of biofilms using only disinfection treatments cannot be completely achieved, a situation that may play a role in biofilm build-up in crevices and scratches, for example, after the cleaning procedure. Mechanical or chemical breakage of the polysaccharide matrix is essential for successful biofilm control.

20.4. DISINFECTING AGENTS

To inactivate or control microbial cells or spores, numerous physical or chemical treatments are used.

20.4.1. Physical Agents

Heat. Exposure of microbial cells to elevated temperatures may lead to destruction. One of the commonest and most useful forms of disinfection is the application of moist heat to raise the surface temperature to at least 70°C. However, high temperatures will denature protein residues and bake them onto the surface of the equipment. Therefore, it is essential that all materials, such as residual food, is removed by thorough cleaning before the application of heat for disinfection.

Hot Water. Using hot water is the method of choice and is common throughout the food industry. Removable parts of machinery and smaller items of equipment can be submerged in a sink or tank containing water at disinfection temperature for a suitable period of time (e.g., 80°C for 2 min). The disinfectant rinse in mechanical washing machines should reach this disinfection temperature, and the period of immersion should be sufficient to allow the equipment surfaces to reach this temperature. Water at disinfection temperature will scald bare hands, so basket racks or some other types of receptacle will have to be used where the process is carried out manually.

Steam Disinfection. It may not be practical to have steam available for disinfection throughout the premises. Steam jets are useful to disinfect areas of machinery and surfaces that are difficult to reach or that must be disinfected in situ on the factory floor. The heating of surfaces during the application of high-temperature steam promotes their subsequent drying. The use of steam can present problems by creating condensation on treated surfaces. High-pressure steam can strip paint from painted surfaces and lubricants from the working parts of machinery. Moreover, some types of material, such as plastics, are unsuitable for treatment with live steam. Steam jets should only be used by trained personnel, as they can be dangerous in a cleaning process.

20.4.2. Chemical Agents

It is important to point out that chemical disinfectants that are liable to taint food, such as phenolic compounds, should not be used in food premises. Thus, care should be taken that chemical disinfectants do not cause harm to personnel.

Chlorine and Chlorine-Based Products Including Hypochlorite Compounds. Properly used, chlorine compounds are among the most suitable for food-processing plants. They can be obtained as liquid hypochlorite solutions or can be combined with a detergent in a chlorinated crystal form. These disinfectants act rapidly against a wide range of microorganisms and are relatively cheap. They are the most suitable for general-purpose disinfection in food premises. They should be used at concentrations of 100–250 ppm of active chlorine per liter. This group of disinfectants is corrosive to metals and also has a bleaching action. Surfaces disinfected with them should therefore be subjected to a final rinsing with potable water of quality after an adequate contact time. Chlorine disinfectants, with the exception of chlorine dioxide, are readily inactivated by the presence of organic matter. Hypochlorite is an effective disinfectant because most microorganisms do not possess specific enzymatic mechanisms for detoxification of HOCl. Chlorine can also damage the cell membrane, resulting in leakage of cell components, and forms substitution products with proteins and amino acids.

Iodophors. Always blended with a detergent in acid medium, iodophor substances are particularly suitable in circumstances where an acid cleaner is required. They have a rapid action and a wide range of antimicrobial activity. A solution of about 25–50 ppm active iodine is required for the disinfection of a clean surface. Like chlorine, iodophors have a corrosive action on metals, and a thorough rising with water of drinking quality is required after a suitable contact time. They are readily inactivated by organic matter. Iodophors give visual indication of their effectiveness since they lose their color when the residual iodine has fallen to ineffective levels. They are not toxic when used in normal concentrations but may add to the total dietary iodine load. They have little taste or smell but may combine with substances in the food to cause taint. For these reasons, they should be thoroughly rinsed away after use.

Quaternary Ammonium Compounds. The QACs are combined detergents and disinfectants and have good detergent characteristic. They are colorless, relatively noncorrosive to metal, and nontoxic but may have a bitter taste. They are not as effective against gram-negative bacteria as chlorine, chlorine-based disinfectants, and iodophors. Quaternary ammonium compounds react with the cell membrane, denature essential cell proteins, and inactivate cellular enzymes. The solutions tend to adhere to surfaces, and a thorough rinsing is necessary. They should be used at a concentration of approximately 200–1000 ppm. They are readily inactivated by magnesium and calcium compounds in hard water and are not compatible with soaps and anionic detergents.

Acid Anionic Surfactants. Acid anionic sanitizing agents offer distinct advantages over most chemical sanitizing agents: the absence of covalent reactions, chemical stability, convenience of use, freedom from organoleptic properties, environmental safety, and biodegradability. These comparatively recent disinfectants have detergent as well as bactericidal properties. They are low in toxicity, relatively noncorrosive, tasteless, and odorless and are efficient disinfectants when used according to manufacturer's recommendations. They are inactivated by organic matter. The anionic sanitizing agents exhibited bactericidal activity equivalent to that of hypochlorites. This activity was observed against a broad spectrum of microorganisms, including gram-positive and gram-negative bacteria.

Strong Acids and Alkalis. In addition to their detergent properties, strong acids and alkalis have considerable antimicrobial activity. Particular care should be taken to ensure that they do not contaminate food.

20.5. CRITERIA FOR DISINFECTANT USAGE

After a thorough cleaning program, the disinfectant must satisfy the following criteria:

1. Suitable for typical food factory applications (be active under the temperature regimens of the food-processing facilities, offer a recognized performance level at ambient temperature within a contact times of 5 min or less, still be active after use and prolonged storage under a variety of conditions)
2. Non-toxic (even at low concentrations, disinfectants must present no health risk to the food consumer)

3. Safe for cleaning operatives (product description, physical data such as appearance, odor, pH, boiling point, and solubility; health hazard, first aid in case of inhalation, eye and skin contact, or ingestion; fire hazards, hazardous reactions with other chemical, spillage or leakage procedure and disposal, transport storage and handling precautions, operator safety aspects and form of application)
4. Nontainting
5. Effective (widest possible spectrum of activity against microorganisms such as viruses, bacteria, fungi, spores, parasites)
6. Effective under environmental conditions (e.g., temperature, pH, interfering debris, water hardness, contact time, items to be treated)
7. Cost (be reasonable)

20.6. EFFICIENCY OF SANITIZING AGENTS

In the food industry, the performance of chemical disinfecting agents against microorganisms is greatly influenced by many factors, including physical, chemical, and biological.

20.6.1. Physical

Temperature. In general, the kinetics of many sanitizing agents against cell growth is greatly affected by the temperature at which the sanitizing is done. A change from low temperature (4–20°C) to high temperature (20–37°C) can reduce by a factor of 5–60 the concentration of product needed to disinfect some surfaces such as stainless steel. Thus, between 4 and 20°C, the product activity is enhanced and, generally, the concentration of disinfectant necessary to sanitize decreases by half, especially for 10 min of exposure time. The significant interval is observed between 20 and 37°C for an exposure time of 20 min. It is interesting to note that an increase in temperature increases the sporicidal activity of sanitizing agents. However, it is also important to remember that at the hot temperatures (50°C) likely to impair growth through protein denaturation, which injures cellular structures such as bacterial membranes, some disinfectants such as iodine compounds are unstable and decay rapidly, resulting in staining material. On the other hand, refrigeration temperature, such as 4°C or below, might also have a lethal effect by interfering with cell permeability and allow greater penetration of substrate, including greater uptake of bactericidal agents such as QACs and glutaraldehydes. In general, products that act via metabolic interference or by inhibition of growth become more effective when the temperature goes to 4 from 50°C. However, in many cases with hot hypochlorite solutions, the corrosive action of chlorine become a problem. Thus temperature is also another important variable affecting the efficacy of the sanitizers. An increase in temperature greatly enhances sanitizer activity profile, whereas at low temperature (4°C) higher concentrations of disinfectants are often needed to inactivate cells of the microorganisms. Lastly, in most cases, the disinfectants are much less effective against microorganisms at refrigeration temperature (4°C) than at ambient (20°C), and among many sanitizing agents used on surfaces contaminated with microbial cells, iodophor and phosphoric acid appear ineffective.

Exposure Time. Regardless of the temperature, the microbial resistance to sanitizing agents is also influenced by the length of exposure times, which can vary from less than 1 min to several minutes for surface disinfection. Depending on the stability of the disinfectant, a similar product with bacteriostatic action within 10 min exposure can have a bactericidal action at 20 min. It is therefore desirable that for a sanitizing agent to be effective, it must produce its effective inactivation of the cells on clean surfaces after minimum contact times that vary, in many cases, between 2 and 5 min. The longer the contact time, the greater the effectiveness of the product in significantly reducing the contamination on porous than on nonporous surfaces. The overall efficacy of disinfection on this surface is a function of the chemical agent used, especially on the attached cells. In several reports, a two- to threefold product increase in contact time results in a corresponding two- to threefold decrease in the critical disinfection point (CDP) value, which greatly enhances the efficacy of sanitizing agents. Also, in many situations, the increase of exposure time (10–30 min) can decrease by a factor of 2–3 the concentration of product needed to disinfect surfaces such as stainless steel or glass.

Concentration of Product. The effective concentration of the chemical solution will vary according to the intended use and must be adequate for the particular purpose. The solutions should therefore be made up strictly according to the manufacturer's instructions.

Product Stability. The choice of agents and procedures to be used for industrial environmental sanitation depends on a variety of factors, and no single agent or procedure is adequate for all purposes. These factors include the nature of treatment, the degree of microbial killing required, and the cost and ease of use of the available agents. To assure the stability of the product, all disinfectant solutions should be freshly made using clean utensils. Topping-up existing solutions or prolonged storage of ready-to-use dilute solutions may render the disinfectant solution ineffective or may allow it to become a reservoir of resistant microorganisms. If mixed with detergents or other disinfectants, sanitizer products may be inactivated.

pH. The pH acts on microorganisms, modifying their surface charges, and on the disinfectant, changing the degree of ionization. Thus, iodophor compounds possess an optimal activity in acid medium rather than in alkali medium where their activity is reduced.

Contact Surfaces. Two of the most important factors that influence greatly the efficacy of disinfectants and the rate of microbial deposition are the physical and chemical nature of the surface. The characteristics that affect the disinfectant include surface finish, porosity, hardness, wettability, geometry, and chemical reactivity with the disinfectant. Also, with continued use, a surface may become physically abraded or corroded, which impairs its smoothness and renders it unhygienic. For this reason, many commercial preparations include a wetting agent in disinfectants.

In several studies using a variety of surfaces, various sanitizing products demonstrate similar action on hard surfaces such as steel, painted plasterboard, and ceramic tile, but the agents differed in their action on porous surfaces such as wood, plastic, and rubber. Thus, among the surfaces, glass is a very good surface for food equipment because its hardness, smooth surface, corrosion resistance, transparency, and cleanability result in more efficient removal of dirt and attached cells. Stainless steel, although lacking the transparency and corrosion resistance of glass, also has cleanability similar to glass except that it is much easier to fabricate and resistant to impact damage. Rubber and rubber-like materials, used in tubing and gaskets, have initially a smooth surface that can be mechanically or chemically abraded and lose their elastic properties and dimensional stability. In many studies, the dirt or cells attached to rubber surfaces were approximately 10 times greater than that on glass or steel. The surface cracks that eventually occur accumulate dirt and provide microorganisms with sites protected from the sanitizing agents, which makes microbial control of damaged rubber surfaces difficult to achieve.

20.6.2. Chemical

Organic Compounds. Inadequately cleaned food contact surfaces may result in residual food particles, and many foods leave protein residues such as blood, food debris, fat, dust, and milk stone that may occlude microorganisms and reduce the effectiveness of several chemical disinfectants by fixation or oxidation of particles that render antimicrobial agents unavailable to react with microorganisms. Disinfectants will not act at all where there is gross soiling. This effect is lesser for phenolic products than quaternary ammonium and halogen compounds. Disinfection with chemicals must, therefore, always be combined with a cleaning process.

Water Hardness. Water quality plays an important role in the efficacy of sanitizing agents. However, the effectiveness of disinfectants may be affected by the degree and nature of the hardness of water used for dilution. Generally, hard water contains calcium, magnesium, and less frequently other ions such as iron, manganese, aluminum, strontium, or zinc that can react, causing acids and alkalis to precipitate and thus reducing disinfectant activity. Also, hard water reacts with soaps, other anionic detergents, bicarbonate cleaners, and a variety of disinfectants to affect their sanitizing actions. Hard water can be softened by addition of a sequestering agent, such as sodium tripolyphosphate, by the lime-soda process, which precipitates and removes the hardness, or preferably by the so-called zeolite process, which uses ion exchange resins to replace the alkaline metal ions with sodium. It must also remembered that potable water should be used in all cleaning regimens. Soft water that is free of minerals is necessary for proper detergent emulsification.

20.6.3. Biological

Nature of Microorganisms. Microorganisms vary widely in their responses to physical and chemical stresses. The differences in chemical resistance exhibited by various vegetative bacteria are relatively minor except for tubercle bacilli, which, presumably because of their hydrophobic cell surfaces, are comparatively resistant to a variety of low-level disinfectants. Among the ordinary vegetative bacteria, staphylococci and enterococci are somewhat more resistant than most other gram-positive bacteria. A number of gram-negative bacteria such as *Pseudomonas*, *Klebsiella*, or *Enterobacter* may also show somewhat greater resistance to some disinfectants. This resistance is more important when the microorganisms are contained in a biofilm.

It is important to note that gram-positive bacteria such as *Listeria*, *Bacillus*, or *Staphylococcus* can produce extracellular lipoteichoic acids that, being lipophilic, may prevent penetration of sanitizer agents. Also, adherent growth can stimulate the amount of teichoic acid and increase the lipid content of the outer membrane of gram-negative bacteria such as *Yersinia*, *Salmonella*, *Aeromonas*, or *Campylobacter* to increase resistance to chlorine. The quaternary ammonium compound sanitizers were shown to be more effective on gram-positive than on gram-negative bacteria. Finally, spore-forming foodborne bacteria have high heat and chemical resistance, and some adhere strongly to equipment surfaces. There they form biofilms in which the vegetative cells also show increased resistance to detergents and heat.

Presence of Biofilm. The potential contamination of food from biofilms in the food industry is of major health concern because microbial cells living within biofilms are inherently more resistant to antimicrobial treatments than those suspended in solution. Disinfectants showing activity in suspension tests do not necessarily do so on contaminated inert surfaces because microbial cells adhering to surfaces might behave differently than cells in suspension when exposed to hostile environmental conditions. Therefore, determination of disinfectant effectiveness should involve testing against adhering cells. Moreover, the concentration of sanitizer agent needed to destroy microorganisms could to be multiplied 10- or 100-fold to obtain the same results against cells in a biofilm compared to cells in suspension. Explanations for the increased resistance of biofilm microbial cell may include limitations to the free diffusion of antimicrobial agents through the biofilm matrix, interaction of the antimicrobial agents with the biofilm matrix (cells and extracellular matrix), and variability in the physical and chemical environment associated with individual biofilm bacteria or regions of the biofilm. It is important to know that this resistance increases with the age of the biofilm. The greater the amount of biofilm remaining on the contact surfaces, the greater is the resistance to sanitizer agents and hostile environments. Recent research using various materials including stainless steel, rubber, polyester, and/or polyurethane and Teflon demonstrated that the type of attachment also influences sanitizer efficacy.

Adequate and frequent hygiene operations are generally assumed to control biofilm formation on food contact surfaces, but it seems that no routine cleaning system is completely effective. It is becoming clear that the smoothness of surface and frequent washing of the processing equipment constitute important factors in preventing biofilm formation. Also, it is usually easier to remove a recently formed biofilm from a smooth surface than from a rough one. Surface roughness may hinder the penetration of antimicrobial agents or high-pressure washes, thereby protecting bacteria. Microbial cells present within cracks or crevices likely acquire a significant degree of physical protection against sanitizing treatments. Consequently, surface roughness likely favors bacterial survival under conditions of stress.

20.7. TESTING OF SANITIZING AGENTS

20.7.1. Direct Swabbing Techniques

This technique, used for the assessment of disinfection efficacy, involves the collection of samples from a given area (e.g., 100 cm^2) by swabbing and a subsequent quantitative (or semiquantitative) assessment of surviving organisms. Even after proper sanitation, 1 or 2 of 10 swabs may be positive in either coliform or staphylococci tests (e.g., in speed tests).

20.7.2. Agar Cylinder Techniques

A cylinder of stainless steel (e.g., 20 $\times$ 8 cm) is filled with an agar medium. The cylinder is fitted with a piston, enabling a 3- to 4-mm layer of medium to be pushed out to contact the surface to be checked. This layer is then cut and placed in a petri dish. Incubation of the agar slice provides a quantitative assessment of surviving microorganisms within a given area.

20.7.3. Agar Sausage Method

One end of a Teflon tube is sealed and filled with agar medium to produce an agar sausage, which is removed from the tube. One end of the agar sausage is cut before touching the surface to be investigated. A slice is then cut and placed in a petri dish, providing a quantitative assessment of surviving microorganisms.

20.7.4. Agar-Carrying Linen

A circle of linen is cut to the dimension of a petri dish, with a small flap extending beyond the dish when the linen is positioned inside the base of the dish. The linen circle and the dish are sterilized. The petri dish is then filled with agar medium. The cover of the petri dish is removed, and the flap of linen can be used to lift the agar medium out of the dish and place it on the surface to be analyzed. The flap should again be used to return the agar to the petri dish. This technique is considered to be more practical than the other tests described above.

20.7.5. Ready-to-Use Test

Although all the above techniques are more or less feasible and can provide very useful information on the efficacy of disinfectant agents, they cannot be easily integrated into the disinfection practice of the premises, as they all require a laboratory background and skill in practical microbiology. Therefore, some ready-to-use sets, which do not require such skills, have been developed. The use of one of these sets is described below.

A piece of linen (2 cm^2) is impregnated with agar medium, mixed indicators, and catalysts; this is placed in a small plastic bag. The set also contains a wetting agent, tweezers, an incubation bag (all are sterilized), and a strip of glue. The bag containing the medium should be cut open, and the pieces of linen (with medium) should be placed on, for example, the floor using the tweezers. The pieces of linen should be wetted and then lightly rubbed on the floor. The samples should be placed in the incubation bag, which should be closed with the strip of glue and stored at room temperature in a dark place. The samples should be investigated after 4–6 h. If the violet color of the medium has not changed or has changed only one of five pieces, this demonstrates the efficacy of the disinfection (representing a count of <100 viable bacteria/100 cm^2). If the violet color of the medium has changed from violet to yellow/brown in 6 h, this means that the disinfection has not been effective. In all cases, when sampling for microbial monitoring of equipment and food contact surfaces, the use of quenching (neutralizing) agent is required to eliminate any residual disinfectant.

20.7.6. Impedimetric Techniques

This technique measures microbial growth by detecting electrical charge in the surrounding incubation medium. For instance, the rate of change in conductivity can be related to the number of viable microorganisms present. The electrical charge can thus be correlated to the number of microorganisms. The advantage of impedance techniques is that they may be employed whether organisms are in suspension or attached to surfaces, provided that the organisms are viable and can change the electrical properties of the incubation medium. This makes the technique particularly suitable for testing surface disinfectants, as there is no need to remove the organisms from the test coupon prior to enumeration. This leaves the biofilm or attached cells in the same morphological state and does not subject them to removal stresses that may influence the degree of disinfectant efficacy recorded.

20.7.7. Bioluminescence Technique

Determination of intracellular adenosine triphosphate (ATP) content by bioluminescence, a time-saving method (5–10 min), is proposed for estimating the number of organisms attached on surfaces and those in suspension. Using a luminometer in combination with suitable reagents enables measurement of the level of ATP. This technique is based on the assessment of level of ATP present in microbial cell samples via an enzyme-linked system that produces light in proportional to the concentration of ATP present. A correlation between bioluminescence and viability of cells is reported for a range of antimicrobial agents on solid surfaces and in suspension tests. As with impedimetric techniques, the use of bioluminescence not only measures the total loss of viability of an attached population after sanitizing disinfection but can also be used to give some idea of biostasis and viability recovery of microorganisms immediately after the biocide application. This may be useful in determining and controlling the growth of organisms on food surfaces in the period between disinfectant application and the subsequent start of production. Thus, bioluminescence appears to be a useful tool for detecting attachment of foodborne organisms, biofilm formation on food contact surfaces, and antimicrobial effects on microbial cells. Nevertheless, this method possesses certain limitations because it is unable to distinguish between microbial and nonmicrobial (residues) sources of contamination on a surface.

20.8. CONCLUSIONS

In the present chapter, we have looked at several important aspects of sanitation or sterilizing agents and the control of food pathogen microorganisms involved in the food-processing industry. In the food industry, cleaning and disinfection are two of the more important measures necessary to assure the production of safe foods. Incorrectly selected, disinfectants or sterilizing agents could be ineffective or may give rise to product taint, health hazards to sanitation operatives, or toxicity to food product consumers. Therefore, the advice of the disinfectant manufacturer should always be sought if the suitability of disinfectant is in doubt.

However, the physical and chemical nature of the cell wall of several microorganisms does not especially predispose it to resist sanitizing agents. The nature of the food contact combined with the secretion of extracellular material called "glycocalyx" would therefore play a role on the effect of antimicrobial agents in the destruction of pathogenic organisms. It should be borne in mind that in the food-processing industry, microorganisms are present everywhere and the prevention of contamination depends on the nature of the organism, the type of disinfectants, the choice of surface materials and their finish, good plant design, control of the process, and staff training.

In the development of advanced process hygiene regimens, we must not forget that under prolonged continuing use, surfaces may become physically abraded or corroded, which impairs their smoothness and makes them unfit for hygienic use. The cracks and crevices that eventually occur on a surface may provide microorganisms with sites where, protected from the cleaning agents, they make microbial control difficult to achieve. With regard to the resistance of attached cells to sanitizers, it is important to know that it was assumed that several organisms such as *Aeromonas*, *Campylobacter*, *Listeria*, *Salmonella*, *Vibrio*, and *Yersinia* have the same range of sensibility as other nonsporulating organisms. However, sanitizers appear less effective at removing microbial cells on porous surfaces, especially at refrigeration temperature. Also, the use of some surfaces in the food industry poses the most important limitation, with plastic surfaces such as polypropylene and polyester or rubber materials being much more difficult to clean and sanitize than glass or stainless steel. Thus, the resistance of attached microorganism cells to sanitizing agents is related to the surfaces and the temperature at which attachment is achieved.

Industry must follow strict disinfection standards, particularly for plastic or rubber equipment. "Clean in place" (CIP) cleaning and disinfection methods, if well used, are very efficient against most types of attached bacteria cells on equipment surfaces. Therefore, it is important to verify that cleaning and sanitizing regimens are used at proper temperatures, pressures, flow rates, product strength, and contact time. However, there should be no hesitation in making changes to these procedures in order to manually clean and soak the pieces of equipment to which bacteria are more likely to adhere. Lastly, while sanitary measures and improved methods of detection have helped in reducing the frequency of food contamination by food pathogens, identification of the factors that influence the survival and growth of pathogenic microorganisms in the food industry remains crucial for better control of these agents.

To reduce the potential for survival and buildup of bacteria food pathogens such as *Yersinia* spp., *Campylobacter* spp., *Aeromonas* spp., *Shigella*, or *Salmonella* spp., the attention of the food industry must be focused on general plant conditions for the prevention of contamination at its source and during processing by using the HACCP concept. These points include surfaces in direct and indirect contact with product, air flow patterns, especially going from dirty to clean areas, and process fluids in direct or indirect contact with the product.

Since processing plants have a wide variety of contact surface, such as different types of stainless steel and plastic or rubber material in various conditions (e.g., new, pitted, frayed, cracked), it is not surprising that sometimes sanitizers have little effect on attached cells even when specific hygiene practices are respected. Also, there is much evidence to indicate that the main source of postprocess contamination comes from microorganisms on surfaces in contact with the food. Thus, processing rooms and any equipment coming in contact with food products should continually be monitored for crevices and cracks that could serve as niches for microorganisms. Improper welds and similar irregular surfaces that may lead to ineffective cleaning and sanitizing should be eliminated.

In addition, opportunities for contamination must be limited by controlling the movement of raw material, final product, and staff and by preventing as much as possible the entry of food pathogen agents inside a plant. Despite the many techniques available for studying the microbial–surface association, the problem of microbial resistance to disinfection is a subject of increasing interest for both the public health and food sectors.

Lastly, it is important to know that, in all cases, the control of food contamination by pathogenic foodborne microorganisms requires the judicious use of approved chemical sanitizing agents in clean areas followed by good manufacturing practice (GMP) to reduce the potential for product contamination, especially in an environment where refrigeration temperatures and porous surfaces are used.

Clearly, the most effective strategy for the control of microbial contamination and subsequent biofilm formation is to keep the material clean, without any cracks or openings, as pathogenic microorganisms can become established. Also, information and educational programs on food safety need to be aimed at increasing the awareness of industry.

Basic research into factors promoting microbial resistance is still needed in order to control and prevent biofouling and provide basic knowledge on surface–bacteria relationships and the efficacy of sanitizing agents against attached cells. In this way, the financial loss due to recall or condemnation of food products can be minimized and at the same time protection of consumer health can be achieved.

Finally, the action of disinfectants or sterilizing agents against microorganisms in the food industry is an excellent biochemical engineering problem. It requires integration of concepts from microbiology, engineering, and chemistry and encompasses phenomena ranging from microbial growth and disinfection kinetics to mass transport and reactor dynamics. The problem is especially interesting in that attached microorganisms are universally known to be more resistant to antimicrobial agents than their planktonic counterparts. Thus, with their long-standing involvement in the design and analysis of surface biocontamination, one of the most exciting prospects and challenges for engineers in the coming century is to perfect materials and agents that could prevent bacterial attachment and biofilm formation.

BIBLIOGRAPHY

Busscher, H. J., A. H. Weerkamp, H.c. van der Mei, A. W. J. van Pelt, H. P. de Jong, and J. Arends. 1984. Measurement of the surface free energy of bacterial cells surfaces and its relevance for adhesion. *Appl. Environ. Microbiol.* **48**, 980–983.

Carpentier, B., and O. Cerf. 1993. Biofilms and their consequences, with particular reference to hygiene in the food industry. *J. Appl. Bacteriol.* **75**, 499–511.

Dunsmore, D., G. Twomey, A. Whittlestone, W. G. Whittlestone, and H. W. Morgan. 1981. Design and performance of systems for cleaning product contact surfaces of food equipment: A review. *J. Food Prot.* **44**, 220–240.

Mafu, A. A., D. Roy, J. Goulet, and D. Montpetit. 1998. Attachment of *Aeromonas hydrophila* to water distribution system pipes after different contact times. *J. Food Prot.* **61**, 1321–1329.

Mafu, A. A., D. Roy, and C. Machika. 1996. Efficiency of disinfecting agents for destroying *Listeria monocytogenes, Yersinia enterocolitica* and *Staphylococcus aureus* on contaminated surface. *Dairy Food Environ. Sanit.* **16**, 426–430.

Mafu, A. A., D. Roy, L. Savoie, and J. Goulet. 1991. Bioluminescence assay for estimated the hydrophobic properties of bacteria as revealed by hydrophobic interaction chromatography. *Appl. Environ. Microbiol.* **57**, 1640–1643.

Mafu, A. A., D. Roy, J. Goulet, L. Savoie, and R. Roy. 1990. Efficiency of sanitizing agents for destroying *Listeria monocytogenes* on contaminated surfaces. *J. Dairy Sci.* **73**, 3428–3432.

Mafu, A. A., D. Roy, J. Goulet, and L. Savoie. 1991. Characterization of physicochemical forces involved in adhesion of *Listeria monocytogenes* to surfaces. *Appl. Environ Microbiol.* **57**, 1969–1973.

Mafu, A. A., D. Roy, J. Goulet, and P. Magny. 1990. Attachment of *Listeria monocytogenes* to stainless steel, glass, polypropylene and rubber surfaces after short contact times. *J. Food Prot.* **53**, 742–746.

Mosley, E. B., P. R. Elliker, and H. Hays. 1976. Destruction of food spoilage, indicator and pathogenic organisms by various germicides in solution and on a stainless steel surface. *J. Milk Food Technol.* **39**, 830–836.

Mosteller, T. M., and J. R. Bishop. 1993. Sanitizer efficacy against attached bacteria in a milk biofilm. *J. Food Prot.* **56**, 34–41.

Mustapha, A., and M. B. Lewen. 1989. Destruction of *Listeria monocytogenes* by sodium hypochlorite and quaternary ammonium sanitizers. *J. Food Prot.* **52**, 306–311.

Stewart, P. S., L. Grab, and J. A. Diemer. 1998. Analysis of biocide transport limitation in artificial biofilm system. *J. Appl. Microbiol.* **85**, 495–500.

Tamasi, G. 1993. Testing disinfectants for efficacy. *Rev. Sci. Technol.* **14**, 75–79.

Van Loosdrecht, M. C. M., J. Lyklema, W. Norde, and Z. A. J. b Zehnder. 1990. Influence of interfaces on microbial activity. *Microbiol. Rev.* **54**, 75–87.

Zottola, E. A., and K. C. Sasahara. 1994. Microbial biofilm in the food industry—Should they be a concern? *Int. Food Microbiol.* **23**, 125–148.

CHAPTER 21

Role of HACCP in Control of Foodborne Illnesses

ROBERT B. GRAVANI

Previous chapters of this book have addressed the epidemiology of foodborne illnesses and described a number of important pathogenic microorganisms, including bacteria, viruses, molds, and parasites as well as seafood toxins, that cause foodborne illness. This chapter will highlight a systematic, preventive approach to reducing the risks of hazards associated with foods.

21.1. TRADITIONAL APPROACHES

For decades, the food industry has used traditional approaches in controlling biological hazards associated with foods. These approaches include the simultaneous use of employee education and training programs, frequent inspections of facilities and operations, as well as microbiological testing of raw ingredients, in-process foods, and finished products.

Employee education and training programs are often structured to provide a thorough understanding of good manufacturing practices (GMPs), including personal hygiene, product flow and cross-contamination, cleaning and sanitation procedures, and the various causes of microbial contamination.

Another method used to control biological hazards involves frequent inspections of facilities and operations to ensure that GMPs are being followed. The GMPs for specific foods are available in advisory and regulatory documents such as the current GMPs in the code of federal regulations and in the various codes for hygienic practice developed by the Codex Alimentarius Committee on Food Hygiene.

Still another measure to control biological hazards in finished products is the thorough microbiological analysis of raw ingredients, in-process products, as well as finished foods. The analysis of samples for indicator organisms (e.g.,

Guide to Foodborne Pathogens, Edited by Ronald G. Labbé and Santos García
ISBN 0-471-35034-6

coliforms and fecal streptococci), spoilage organisms, and pathogens provides some evidence that the product has been manufactured, handled, and distributed according to appropriate practice. It should be noted that end-product sampling and testing is not a perfect system for detecting pathogens in a food and can lead to a false sense of security. For example, if *Salmonella* was present in a batch of product at a rate of one out of every 1000 units of product (defect rate 0.1%), a sampling plan that analyzed 60 units from the batch would have a greater than 94% probability of approving the batch and missing the *Salmonella*-contaminated product. If a company is trying to detect hazards by just taking random samples, there is a high probability that they will not be detected.

Although the traditional approaches for controlling biological hazards are being used by many food companies around the world, foodborne outbreaks still occur. In the United States alone, the Centers for Disease Control and Prevention in Atlanta, Georgia, estimates that there are over 76 million cases of foodborne illness each year, resulting in 325,000 hospitalizations and about 5000 deaths. It has been estimated that each year 130 million Europeans [15% of the total population of the World Health Organization (WHO) European Region] are affected by episodes of foodborne diseases, ranging from mild gastrointestinal infections to severe gastroenteritis. Foodborne disease surveillance programs in countries around the world also indicate that foodborne illnesses are prevalent and this issue is a global concern.

21.2. HAZARD ANALYSIS CRITICAL CONTROL POINT CONCEPT

There is a critical need for a preventive, systematic approach to food safety assurance. This need for a modified approach led to the development of the hazard analysis critical control point (HACCP) concept, which can be used to identify and control biological, chemical, and physical hazards in foods from raw material, production, procurement, and handling to manufacturing, distribution, and consumption of the finished product. The HACCP concept, along with GMPs and prerequisite programs, can be used to reduce the risks of hazards associated with foods.

21.3. ORIGINS OF HACCP

The HACCP concept was developed by the Pillsbury Company with the cooperation and participation of the National Aeronautics and Space Administration (NASA), the U.S. Army Natick Laboratories, and the U.S. Air Force Space Laboratory Project Group. The development of the HACCP system began in 1959 when the Pillsbury Company was asked to produce foods for the manned space flights. There needed to be assurance (as close to 100% as possible) that the food produced for the space program would not be contaminated with pathogens that would cause illness. The project scientists quickly realized the traditional methods of quality control were not adequate to assure the safety of foods being produced. After much research and evaluation, the HACCP concept was developed and was first presented to the scientific community at the 1971 Conference for Food Protection.

The HACCP concept was first used in the acidified and low-acid canned food industry and was then adopted by a few large companies during the mid-1970s and early 1980s. It was not, however, widely used in the food industry during this time. After an important 1985 U.S. National Academy of Science (NAS) publication strongly endorsed and recommended the HACCP concept, the food industry expressed considerable interest in the application of HACCP.

As a result of the NAS report, the National Advisory Committee on Microbiological Criteria for Foods (NACMCF) was established and embraced the HACCP concept. In 1989, the committee developed a HACCP document and guide for maintaining uniformity of the principles and definitions of terminology. Since then, the NACMCF has made several refinements and improvements in the HACCP concept and published revisions in 1992 and 1997. The 1997 document, entitled "HACCP Principles and Application Guidelines," contains many additions and provides a revised and more detailed explanation of the HACCP principles. The Food Hygiene Committee of the United Nations FAO/WHO Codex Alimentarius Commission has also been actively involved in the development of HACCP guidelines for use in international trade. In 1997, Codex adopted their latest version of a HACCP guidance document, which is very similar to the one developed by NACMCF. Today, HACCP is widely used throughout the food processing industry and is also being applied to foods processed and prepared in retail food stores and food service operations. Many companies in countries throughout the world are also applying HACCP in their operations.

The HACCP concept is a systematic approach to the identification, evaluation, and control of biological, chemical, and physical hazards associated with foods. It is a procedure used to design safety into the product and the process by which it is produced.

21.4. PREREQUISITE PROGRAMS

As the HACCP concept continues to evolve, companies have recognized the importance of providing a solid foundation on which a HACCP system can be built. The term *prerequisite programs* describes a wide variety of programs that not only provide the foundation for HACCP but also support the system. Prerequisite programs are vital to the successful development and implementation of effective HACCP plans.

Many of the prerequisite program components are based upon the current GMPs in the Code of Federal Regulations and in the Codex Alimentarius General Principles of Food Hygiene for foods traded internationally. Prerequisite programs also include other important systems such as ingredient specifications, supplier approval programs, ingredient-to-product traceability programs, product recall plans, and consumer complaint management programs. Prerequisite programs provide the basic environmental and operating conditions that are necessary for the production of safe, wholesome food. These programs are intended to keep low-risk, potential hazards from becoming serious and affecting the safety of the product.

Common prerequisite programs include a wide variety of activities. There are about eight major areas where prerequisite program activities should be focused, and these include raw material controls, education and training, facilities, sanitation, production equipment, production controls, storage and distribution, and product controls. An outline of important details in each of these areas is presented below.

Raw Material Controls. Each facility should assure that its suppliers have effective GMPs and food safety programs in place. These may be the subjects of continuing supplier guarantee and supplier HACCP system verification. Other important considerations include:

- Purchasing specifications (e.g., ingredients, products, packaging materials)
- Verification of suppliers
- Letters of guarantee and certificates of analysis
- Sample results accompanying shipments
- Raw material inspection before acceptance
- Acceptance testing procedures
- Receiving and storage standard operating procedures (SOPs)

Education and Training. Food company employees need to understand their important role in assuring the safety of foods and perform their duties with food safety in mind. Some components of a continuing education and training program include:

- Personal safety
- Personal hygiene
- GMPs
- Ingredient and food handling
- Sanitation practices
- HACCP

Facilities. The establishment should be located, constructed, and maintained according to sanitary design principles. There should be linear product flow and traffic control to minimize cross-contamination from raw to cooked foods. Items that need attention are:

- Sanitary design and construction
- Adjacent properties

- Building exterior
- Building interior
- Employee and product flow patterns
- Ventilation (condensation)
- Waste disposal and management
- Sanitary facilities
- Handwash stations
- Water, ice, and culinary steam
- Lighting

Sanitation. All procedures for cleaning and sanitizing equipment and the facility should be written and followed. A master sanitation schedule should be in place. Documented procedures are needed to assure the segregation and proper use of nonfood chemicals in the plant. These include cleaning chemicals, fumigants, and pesticides or baits used in or around the facility. Areas of concern are:

- Personal hygiene
- Cleaning and sanitizing SOPs
- Master sanitation schedule
- Good housekeeping practices
- Effective pest control programs
- Environmental surveillance
- Chemical control programs

Production Equipment. All equipment should be constructed and installed according to sanitary design principles. Preventive maintenance and calibration schedules should be established and documented. Checklists of activities should include:

- Sanitary design and installation of equipment
- Cleaning and sanitizing SOPs
- Preventive maintenance programs
- Equipment calibration programs

Production Controls. A variety of production control programs should be in place to protect against foreign materials and allergens. These include:

- Product zone controls
- Foreign material control program
- Metal protection program

- Glass control program
- Allergen control program

Storage and Distribution. All finished products should be stored and distributed under sanitary conditions and the proper environmental conditions such as temperature and humidity to assure their safety and wholesomeness. Areas of concern are:

- Shipping and distribution SOPs
- Temperature controls
- Transport vehicles

Product Controls. All raw materials and products should be lot-coded and a recall system in place so that rapid and complete tracebacks and recalls can be done when a product retrieval is necessary. A crisis management plan should be developed and periodically updated. Important considerations include:

- Product coding and labeling
- Product traceability and recall procedures
- Crisis management plan
- Complaint investigations

Prerequisite programs are developed and managed separately from HACCP systems, and every area should have its own written SOPs, training, documentation, validation of the program's adequacy, and verification. The day-to-day management of prerequisite programs is normally the responsibility of a number of departments in the processing facility. There may be some situations where certain activities that are normally addressed in prerequisite programs, may be included in the HACCP plan. For example, although sanitation procedures are usually part of a prerequisite program, some processors manage selected sanitation procedures as critical control points (CCPs) in their HACCP systems. This has been done frequently in the meat and dairy industries where sanitation procedures for meat slicers, ice cream fillers, and other pieces of equipment were established as CCPs to help prevent recontamination of processed products with *Listeria monocytogenes.*

To clearly illustrate the relationship of prerequisite programs to HACCP, the HACCP Food Safety Assurance Pyramid is provided in Figure 21.1. To be effective, HACCP depends on the commitment and leadership of top management in a company, a well-developed and effective employee education and training program, a thorough understanding of biological, chemical, and physical hazards, and a good working knowledge of the seven HACCP principles. This HACCP Pyramid is built upon a strong foundation of key food safety and sanitation components (GMPs) and prerequisite programs that were previously discussed.

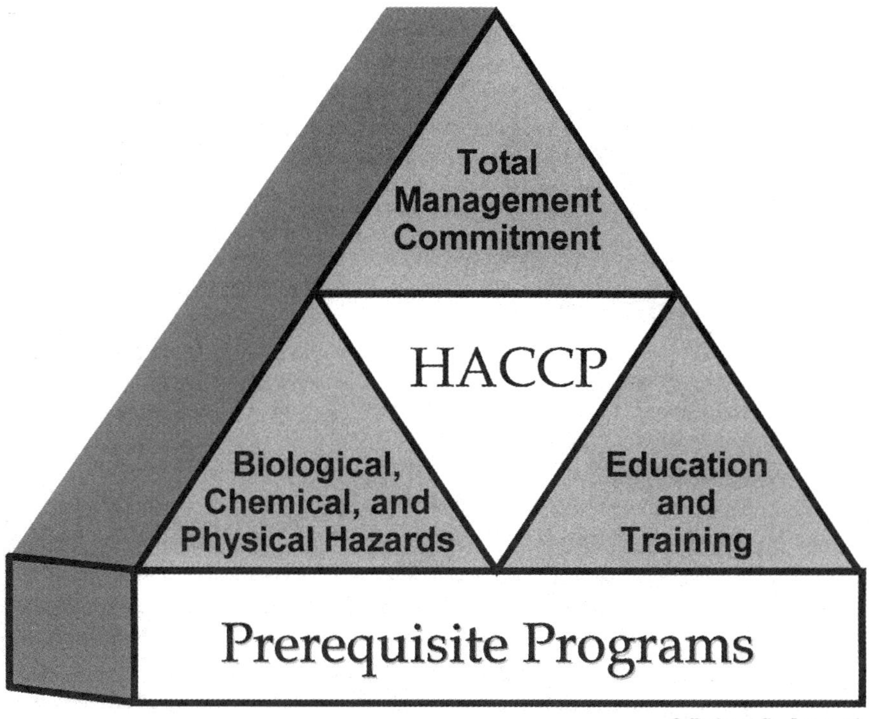

Figure 21.1. HACCP Food Safety Assurance Pyramid.

Well-developed and consistently implemented prerequisite programs can simplify a HACCP plan, so it is very important that food processors develop, maintain, and periodically review the effectiveness of prerequisite programs. Before developing a HACCP plan, processing plant managers should:

- Survey and evaluate present operating conditions
- Assess GMPs
- Strengthen prerequisite programs

Without effective prerequisite programs, a HACCP plan becomes complicated and very difficult to manage.

21.5. HACCP PLAN DEVELOPMENT

In the development of a HACCP plan, five preliminary tasks need to be accomplished before the application of the HACCP principles to a specific product and process. These tasks are listed in Figure 21.2.

Assemble a HACCP team

↓

Describe the food and its distribution

↓

Describe the intended use and consumers of the food

↓

Develop a flow diagram, which describes the process

↓

Verify the flow design

Figure 21.2. Preliminary tasks in development of HACCP plan.

The first task in developing a HACCP plan is to assemble a HACCP team. This multidisciplinary team will have the responsibility of developing the HACCP plan, and it should include individuals who have expertise and specific knowledge of the product and process. Individuals from areas such as engineering, production, sanitation, and food microbiology should be team members.

The HACCP team then describes the food that is being manufactured. This consists of a general description of the product, its ingredients, and processing methods. The method of distribution (refrigerated, frozen, or at ambient temperature) should then be described.

The normal expected use of the food is then described and the intended consumers of the product should be identified. The consumers may be the general public or a particular segment of the population such as infants, immunocompromised individuals, or the elderly.

After these three tasks are completed, the HACCP team should develop a flow diagram, which describes a clear, simple outline of the steps involved in the entire process.

The HACCP team should then perform an on-site review of the operation to verify the accuracy and completeness of the flow diagram. Modifications should be made to the flow diagram as necessary.

After these five preliminary tasks have been completed, the seven HACCP principles should be used to develop the HACCP plan.

21.6. SEVEN HACCP PRINCIPLES

HACCP is a management system that is designed for use in all segments of the food industry and is product and process specific. The HACCP concept is based on seven universally accepted principles, which are as follows:

1. Conduct a hazard analysis.
2. Determine critical control points.
3. Establish critical limits.
4. Establish monitoring procedures.
5. Establish corrective actions.
6. Establish verification procedures.
7. Establish record-keeping and documentation procedures.

Principle 1: Conduct a Hazard Analysis. The purpose of the hazard analysis is to determine which of the potential biological, chemical, or physical hazards associated with a food or a manufacturing process presents a significant risk to consumers. The hazard analysis is the key to preparing an effective HACCP plan. If the hazard analysis is not done properly, and hazards that need to be controlled within the HACCP system are not identified, a serious food safety incident (foodborne illness or injury) can result.

The hazard analysis involves two stages. The first stage, hazard identification, is actually a brainstorming session that involves the review of ingredients used in the product, the activities conducted at each step in the process and the equipment used, the final product and its method of storage and distribution, and the intended use and consumers of the product. Hazard identification focuses on developing a list of potential food hazards associated with each process step. In stage 2, the hazard evaluation, each potential hazard is evaluated based on the severity of its effects (e.g., illness, injury, or death) and its likelihood of occurrence. The potential hazards that present a significant risk to consumer health are the hazards that should be addressed in the HACCP plan.

It is important that food processors conduct a hazard analysis on all existing and any new products to be manufactured, since ultimate microbiological safety of nonthermally processed foods is directly related to the quality of raw materials. Any hazard analysis must begin with identification of hazards associated with raw materials, with particular attention being given to raw products of animal origin (i.e., milk, meat, poultry, and seafood), all of which may harbor foodborne pathogens. Although heat treatments, acidulation, fermentation, salting, and drying are designed to destroy or inhibit growth of pathogenic and spoilage microorganisms, other operations such as slicing and dicing, cooling of cooked products, and filling and packaging may allow pathogenic organisms to contaminate the final product. Therefore, all hazards associated with manufacturing procedures and postprocessing contamination must be fully understood along with the consequences of processing failures and/or errors. Food processors should also be familiar with the effect of various physicochemical factors (i.e., pH, water activity, preservatives, and type of packaging with or without modified atmosphere) on the behavior of pathogenic organisms in the product during processing, distribution, storage, and use by the consumer. The National Advisory Committee HACCP document contains a series of questions regarding ingredients, intrinsic factors of the food during and after processing,

processing procedures, microbial content of the food, facility design, equipment design and use, packaging, sanitation, employee health, and others that can be used when conducting a hazard analysis. It should be noted that any change in raw materials, product formulation, processing, packaging, distribution, or intended use of the product should prompt an immediate reassessment of hazards, since these changes have the potential to affect product safety adversely.

Principle 2: Determine Critical Control Points. The HACCP team determines CCPs based upon the results of the hazard analysis. The hazards that were identified as being reasonably likely to cause illness or injury if not controlled need to be addressed in the HACCP plan. A CCP is a step at which control can be applied and is essential to prevent or eliminate a food safety hazard or reduce it to an acceptable level. While there are many control points in every process, CCPs are those where hazards can be properly controlled. If control at these CCPs is lost, a foodborne illness or injury will result. The complete and accurate identification of CCPs is fundamental to controlling food safety hazards, so CCPs must be carefully identified and documented. Examples of CCPs may include thermal processing, chilling, product formulation control, and checking products for metal contamination. Each food processor needs to determine the best location for CCPs to control identified hazards.

Principle 3: Establish Critical Limits. A critical limit is a boundary of safety and is used to distinguish between safe and unsafe operating conditions at a CCP. Each CCP will have one or more control measures to assure that the identified hazards are prevented, eliminated, or reduced to acceptable levels. Critical limits may be based on factors such as temperature, time, moisture level, pH, water activity (a_w), physical dimensions, humidity, titratable acidity, flow rate, salt concentration, available chlorine, viscosity, or preservatives. Critical limits must be scientifically based and may be obtained from a variety of sources, including regulatory standards and guidelines, the scientific literature, experimental results, equipment manufacturers, and experts. An example of a critical limit for the cooking of ground beef patties in a continuous process would be:

- Oven temperature
- Time in oven (rate of heating and cooling)
- Belt speed
- Patty thickness
- Patty composition
- Oven humidity

Principle 4: Establish Monitoring Procedures. Monitoring is a planned sequence of observations or measurements to assess whether a CCP is under control and to produce an accurate record for future use in verification. Monitoring facilitates the tracking of an operation and is used to keep the process in control. Monitoring is also used to determine when there is a loss of control and a deviation occurs at a CCP (exceeding or not meeting a critical limit). Corrective actions must then be taken to get the process back in control. To properly establish monitoring procedures, the HACCP team must describe:

- What will be monitored
- When it will be monitored (the frequency or maximum lapsed time between monitoring procedures)
- Who will perform the monitoring procedures.

Most monitoring procedures need to be rapid and often include visual observations and measurement of items such as temperature, time, pH, and moisture level. Microbial tests are seldom effective for monitoring owing to their time-consuming nature and problems with assuring detection of contaminants. Individuals who are involved in monitoring should receive proper training and fully understand the purpose and importance of the monitoring, be unbiased in their monitoring and reporting, and accurately record the results that they observe. All records and documents associated with CCP monitoring should be signed (or initialed) and dated by the person doing the monitoring.

Principle 5: Establish Corrective Actions. When a deviation from an established critical limit at a CCP occurs, there will be an unsafe food (that will cause illness or injury), and corrective actions must be taken to address the problem. Through the establishment of corrective actions, foods that may be hazardous are prevented from reaching consumers. Ideally, data gathered from monitoring at each CCP should detect trends that are moving toward potential deviations, enabling the operator to make adjustments before there is a need to take corrective actions. Due to the complexity of many food-processing operations, complete control is often not possible, so some deviations will occur. When there is a deviation from critical limits, corrective actions are needed to:

- Determine and correct the cause of noncompliance
- Determine the disposition of noncompliant product
- Record the corrective actions that are taken

The HACCP plan should specify what is done when a deviation occurs, who is responsible for implementing the corrective actions, and that a record of the actions taken will be developed and maintained. Specific corrective actions should be developed for each CCP and included in the HACCP plan.

Principle 6: Establish Verification Procedures. Verification determines the validity of the HACCP plan and is used in evaluating whether the facility's

HACCP system is functioning according to the HACCP plan. An effective HACCP system requires little end-product testing, since sufficient validated safeguards are built in early in the process. Firms should rely on frequent reviews of their HACCP plan, verification that the plan is being correctly followed, and review of CCP monitoring and corrective action records.

Another important aspect of verification is the initial validation of the HACCP plan to determine that the plan is scientifically and technically sound, that all hazards have been properly identified, and that if the HACCP plan is properly implemented, these hazards will be effectively controlled. Subsequent validations are performed and documented by the HACCP team or independent expert as needed. Validations are conducted when there is an unexplained system failure, when a significant product, process, or packaging change occurs, or when new hazards are recognized.

Principle 7: Establish Record-Keeping and Documentation Procedures. The establishment of an effective record-keeping system is an integral part of a HACCP system. Records are the only reference available to trace the production history of a finished product. If questions arise concerning the safety of a product, a review of records may be the only way to prove that the product was prepared and handled in a safe manner in accordance with the company's HACCP plan. Records include a summary of the hazard analysis, the HACCP plan, supporting documentation, and daily operational records. Company management, supervisors and inspectors play a primary role in assuring that all HACCP records are accurate and complete and reflect the actual operating conditions in the establishment. Records and record review assist in assuring product safety.

Through the development and effective implementation of HACCP in food-processing operations, many benefits result, and these include:

- Demonstrating management commitment to food safety assurance
- Producing safe products
- Providing evidence of safe processing and handling of foods
- Increasing the level of confidence in product safety
- Improving the level of customer confidence in product safety
- Using resources effectively
- Complying with regulatory requirements
- Receiving a product quality dividend

A well-designed, developed, and implemented HACCP plan, built on a strong foundation of GMPs and prerequisite programs, can reduce the risk of foodborne illness and provide greater safety assurance than traditional approaches.

BIBLIOGRAPHY

Bauman, H. E. 1992. Introduction to HACCP. In: M. D. Pierson and D. A. Corlett, Jr. (Eds.), *HACCP: Principles and Applications*. New York: Van Nostrand Reinhold, pp. 1–5.

Corlett, D. A., Jr. 1998. *HACCP Users Manual*. Gaithersburg, MD: Aspen.

FAO/WHO Collaborating Centre for Research and Training in Food Hygiene and Zoonoses. 1995. WHO Surveillance Programme for Control of Foodborne Infections and Intoxications in Europe. Sixth Report 1990–1992.

Food and Drug Administration. 1999. Current Good Manufacturing Practices in Manufacturing, Packing or Holding Human Food. Title 21, Code of Federal Regulations. Part 110. Washington, DC: U.S. Government Printing Office.

Gravani, R. B. 1999. Incidence and control of *Listeria* in food processing facilities. In: E. T. Ryser and E. H. Marth (Eds.), *Liseria, Listeriosis and Food Safety*, 2nd ed. New York: Marcel Dekker, pp. 657–709.

Gravani, R. B. 1999. Prerequisite programs are vital to HACCP success. *Sanitarian* **3**(1), 1, 3.

Joint FAO/WHO Food Standards Programme, Codex Alimentarius Commission. 1997. *Recommended International Code of Practice, General Principles of Food Hygiene in CAC/RCP*, rev. 3, Suppl. to Vol. 1B, 2nd ed. Codex Alimentarius General Requirements (Food Hygiene). Rome: FAO/WHO, pp. 1–126.

Joint FAO/WHO Food Standards Programme, Codex Alimentarius Commission. 1997. *Hazard Analysis and Critical Control Point (HACCP) System and Guidelines for Its Application*. Alinorm 97/13A. Codex Alimentarius Committee on Food Hygiene.

Mead, P. S., L. Slutsker, V. Dietz, L. F. McCaig, J. S. Bresee, C. Shapiro, P. M. Griffin, and R. V. Tauxe. 1999. Food-related illness and death in the United States. *Emerg. Infect. Dis.* **5**(5), 607–625.

Mortimore, S., and C. Wallace. 1998. *HACCP: A Practical Approach*, 2nd ed. Gaithersburg, MD: Aspen.

National Academy of Sciences. 1985. *An Evaluation of the Role of Microbiological Criteria for Foods and Food Ingredients*. Washington, DC: National Academy Press.

National Advisory Committee on Microbiological Criteria for Foods. 1998. HACCP principles and application guidelines. *J. Food Protection* **61**(9), 1246–1259.

Sperber, W. H., K. E. Stevenson, D. T. Bernard, K. E. Deibel, L. J. Moberg, L. R. Hontz, and V. N. Scott. 1998. The role of prerequisite programs in managing a HACCP system. *Dairy Food Environ. Sanit.* **18**(7), 418–423.

Stevenson, K., and D. Bernard (Eds.). 1999. *HAACP: A Systematic Approach to Food Safety*. Washington, DC: Food Processors Institute.

CHAPTER 22

Prion Diseases

JUDD AIKEN

22.1 INTRODUCTION

The outbreak of the foodborne epidemic, bovine spongiform encephalopathy (BSE) in Great Britain, its spread to Europe and the link between BSE and a new human form of the disease, variant Creutzfeldt-Jakob disease (vCJD), have focussed considerable attention on prion diseases. These inevitably fatal neurological disorders, also referred to as transmissible spongiform encephalopathies (TSEs), share several hallmark characteristics including spongiform degeneration in the central nervous system, accumulation of a structurally abnormal form of a brain protein (the prion protein, PrP) in infected animals, and lack of an immune response. Uncertainty over the number of humans currently infected with vCJD, extreme resistance of the infectious agent to inactivation, lack of a cure or even a preclinical diagnosis, and uncertainty over the mode of transmission of BSE to humans makes these diseases particularly vexing.

The biology of these disease agents is different from other infectious agents. These differences include an extended preclinical phase, their ability to resist traditional sterilization methods, and difficulties in diagnosis of the disease. These characteristics have had tragic consequences including the exposure of the cattle population in Great Britain to contaminated feed and the transmission of the resulting bovine disease to humans. One somewhat paradoxical trait is that (with perhaps the notable exception of chronic wasting disease of deer and elk) prion diseases are not readily transmissible. Ingestion of contaminated food is the most common means of transmission, although, in the case of vCJD, the source of infection (meat, milk, processed bovine products) is not yet clear.

TSEs have been identified in a number of species (Table 22.1) and include scrapie in sheep and goats, BSE, transmissible mink encephalopathy (TME), feline spongiform encephalopathy (FSE), and chronic wasting disease of deer

Guide to Foodborne Pathogens, Edited by Ronald G. Labbé and Santos García
ISBN 0-471-35034-6

TABLE 22.1. Animal and Human Prion Diseases

Species	Prion Disease	Source of Infection
Sheep	Scrapie	Acquired, maternal
Cattle	Bovine spongiform encephalopathy	Contaminated feed
Mink	Transmissible mink encephalopathy	Contaminated feed
Cats	Feline spongiform encephalopathy	BSE infected tissue or meat-and-bone meal
Deer and elk	Chronic wasting disease	Origin unknown; self-Sustaining
Human	Kuru	Ritualistic cannibalism
	Creutzfeldt-Jakob disease	
Human	Iatrogenic	Infection
	Sporadic	Unknown
	Familial	PrP gene mutation
	Variant CJD	Infection—source BSE
Human	Gerstmann-Straussler-Scheinker syndrome	PrP gene mutation
	Fatal familial insomnia	
Human	Familial	PrP gene mutation
	Sporadic	Unknown

and elk. The human diseases include Kuru maintained by ritualistic cannibalism, CJD that has sporadic, acquired and familial manifestations, and two familial forms, Gerstmann-Straussler-Scheinker syndrome (GSS) and fatal familial insomnia (FFI). All TSEs have been experimentally transmitted to a number of species from nonhuman primates to rodents. Each of these diseases has its unique set of characteristics, including range of species that can be infected. These characteristics differ between species and there are varying degrees of host restriction.

Scrapie is a disease of sheep and, rarely, goats that has been recognized for at least 250 years. The term "scrapie" is derived from the pronounced rubbing and scratching of the skin, which occurs in infected sheep three to four years of age. Clinical manifestation of scrapie is characterized by ataxia and recumbancy. Scrapie has a worldwide distribution with the notable exception of Australia and New Zealand due to an aggressive scrapie eradication program. Scrapie is the prototypic TSE but relatively little is known about it. Even the etiology of scrapie is unclear.

Transmissible mink encepahlopathy is a rare disease of ranch-raised mink. It was first described in Wisconsin in 1947 and has since been observed in Ontario, Finland, Germany, and Russia. The incubation period of natural TME is 7–12 months with clinical symptoms including hyperexcitability and, ultimately, motor incoordination. Exposure is via contaminated foodstuffs, although the source, once believed to be sheep scrapie, is not clear.

Chronic wasting disease (CWD), an emerging TSE in wild deer and elk in the western U.S. and Canada, was originally described in captive mule deer and elk in Wyoming and Colorado in 1967. Clinical signs of CWD include emaciation and a reduced fear of humans. The origin and mode of transmission of CWD is unknown, but the mortality rates within a given captive population can be very high (>90% of all animals in residence at one facility). Transmission appears to occur via both lateral and maternal pathways.

Bovine spongiform encephalopathy was first identified in the United Kingdom in 1985. BSE is a foodborne infection believed to be caused by the survival of infectivity in cooked animal offal that was incorporated into commercial diets of cattle in Great Britain. BSE reached epidemic proportions in the UK during the late 1980s and early 1990s, resulting in the death of almost 200,000 cattle. In addition to being devastating to British agriculture, BSE appears to have been the source of a novel feline form of the disease (FSE), and a new human disease of unknown scope, variant CJD.

Human TSEs are primarily sporadic in origin. Sporadic CJD accounts for 90–95% of the reported cases of CJD, affecting one person in a million per year. The disease generally occurs in people older than 50 years of age, although cases have been reported in individuals in their early teens to late eighties. The etiology of sporadic CJD is unknown and there is no link between scrapie in sheep and sporadic CJD. Familial TSEs (GSS, familial CJD and FFI) are autosomal dominant disorders that have been linked to specific mutations in the PrP gene and that occur at an incidence of 1 per 10 million per year. Two factors have contributed to iatrogenic CJD: the presence of CJD titer in preclinical patients and the resistance of these disease agents to inactivation. Iatrogenic CJD has been documented primarily by exposure to central nervous system tissue from infected individuals, specifically dura mater, corneal transplants, and cadaveric putuitary growth hormone treatment.

Kuru is a human TSE of the Fore cultural group of Papua, New Guinea. The disease was perpetuated by ritualistic cannibalistic practices with, at its height, approximately 1% of the population being infected. The incidence of Kuru has declined dramatically since the cessation of the cannibalism in the late 1950s; however, due to the long incubation periods that characterize all TSEs, a few cases still occur.

Variant CJD (vCJD) is an emerging TSE with the first cases occurring in 1995. As will be described below, vCJD can be distinguished from CJD based upon clinical, biochemical, and rodent transmission studies. Whereas sporadic CJD affects individuals in their fifties or sixties, vCJD has to date been disease of teens and young adults. Biologic and biochemical studies have tightly linked vCJD with BSE. It is assumed that consumption of beef products is the source of the infection. Uncertainties over the length of the incubation period, route of infection and number of individuals exposed to contaminated bovine products have resulted in very disparate estimates of the future course of this TSE.

22.2 NATURE OF ILLNESS CAUSED

Prion diseases are inevitably fatal inducing a progressive neurologic dysfunction after a long incubation period. Typical pathological features of prion diseases include spongiform vacuolation, accumulation of PrP^{Sc}, and astrocytosis, and are often accompanied by the accumulation of amyloid plaques. A unique and very pertinent characteristic of all prion diseases is the extended preclinical stage of the disease. TSEs have long incubation periods, taking months to develop in mink, years to develop in sheep, deer, and cattle, and years to decades in humans. The consequences of the long preclinical stage include an inability to diagnose animals and humans in the early stages of the disease. Concerns of potential prion contamination of the blood supply are exacerbated by the need of (and current lack of) a test to identify CJD-infected humans.

The incubation period of BSE ranges from 2–8 years. Clinical signs of BSE initially involve a change in the animal's temperament, with the animal becoming nervous or aggressive. During the 2–6 month clinical period the disease develops into an obvious lack of coordination, difficulty in rising, and loss of weight. The initial clinical signs of scrapie occur 2–5 years after infection, with changes in temperament often followed by rubbing against enclosures. As the disease progresses, the affected animals exhibit a loss of coordination, weight loss, and gait abnormalities.

Human prion diseases can be distinguished clinically. GSS is typified by chronic progressive ataxia and terminal dementia and has a clinical duration of 2–10 years. Fatal familial insommnia initially presents as insomnia followed by ataxia and dementia. Sporadic CJD affects individuals in their fifties and sixties. Death occurs within six months of the onset of the clinical stage that presents as a rapidly progressive multifocal dementia and often ataxia. vCJD clinically presents as a psychiatric disturbance with depression being a predominant feature. The clinical course that develops includes ataxia and dementia and is primarily observed in teenagers and young adults. The clinical course is more extended than classical CJD, with vCJD having a clinical course of approximately 1 year.

Pathogenesis

The pathogenesis of prion diseases can vary depending upon the host species and strain of the agent. All TSEs replicate in nervous tissue and exhibit the highest levels of titer and PrP^{Sc} accumulation in the brain and spinal cord. For example, hamster-adapted strains of scrapie and TME reach levels of 10^9 LD_{50} per gram in the brains in infected animals at the terminal stage of the disease process. Infectivity accumulates throughout the course of the disease with the outcome that considerable amounts of infectious agent are present long before the onset of clinically recognizable disease. There have, for example, been approximately 180,000 documented cases of BSE in Great Britain. Undoubtedly many more cattle harboring high levels of infectivity were slaughtered at a preclinical stage of infection.

It is fortunate that BSE, which does not exhibit a typical strong TSE species barrier, has limited infectivity outside the bovine nervous system. This contrasts with scrapie (when sheep are orally infected) in which infectivity is identified in the digestive tract and lymphoreticular system prior to progressing through the autonomic nervous system, to the spinal cord and brain. Studies characterizing the disease-associated PrP isoform in various tissues from CJD and vCJD patients determined that PrP^{Sc} is readily detectable in lymphoreticular tissues from vCJD and not detectable in sporadic CJD. This is likely due to the oral source of the infection and suggests a greater potential of iatrogenic transmission of vCJD.

22.3 CHARACTERISTICS OF THE AGENT

The unusual biology of these diseases has been long recognized, and has influenced how these disorders have been characterized. Based upon their long incubation periods and transmissibility, TSEs were originally described as unconventional or slow viruses. The extreme resistance of these agents to ionizing and gamma irradiation combined with the inability to isolate a TSE-specific microorganism prompted speculation that there existed a non-nucleic acid mode of replication. In 1968, a mathematician, J. S. Griffith, proposed three means by which a protein could have self-replicating properties. One of Griffith's models involved the interaction of two proteins having the same primary amino acid sequence yet differing structurally. Infectivity purification studies in the late 1970s identified a brain homogenate fraction that was enriched for infectivity. This purification was performed independently by two groups. The detergent extraction and centrifugation steps resulted in the formation of fibular structures, referred to as scrapie-associated fibrils in 1981 or the similarly structured prion rods in 1982.

Biochemical characterization of the highly infectious preparations identified a protease resistant protein termed the prion protein. This glycoprotein had a molecular weight of 33–35 kDa (in the absence of protease treatment). Treatment with mild protease (50–100 ug/ml of proteinase K) reduced the protein to 27–30 kDa. Characterization of the gene encoding the prion protein quickly led to the realization that the prion protein was not unique to an undiscovered microorganism but rather was expressed in uninfected animals and encoded by a single-copy gene in the nucleus. The difference between the infection-associated and uninfected forms of the protein involved the structure of the two otherwise identical proteins (Table 22.2). The disease-associated form, in addition to being resistant to proteinase digestion, was found to have more beta sheet structure than the form of the protein expressed in uninfected animals. In 1982, Stanley Prusiner formally proposed the prion hypothesis that identified Griffith's hypothetical protein as the prion protein and defined prions as "small

TABLE 12.2. Prion Protein Nomenclature

	Protease Sensitivity	Description
PrP^{C}	Sensitive	Normal isoform of the prion protein
PrP^{Sc}	Resistant	Disease-associated isoform of the prion protein
PrP-sen	Sensitive	Refers to protease digestion characteristics of PrP, often in the absence of transmission data
PrP-res	Resistant	Refers to protease digestion characteristics of PrP, often in the absence of transmission data
Prion rods	Resistant	Structures produced upon detergent extraction of infected tissue. Highly infectious, comprised primarily of PrP^{Sc}
Scrapie associated fibrils (SAF)	Resistant	Very similar to prion rods

proteinaceous particles which are resistant to inactivation by most procedures that modify nucleic acids. Prusiner proposed that the interaction of the prion protein (disease associated form) with the normal cellular form resulted in the conversion of the normal form to the disease form, increasing the amount of abnormal form and, thus, the level of infectious agent. Prusiner was awarded the Nobel Prize for Medicine in 1997, although formal proof of the hypothesis has remained elusive.

The PrP gene is present and expressed in the genomes of all mammals studied. The human PrP is a glycoprotein of 253 amino acids. It is a cell surface glycoprotein expressed primarily in neurons but also in astrocytes and other cells. Knockout experiments in the mouse have demonstrated that it is not an essential gene and that KO mice, when infected with mouse-adapted scrapie, do not replicate the abnormal form of the protein, nor develop spongiform lesions characteristic of the disease.

PrP^{C} is a cell surface sialoglycoprotein of unknown function. Under physiological conditions, PrP^{C} is synthesized in the endoplasmic reticulum and transported through the Golgi towards the cell surface. Like other GPI-anchored proteins, PrP^{C} is located primarily in cholesterol-rich, detergent-resistant microdomain complexes of the plasma membrane (rafts). Cell culture studies have demonstrated that, once in the membrane, some PrP molecules are released into the extracellular space, while most PrP is internalized into an endocytic compartment. The normal function of the protein is not known. There is some evidence, based upon a metal-binding domain present in the N-terminal region of the polypeptide and the binding of copper to synthetic peptides, that PrP^{C} is a metalloprotein. The transition metal induces a dramatic effect on PrP biochemical properties resulting in an increase aggregation, beta-sheet content and acquisition of protease-resistance.

PrP^{Sc} represents a conformational variant of PrP^{C}. In contrast to PrP^{C}, PrP^{Sc} forms insoluble aggregates with B sheet content characteristic of an amyloidogenic protein polymer. PrP^{Sc} assembles into fibrils both in vivo and in vitro, is resistant to heat, radiation, conventional disinfectants such as alcohol and formalin, and is partially resistant to digestion with proteinase K (PK). PK digestion, in most cases, removes 60–70 amino acid residues from the N-terminus, generating PrP27–30, the protease resistant core of PrP^{Sc}.

Given the lack of evidence of an essential, scrapie-specific nucleic acid, hypotheses of the generation of PrP^{Sc} have suggested an autocatalytic process involving the interaction of PrP^{C} with PrP^{Sc}. Two autocatalytic models have been proposed and generated considerable debate: the template-assisted conversion and nucleation-dependent polymerization. Both are based upon an autocatalytic process in which the interaction of misfolded PrP with PrP^{C} results in the conversion of PrP^{C} to the pathologic form. The template-assisted conversion model hypothesizes that the conformational transition of monomeric PrP^{C} is induced by a monomeric PrP^{Sc} through a cycle of unfolding and refolding reactions. PrP^{C} combines with PrP^{Sc} by formation of a heterodimer at an intermediate stage. Following the conversion of PrP^{C} into PrP^{Sc} in the heterodimer, the two PrP^{Sc} molecules dissociate from the complexes and restart the reaction. The nucleation-dependent polymerization model argues that infectious PrP^{Sc} is an ordered aggregate that acts as a seed. PrP^{C} upon binding to the seed acquires the conformation of the PrP^{Sc} subunits in the oligomer.

TSE Strains

TSE strains have the unique distinction of being historically employed as evidence of the existence of an essential nucleic acid as the infectious agent and, more recently, also being used to support the protein only (prion) hypothesis.

Different strains of TSEs can adapt and/or exist in a given species. The first TSE strains identified were in goats infected with scrapie. It is estimated that twenty different prion strains have been produced upon transmission of TSEs to rodents. Strains are defined by a number of characteristics, the most easily identified being incubation period and clinical symptoms. Strain-specific histopathological differences (brain location of spongiform changes, number and size of spongiform changes) have led to the development of the lesion profile analysis, which measures the extent and distribution of spongiform degeneration in the central nervous system. Lesion profiling, in which nine standard areas of the brain are assigned a score based on the intensity of vacuolation, provides a quantitative assessment of spongiform degeneration. The migration pattern of PrP^{Sc} on PAGE has also proven to be a useful tool to distinguish TSEs. Human TSEs can be classified based upon PK digestion products of PrP^{Sc} as the migration of PrP^{Sc} bands representing different degrees of PrP glycosylation. Defining the ratio of di-, mono-, and nonglycosylated forms of PrP^{Sc} is referred to as a glycoform profile.

A strong link between TSE strain and PrP^{Sc} structure was identified by Richard Marsh and colleagues. The investigators passaged TME into hamsters and, after numerous passages, identified two stable hamster-adapted strains. The strains, referred to as Hyper (HY) and Drowsy (DY) were easily distinguished based upon incubation period as well as clinical characteristics. HY exhibited hyperexcitability and cerebellar ataxia at 65 days post inoculation while DY exhibited the predominant clinical sign of lethargy with no hyperexcitability or cerebellar ataxia 168 days post inoculation. The proteinase K-resistant forms of PrP^{HY} and PrP^{DY} migrate differently on SDS-PAGE. All three isoforms (diglycosylated, monoglycosylated and nonglycosylated) of PrP^{DY} migrate with a 1–2 kDa lower molecular weight than the isoforms generated by a HY infection. This molecular mass difference after PK digestion was shown by protein sequence analysis and PrP antibody mapping to be due to differences in the N-terminal region of PrP. These observations indicated that the faster gel migration of PrP^{DY} after PK digestion was due to additional truncation at the N-terminal end. These qualitative differences in the proteolytic degradation pattern of the two proteins having identical primary sequences strongly suggest that different second or tertiary conformations exist between HY and DY.

Species Barrier Effect

A characteristic of prion diseases that has had an impact both on the origin of BSE and its subsequent transmission to humans is the species barrier effect. Transmission of a TSE to a new host species is an inefficient process, resulting in an increase in the length of the incubation period in the donor host species. Subsequent passage in the new host species results in the reduction and eventual stabilization of the incubation. Some TSEs, however, are not pathogenic for some host species. For example, the TME agent does not establish an infection in several strains of mice and passage of CWD to hamsters has been unsuccessful. Experimentally, strong species barriers can be overcome through repeated passaging of the agent in the new host. For example, TME has limited pathogenicity in ferrets, initiating clinical symptoms after an extended incubation period of 24 months. Second passage (ferret to ferret) results in a reduction of incubation period to 18 months while a third passage produces a 4 month incubation that is stable upon further ferret passage. Ferret-adapted TME (4 month incubation period) has limited pathogenicity in mink, producing an almost 24 month incubation period.

The apparent strong sheep-to-bovine-species barrier may have been overcome agriculturally in a similar manner. Scrapie-infected sheep were rendered and the meat and bone meal by-products included as a supplement to cattle rations. The physico-chemical stability that characterizes these infectious agents resulted in the scrapie agent surviving the heat treatment present in the rendering process. Cattle fed scrapie-infected meat and bone meal were then rendered and included in meat and bone meal supplements, thus recycling infectivity in the cattle population of Great Britain.

It is a rather unique, and certainly unfortunate, characteristic of the BSE agent that it transmits readily to numerous other species. Experimentally BSE has been transmitted, in addition to cattle, to a number of species including mice, mink, sheep, goats, marmosets, and macaque monkeys. It is this weak species barrier that has led to the emerging vCJD epidemic. VCJD is, in essence, human BSE.

22.4 EPIDEMIOLOGY

The TSE landscape has shifted considerably over the past 20 years, from CJD being an extremely rare and relatively unknown disease and scrapie being an agricultural nuisance and of veterinary interest, to the outbreak of the agriculturally disastrous BSE epidemic and the realization that BSE can and has been transmitted to humans. The extent of the bovine to human transmission is unknown and is of considerable, and justifiable, public concern.

BSE—Great Britain

First documented in 1986, the initial cases of BSE occurred in 1985, although it likely was cycling in cattle prior to that time. The disease peaked in January of 1993 with approximately 1000 new cases documented per week. There have been almost 181,000 documented cases of BSE in Great Britain. This is certainly a considerable underestimate of the total number of infected cattle, as infected preclinical animals would be excluded. The decline in cases beginning in 1993 is attributed to the 1988 ban on the inclusion of meat-and-bone meal in cattle feed. Due to the approximate 5 year incubation period, the effects of the ban were not observed until 1993 and 1994.

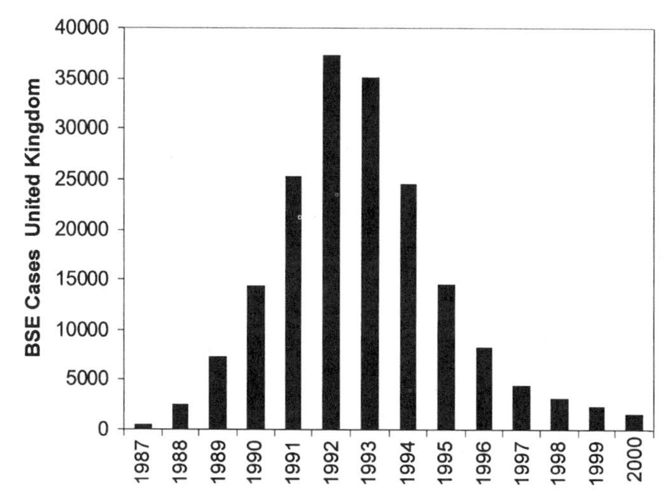

Figure 22.1. The bovine spongiform encephalopathy epidemic in the United Kingdom. The decline in cases that began in 1993 is attributed to the 1988 ban on mammalian meat and bone meal supplements being included in feed. Monthly updates of BSE in Great Britain can be obtained at http://www.maff.gov.uk/animalh/bse/

BSE—Europe

BSE is present in Europe, being reported in Belgium, the Czech Republic, Denmark, France, Germany, Ireland, Italy, Netherlands, Portugal, and Switzerland. Although in a few countries including Portugal and Germany, the number of cases of BSE have been increasing, the number of infected animals is small (hundreds of cases per year).

BSE—North America

There have been no cases of BSE reported in the United States. A single case was reported in Alberta, Canada, an import from Great Britain. In August 1997, a feed ban was instituted in the United States prohibiting the feeding of mammalian protein to ruminant animals.

Other Animal TSEs

Scrapie has a worldwide distribution with the notable exceptions of Australia and New Zealand. The latter countries are scrapie-free. Epidemiologic studies do not provide any support of a link between scrapie in sheep and CJD in humans.

CWD is present in wild populations of deer and elk in a few counties in three western states, Colorado, Wyoming, and Nebraska. In addition, CWD has occurred in ranch-raised elk. In the endemic region, CWD can be present at very high levels. For example, in Larimer County Colorado, 15% of the deer were positive for CWD. There is little doubt that such estimates considerably underestimate the number of infected deer in the region. Animals in the early stages of infection would not have sufficient PrP^{sc} accumulated in their brains to be detected. Although it has been argued that CWD is not expanding its range, the efficient animal-to animal transmission (by as-of-yet-unknown means) suggests the potential for large expansion of the endemic range. It is unclear if CWD is the result of transmission of sheep scrapie to deer and elk or if CWD represents a naturally occurring TSE of cervids. Regardless of the origin, CWD is clearly a self-sustaining disease.

Human TSEs

The epidemiology of human prion diseases encompasses three forms: sporadic, familial, and acquired. CJD (sporadic) has an incidence of approximately one individual per million per year worldwide. It occurs in individuals in their fifth to sixth decades of life and has a worldwide distribution. The familial forms of CJD and GSS are even more rare, affecting 1 per 10 million per year.

vCJD is clearly an emerging disease. Not surprisingly, given the BSE link, almost every case has occurred in the U.K. As of this writing (July 2001), there have been 102 cases in Great Britain (Figure 22.2), one in Ireland and two in France. Given the long incubation periods that characterize prion diseases,

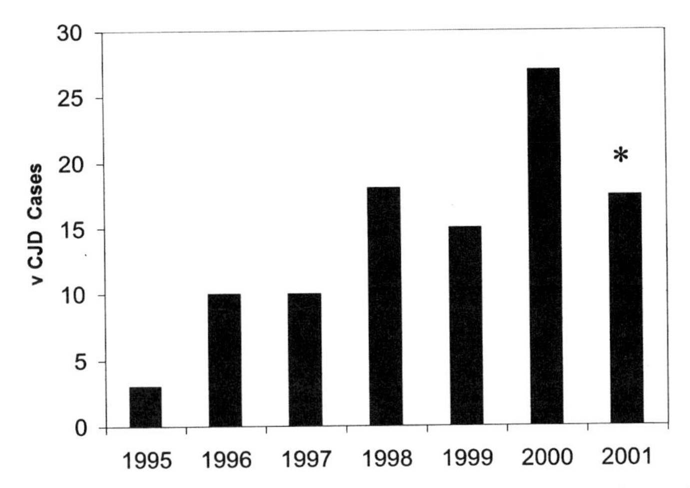

Figure 22.2. The number of vCJD cases in Great Britain. Monthly updates of vCJD in Great Britain can be obtained at http://www.doh.gov.uk/cjd/cjd_stat.htm.
*The 2001 data includes cases only to July.

cases of vCJD can be expected in Great Britain for decades to come. It is not clear, however, how large an epidemic may develop. Although it is assumed, and most likely, to be the result of oral exposure of BSE, the route of infection is not known. It is also not known how many infected cattle were included for human consumption.

22.5 DETECTION OF THE ORGANISM

One of the greatest challenges in this field is the need to develop accurate and highly sensitive methods of assaying for TSEs. Traditional detection and/or verification of a TSE infection involved the histological examination of the brain for evidence of spongiform degeneration typically combined with infectivity studies to determine transmissibility. The identification of the disease-associated form of the prion protein and the generation of PrP antibodies has facilitated Western blot and immunohistochemical approaches for the detection of PrP^{Sc}-containing tissue. It should be emphasized that the most accurate diagnosis occurs in animals and humans during the clinical stages of the disease through the examination of CNS. Brains of infected animals during clinical phase of the disease contain the highest level of spongiform degeneration and the greatest accumulation of PrP^{Sc}. The earlier the stage of infection, the more difficult these diseases are to diagnose.

The Western blot analysis involves the treatment of tissue homogenates with mild levels of proteinase K (50–100 ug/ml for 1–2 hours). PrP^{C}, the only isoform in uninfected tissue, is completely digested while the disease associated form of the protein, PrP^{Sc} exhibits resistance to the digestion (Figure 22.3). A portion of the N-terminal region of the PrP^{Sc} isoform is cleaved during the

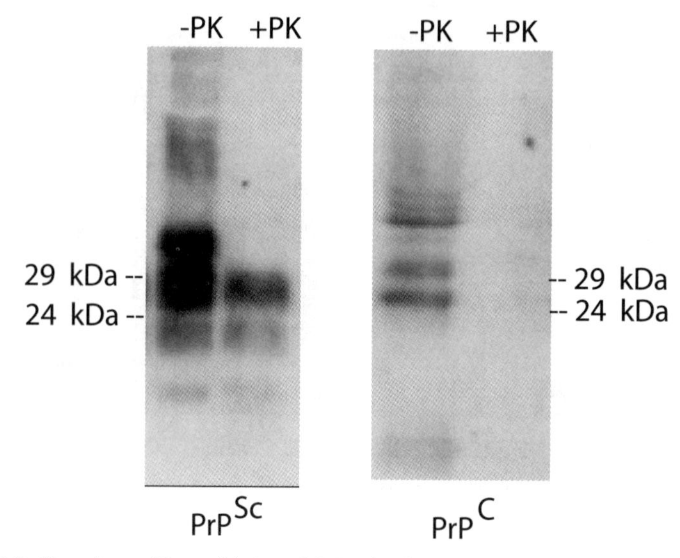

Figure 22.3. Proteinase K sensitivity of PrP obtained from brain homogenate from an infected animal (left panel) and with PrP^C (right panel) which is not associated with infectivity.

digestion resulting in a smaller protease resistant core of approximately 27–30 kDa (Figure 22.3).

22.6 PHYSICAL MEANS OF DESTRUCTION OF THE ORGANISM

One of the most challenging areas of prion disease research involves the identification of processes that result in the sterilization of the infectious agent. The extreme physicochemical stability of the prion disease agent is the underlying cause of the BSE epidemic as well as the iatrogenic transmission of CJD via contaminated surgical instruments. The resistance of these agents to inactivation has been recognized for years. The resistance of scrapie to formalin during the preparation of a vaccine resulted in the accidental transmission of scrapie in the 1930s. Standard autoclaving (121°C for 15 minutes) is ineffective in eliminating infectivity.

Many of the inactivation studies have been performed in the laboratory of David Taylor. Chemical methods of inactivation, including ethanol, formaldehyde, gluteraldehyde, and hydrogen peroxide, all of which exhibit efficacy in the sterilization/decontamination of microorganisms, are of little practical use with prion diseases. One hour exposure to NaOCl solution containing 20000 ppm of CL_2 is suitable for inactivating TSE agents. It should be noted that there are TSE strain differences in inactivation. Richard Kimberlin and colleagues determined that autoclaving one mouse-adapted strain (strain 139A) for 2 hours at 126°C resulted in its inactivation while a second strain (strain 22A) was not.

Chemical denaturation of infectious preparations results in a reduction of infectivity and concomitant decline in the amount of protease resistant PrP. Our group has shown that both infectivity and the abnormal form of the protein can be regenerated upon dilution of the denaturant. It should be emphasized that the preceding experiments were performed under carefully controlled laboratory conditions. The study does emphasize the need to ensure the destruction of the protein during inactivation treatments.

The UK government advisory committee recommends 134–137°C for 18 minutes or a series of successive cycles of 134–137°C for a minimum of 3 minutes per cycle for autoclaving CJD. The Office International Des Epizooties recommends the following treatment for the inactivation of TSEs in meat-and-bone meal containing ruminant proteins: i) reduction of particle size to 50 mm prior to heating, and ii) raw material should be subjected to saturated steam conditions to a temperature of ≥133°C for 20 minutes.

22.7 PREVENTION/CONTROL MEASURES

Prion diseases are always fatal. There are no drug treatments or vaccines that provide a cure. A series of compounds have been identified ("anti-scrapie drugs") that extend incubation period. They are, however, not cures: they can be toxic and do require continuous treatment initiated at the early stages (preclinical) of the infection. As a result, control measures involve eliminating the source of infection and, in the case of humans, reducing the likelihood of iatrogenic transmission.

Animal Diseases

With scrapie, control measures have often involved the destruction of affected animals as well as the flocks. Scrapie remains a self-sustaining disease of sheep throughout the world. Scrapie eradication programs have been successful in Australia but not in the U.S.

The unusual (for TSEs) transmission characteristics of CWD argue that CWD is an emerging disease and that its eradication will prove to be very difficult. As described above, animal transmission studies suggest a "typical" species barrier; thus, CWD would appear not to be health risk to humans.

BSE, being a foodborne disease, has declined tremendously as a result of the ban on feeding ruminant meat-and-bone meal to cattle. There is little evidence of cattle-to-cattle transmission of BSE. It is, therefore, argued that strict adherence to the feed ban will result in its elimination.

vCJD

The lack of treatment for these diseases has focussed efforts to minimize further transmission of vCJD. It is believed that the dramatic decline in BSE coupled with the exclusion of bovine brain or nervous tissue in human food, mandated by the British government in 1995, has tremendously reduced and/or eliminated further BSE to human transmission. The greatest concern is the unknown number of humans currently infected who are at a preclinical stage of the disease. Recent evidence suggests low levels of infectivity may be present in blood from infected animals. This coupled with the data from John Collinge's laboratory, that compared to sporadic CJD there are greater levels of abnormal PrP (and presumably infectivity) in peripheral tissues, the potential for iatrogenic transmission through the blood supply is being considered. In the U.S., there are restrictions on donating blood if the individual has visited or lived in Great Britain.

Estimates of the total number of potential cases of vCJD range from a few dozen to hundreds of thousands of individuals. There are currently too many unknowns (route of infection, number of individuals exposed, effectiveness of CNS exclusion from meat, levels of infectivity present) to provide an accurate assessment of the future incidence of the disease. Given the long incubation periods that characterize these diseases, however, we can expect additional vCJD cases to occur for decades to come.

BIBLIOGRAPHY

Almond, J. and Pattison, J. 1997. Human BSE. *Nature* **389**: 437–438.

Bessen, R. A. and Marsh, R. F. 1994. Distinct PrP properties suggest the molecular basis of strain variation in transmissible mink encephalopathy. *Journal of Virology* **68**: 7859–7868.

Bruce, M. E., Will, R. G., Ironside, J. W., McConnell, I., Drummond, D., Suttie, A., McCardle, L., Chree, A., Hope, J., Birkett, C., Cousens, S., Fraser, H., and Bostock, C. J. 1997. Transmissions to mice indicate that 'new variant' CJD is caused by the BSE agent. *Nature* **389**: 498–501.

Caughey, B. 2001. Interactions between prion isoforms: the kiss of death? *Trends in Biochemical Sciences* **26**: 235–242.

Chesebro, B. 1999. Prion protein and the transmissible spongiform encephalopathy diseases. *Neuron* **24**: 503–506.

Collinge, J. 2001. Prion disease of humans and animals: their causes and molecular basis. *Annual Review of Neurosciences*. **24**: 519–550.

Harris, D. A. 1999. Cellular biology of prion diseases. *Clinical Microbiology Review*. **12**: 429–444.

Kretzschmar, H. 1999. Molecular pathogenesis of prion diseases. *European Archives of Psychiatry and Clinical Neuroscience* **249**: Suppl. 3 III/56–III/63.

McKenzie, D., Bartz, J. C., and Aiken, J. M. 1998. The molecular basis of strains in the transmissible spongiform encephalopathies. *Bulletin de l'Institute Pasteur* **96**: 35–47.

Prusiner, S. B. 1998. Prions. *Proc. Natl. Acad. Sci. USA*. **95**: 13363–13383.

Weissmann, C., Raeber, A. J., Montrasio, F., Hegyi, I., Frigg, R., Klein, M. A., and Aguzzi, A. 2001. Prions and the lymphoreticular system. *Philosophical Transactions of the Royal Society of London B*. **356**: 177–184.

INDEX